Lecture Notes on Multidisciplinary Industrial Engineering

Series Editor

J. Paulo Davim, Department of Mechanical Engineering, University of Aveiro, Aveiro, Portugal

"Lecture Notes on Multidisciplinary Industrial Engineering" publishes special volumes of conferences, workshops and symposia in interdisciplinary topics of interest. Disciplines such as materials science, nanosciences, sustainability science, management sciences, computational sciences, mechanical engineering, industrial engineering, manufacturing, mechatronics, electrical engineering, environmental and civil engineering, chemical engineering, systems engineering and biomedical engineering are covered. Selected and peer-reviewed papers from events in these fields can be considered for publication in this series.

Luis Farinha · Daniel Raposo · João J. Ferreira ·
Marcelo Calvete Gaspar · Maria Leopoldina Alves ·
Luis Serrano
Editors

Sustainable Innovations in Mechanical Engineering and Management Sciences

Selected Contributions to the 8th International Conference on Sustainable Innovation, Entrepreneurship and Technology (RegionalHelix 2024), November 14–16, 2024, Leiria, Portugal

Springer

Editors
Luis Farinha
Instituto Politécnico de Castelo Branco
(Polytechnic University of Castelo Branco)
Castelo Branco, Portugal

João J. Ferreira
Management and Economics Department
University of Beira Interior
Covilhã, Portugal

Maria Leopoldina Alves
Instituto Politécnico de Leiria
Leiria, Portugal

Daniel Raposo
Instituto Politécnico de Castelo Branco
(Polytechnic University of Castelo Branco)
Castelo Branco, Portugal

Marcelo Calvete Gaspar
Instituto Politécnico de Leiria
Leiria, Portugal

Luis Serrano
Instituto Politécnico de Leiria
Leiria, Portugal

ISSN 2522-5022 ISSN 2522-5030 (electronic)
Lecture Notes on Multidisciplinary Industrial Engineering
ISBN 978-3-032-30371-4 ISBN 978-3-032-30372-1 (eBook)
https://doi.org/10.1007/978-3-032-30372-1

This Springer imprint is published by the registered company Springer Nature Switzerland AG
The registered company address is: Gewerbestrasse 11, 6330 Cham, Switzerland

Preface

This Preface presents strategic and sustainable management as a discipline that combines competitive advantage with development paths aligned with planetary boundaries. Classic strategy research explains superior performance through the Resource-Based View (RBV), which asserts that advantage arises from resources and capabilities that are valuable, rare, inimitable, and organised—the VRIO model (Barney, 1991). To maintain an advantage in turbulent environments, the RBV was complemented by the Dynamic Capabilities (DC) framework, which highlights a firm's ability to sense, seize, and transform (Teece, Pisano, & Shuen, 1997). Simultaneously, Stakeholder Theory provided the normative and instrumental basis for creating value with and for multiple stakeholders (Donaldson & Preston, 1995), paving the way for the Natural-Resource-Based View (NRBV), which suggests that environmental skills (pollution prevention, product stewardship, sustainable development) can serve as durable sources of advantage (Hart, 1995). These streams converge with evidence showing that well-designed environmental regulation and the pursuit of eco-innovation can strengthen competitiveness (Porter & van der Linde, 1995; Bansal & Roth, 2000).

From this theoretical core, competitive advantage is no longer limited to internal assets; it now increasingly relies on the orchestration of ecosystems and mission-oriented policies, where universities, firms, governments, and civil society collaborate in Triple/Quadruple/Multiple Helix configurations. Two Springer volumes are central to this agenda: Multiple Helix Ecosystems for Sustainable Competitiveness (Peris-Ortiz, Ferreira, Farinha, & Fernandes, 2016), which expands the Triple Helix towards Quadruple and Quintuple Helix by incorporating users/citizens and the environment as additional helices; and Regional Helix Ecosystems and Sustainable Growth (Farinha, Santos, Ferreira, & Ranga, 2020), which illustrates how dense regional ecosystems aligned with smart specialization strategies promote innovative entrepreneurship, technology transfer, and growth with sustainability. Without discussing the chapters in detail, this introduction offers the conceptual framework that underpins the two main axes: "Innovation Networks, Entrepreneurship and Regional Growth" and "Sustainability, Circular Economy and Smart Cities."

Innovation Networks, Entrepreneurship and Regional Growth

Innovation networks have shifted the centre of knowledge creation. In frontier sectors, innovative performance relies less on isolated internal routines and more on interorganisational collaborations that combine technological complementarities, trust, and learning. The seminal study by Powell, Koput, and Smith-Doerr (1996) showed in biotechnology that relational position within the network, not merely internal R&D effort, explains innovation rates, aligning with the DC view of coordinating distributed assets (Teece et al., 1997). In this view, partnering and alliance management capabilities become VRIO assets when embedded in routines, reputations, and governance mechanisms that are difficult to replicate. The geography of knowledge spillovers remains crucial. Audretsch and Feldman (1996) demonstrated that patenting tends to concentrate where proximity facilitates recombination and lowers coordination costs. The Knowledge Spillover Theory of Entrepreneurship systematises the micro-mechanism: entrepreneurs act as conduits that transform endogenous knowledge into new ventures and regional growth (Acs, Braunerhjelm, Audretsch, & Carlsson, 2009). Hence, regions with similar knowledge bases display different outcomes because they vary in relational density, absorptive capacity, and the rate of innovative entrepreneurship.

A territorial reading of development underscores that advantage is collective and contextual: ecosystems organised by conventions, norms, infrastructures, and support services shape transaction costs, sectoral diversity, and technological trajectories. Within this frame, policy tools such as innovation-oriented public procurement, living labs, and regulatory sandboxes create lead markets for clean and digital solutions while reducing uncertainty and systemic risk conditions critical to crowding in entrepreneurial investment and accelerating diffusion.

Helix models provide a grammar for collaborative governance. On first mention, we highlight Peris-Ortiz, Ferreira, Farinha, and Fernandes (2016), who compile advances from Triple Helix to Quadruple/Quintuple Helix, incorporating users/citizens and the environment as additional helices and making roles, incentives, and trade-offs explicit. On subsequent mentions, Peris-Ortiz et al. (2016) suffice. Complementarily, *Regional Helix Ecosystems and Sustainable Growth* (Farinha et al., 2020) shows, using cases and metrics, that dense regional ecosystems, well-articulated with universities and intermediaries (incubators, technology transfer offices), foster technology entrepreneurship, internationalisation, and high-skill job creation, aligning competitiveness with sustainability goals. Future research should operationalise comparable indicators of relational quality and ecosystem maturity, testing causal links with innovation, productivity, and decarbonization outcomes.

Sustainability, Circular Economy and Smart Cities

Strategic sustainability has moved from a defensive (compliance) stance to a proactive approach centred on value creation. The NRBV (Hart, 1995) argues that environmental skills, from pollution prevention to ecodesign, can be served as sources of advantage when integrated into the core business model and underpinned by dynamic capabilities (Teece et al., 1997). Evidence indicates that well-crafted environmental policies encourage eco-innovation, which can offset costs and enhance productivity (Porter & van der Linde, 1995). Simultaneously, factors like competitiveness, legitimacy, and ecological responsibility help explain why firms "go green" (Bansal & Roth, 2000).

The Circular Economy (CE) has emerged as a paradigm that minimises the use of virgin inputs and reduces value leakage (waste, emissions, energy) through extended product lifecycles, reuse, remanufacturing, and product-as-a-service models. Geissdoerfer, Savaget, Bocken, and Hultink (2017) define CE as a regenerative system and explore its relationship with sustainability; Kirchherr, Reike, and Hekkert (2017), analysing 114 definitions, highlight the emphasis on the 3Rs (reduce-reuse-recycle) while stressing the need for a systemic shift and clear links to the three pillars of sustainable development. Strategically, CE involves sensing material flows and risks, leveraging partnerships for reverse logistics, and transforming value chains and contracts (NRBV + DC), alongside performance metrics that measure both economic returns and environmental and social impacts.

Smart cities act as territorial platforms to accelerate this transition. Caragliu, Del Bo, and Nijkamp (2011) describe them as integrating human and social capital with ICT infrastructure to promote sustainable growth and enhance quality of life; Albino, Berardi, and Dangelico (2015) review their dimensions and metrics, emphasising that smartness is not merely digitisation but the ability to address urban problems inclusively and sustainably. The Multi-Level Perspective (MLP) (Geels, 2002) offers a framework to understand socio-technical transitions: niche innovations (such as electric mobility, positive-energy buildings, and district heating) interact with regimes under the influence of landscape pressures (climate, culture, politics). Cities that combine enabling regulation, open data, innovative procurement, and Multiple Helix governance foster conditions to scale circular solutions across energy, mobility, and the built environment, thereby enhancing resilience and competitiveness.

Integrating these strands (RBV/VRIO, DC, NRBV, CE, and MLP), a resilient competitive advantage arises from well-coordinated ecosystems, with feedback loops among science, entrepreneurship, and public policy. The cited Springer volumes demonstrate that Multiple Helix governance aligns with smart specialisation and sustainability metrics, proposing a research agenda that involves network analysis, policy evaluation, and quasi-experimental designs to measure effects on productivity, emissions, and well-being.

Overview of Chapters

This volume includes 13 chapters arranged in two interconnected sections. The first eight chapters are under *Innovation Networks, Entrepreneurship and Regional Growth* while the final five chapters constitute *Sustainability, Circular Economy and Smart Cities*. Collectively, the contributions present a cohesive view of innovation dynamics, entrepreneurship, and regional development, as well as digital and circular transitions in urban and sectoral contexts.

The chapters included in this volume primarily originate from selected contributions presented at the 8th International Conference on Sustainable Innovation, Entrepreneurship, and Technology (RegionalHelix 2024), held in Leiria, Portugal, from 14 to 16 November 2024. The most relevant conference papers on this topic were invited to submit substantially revised and extended versions for consideration in this volume. All chapters are original, previously unpublished contributions underwent a double-blind peer-review process to ensure their scientific quality and alignment with the book's thematic focus. In addition to the selected conference papers, a limited number of chapters were written by invited authors to expand the volume's conceptual and empirical scope.

Innovation Networks, Entrepreneurship and Regional Growth

Chapter "Smart Enough to Matter: Rebranding the Rural with Innovation-Driven Growth" illustrates how the smart city concept has evolved from a solely technological perspective to encompass broader territorial development models. Using three inland Portuguese municipalities, it demonstrates that innovation, as an economic aspect of smart city frameworks, can enhance place branding: it builds credible, forward-looking territorial identities by combining creativity, technology, and sustainability, resulting in both symbolic and tangible attractiveness. The key message is that peripheral municipalities can use innovation both as a development strategy and as a narrative of competitive distinction.

Chapter "Driving Agricultural Innovation: The Role of Collaborative Networks in the INTERIOR+ Project under the Triple Helix Framework" explores innovation in agriculture through a Triple Helix approach and COINs (Collaborative Innovation Networks). Based on a case study of the Interior+ project, it shows that well-organised collaborative networks speed up the dissemination of practices and technologies by promoting knowledge sharing, resource pooling, and joint problem-solving. The chapter offers practical guidance on how to establish and maintain such networks, emphasising stakeholders' active role in shaping an intersectoral collaboration strategy.

Chapter "Incubators as Catalysts of Dynamic Capabilities: Strategic Contributions to the Sustainability of Entrepreneurial Ecosystems in Low-Density Territories" examines the role of incubators within entrepreneurial ecosystems in low-density regions, using dynamic capabilities as an analytical lens. Interviews with incubators and entrepreneurs in Portugal's North and Centre (inland) regions suggest that incubators serve as coordination nodes for local actors (municipalities, higher education, business associations), stimulating entrepreneurship, innovation, and investment. Structural constraints remain, such as talent shortages and limited access to finance, yet the chapter highlights incubators as catalysts for adaptation and learning, strengthening ecosystem resilience.

Chapter "Knowledge Creation and Innovation: The Key Drivers of Economic Growth According to the Global Innovation Index" explores relationships between the input and output sub-indices of the Global Innovation Index (GII) across 132 economies. Through factor analysis and cluster analysis, it reveals interaction patterns among innovation pillars and identifies performance trajectories that can inform public policy. The outcome is a concise map of innovation capacities on a global scale, useful for understanding what enhances robust and interconnected ecosystems.

Chapter "Sustainable Regional Development through Entrepreneurship Support Policies: Insights from Low-Density Areas in Northern Portugal" assesses entrepreneurship support policies in low-density areas (the districts of Vila Real and Bragança) through interviews with local stakeholders. Findings confirm shared weaknesses but highlight significant differences in policy adequacy and coordination between regions. By comparing existing measures with stakeholder perceptions, the chapter demonstrates that current instruments do not fully address regional needs and provides context-specific advice to enhance attractiveness and entrepreneurial activity.

Chapter "The Relationship Between Open Innovation and Public Funding in Portuguese SMES: Impact on Innovative Capacity and Collaboration within Innovation Ecosystems" examines how open innovation factors relate to publicly funded RDI projects in Portuguese SMEs. Quantitative analysis shows that financial support encourages collaborative practices, bringing firms closer to universities and research centres and strengthening networks that improve access to external knowledge and international competitiveness. The evidence supports targeted policies to reduce collaboration barriers and foster an inclusive innovation ecosystem.

Chapter "Institutionalization of the Entrepreneurship and Innovation Program at Federal University of Technology—Paraná, Brazil" details the institutionalisation of the Entrepreneurship and Innovation Programme (Proem) at UTFPR (Brazil) through documentary research and institutional theory, in dialogue with the Triple Helix approach. Two main aspects emerge: interactions across institutional spheres and the programme's institutionalisation. The analysis finds that Proem's activities align with helix logic and show potential for long-term stability, influencing the development of technology-based start-ups.

Chapter "Interactions between Innovation Capacity and Innovation Ecosystems: A Bibliometric Review" offers a systematic literature review to explore how innovation ecosystems scholarship integrates the concept of innovation capacity. Using keyword co-occurrence analysis (R), it traces an evolution in methods, highlighting themes that emphasise the importance of collaboration in ecosystems and the need to deepen understanding of how organisations can enhance their capacity for innovation. The chapter concludes with a research agenda aimed at this conceptual and empirical integration.

Sustainability, Circular Economy and Smart Cities

Chapter "Technology Parks in Unequal Territories: Evidence and a Framework for Inclusive Innovation" analyses science and technology parks through the lenses of the Triple Helix and regional development. Fieldwork in three parks (Spain, Portugal, Brazil), combined with comparative secondary data, indicates that the most successful operations occur in territories with institutional capacity to internalise knowledge. Parks act as cooperation frameworks that support start-ups and new businesses; further research is recommended to assess the impact of implementation strategies.

Chapter "Digital Pathways Towards Efficient Urban Mobility in Smart Cities" examines digital technologies for urban mobility in smart cities, emphasising three key enablers: artificial intelligence (AI), the Internet of Things (IoT), and mobility-as-a-service (MaaS). Based on expert interviews, it highlights improvements in real-time decision-making, route optimisation, and equitable access, assuming sufficient infrastructure, strong governance, and adaptable policies are in place. Major challenges include fragmented physical systems, data governance issues, rigid regulation, and digital inequalities. Recommendations involve co-design with users, modular technologies, and transparent, ethical data management.

Chapter "EcoXperience Circular Business Models for Agri-Food Waste Valorization: A Case Study" concentrates on the circular economy in the agri-food sector through the EcoX case: a micro-enterprise that upcycles used cooking oil into environmentally friendly detergents using a patented process. The qualitative analysis reveals waste reduction, the replacement of virgin materials, and sustainable consumption (bulk sales, reusable packaging). Its unique contribution is to demonstrate the simultaneous integration of the five circular business model types (circular supplies/design, resource recovery, product life extension, sharing, and product-as-a-service), providing practical lessons for the adoption and expansion of circular strategies.

Chapter "Strategic Communication of Sustainability in the Agri-Food Sector: Challenges and Digital Opportunities" explores the link between sustainability and digitalisation in supporting Portuguese Geographical Indication (GI) products (PDO/PGI). Using GI View and Facebook content from 15 producer groups, it observes inconsistent digital presence, a tendency to emphasise institutional group names over product names, and few references to sustainability. The chapter highlights a strategic opportunity: aligning digital storytelling with transparency, purpose, and origin values, thereby reinforcing GIs' role in sustainable rural development. Chapter

"Development of a Rehabilitation Action Plan to Promote Urban Sustainability in Existing Neighbourhoods" focuses on the sustainable rehabilitation of existing neighbourhoods, presenting a strategic framework that combines physical modifications to the built environment with social dynamics to foster community involvement. Using a mixed-methods approach (spatial analysis, case studies, policy review), it details key elements: diagnostics, sustainable construction guidelines, and participatory mechanisms to promote local ownership. The framework links regulatory planning with residents' lived experiences, aiming to enhance social cohesion, combat urban desertification, and strengthen the sense of place.

Castelo Branco, Portugal — Luis Farinha
Castelo Branco, Portugal — Daniel Raposo
Covilhã, Portugal — João J. Ferreira
Leiria, Portugal — Marcelo Calvete Gaspar
Leiria, Portugal — Maria Leopoldina Alves
Leiria, Portugal — Luis Serrano
2026

References

Acs, Z. J., Braunerhjelm, P., Audretsch, D. B., & Carlsson, B. (2009). The knowledge spillover theory of entrepreneurship. *Small Business Economics*, *32*(1), 15–30.

Albino, V., Berardi, U., & Dangelico, R. M. (2015). Smart cities: Definitions, dimensions, performance, and initiatives. *Journal of Urban Technology*, *22*(1), 3–21.

Audretsch, D. B., & Feldman, M. P. (1996). R&D spillovers and the geography of innovation and production. *American Economic Review*, *86*(3), 630–640.

Bansal, P., & Roth, K. (2000). Why companies go green: A model of ecological responsiveness. *Academy of Management Journal*, *43*(4), 717–736.

Barney, J. (1991). Firm resources and sustained competitive advantage. *Journal of Management*, *17*(1), 99–120.

Caragliu, A., Del Bo, C., & Nijkamp, P. (2011). Smart Cities in Europe. *Journal of Urban Technology*, *18*(2), 65–82.

Donaldson, T., & Preston, L. E. (1995). The stakeholder theory of the corporation: Concepts, evidence, and implications. *Academy of Management Review*, *20*(1), 65–91.

Farinha, L., Santos, D., Ferreira, J. J., & Ranga, M. (Eds.). (2020). *Regional helix ecosystems and sustainable growth: The interaction of innovation, entrepreneurship and technology transfer*. Springer. https://doi.org/10.1007/978-3-030-47697-7

Geels, F. W. (2002). Technological transitions as evolutionary reconfiguration processes: A multi-level perspective and a case study. *Research Policy*, *31*(8–9), 1257–1274.

Geissdoerfer, M., Savaget, P., Bocken, N. M. P., & Hultink, E. J. (2017). The circular economy—A new sustainability paradigm? *Journal of Cleaner Production*, *143*, 757–768. https://doi.org/10.1016/j.jclepro.2016.12.048

Hart, S. L. (1995). A natural-resource-based view of the firm. *Academy of Management Review*, *20*(4), 986–1014.

Kirchherr, J., Reike, D., & Hekkert, M. (2017). Conceptualizing the circular economy: An analysis of 114 definitions. *Resources, Conservation & Recycling*, *127*, 221–232. https://doi.org/10.1016/j.resconrec.2017.09.005

Peris-Ortiz, M., Ferreira, J. J., Farinha, L., & Fernandes, N. O. (Eds.). (2016). *Multiple helix ecosystems for sustainable competitiveness*. Springer. https://doi.org/10.1007/978-3-319-29677-7

Porter, M. E., & van der Linde, C. (1995). Toward a new conception of the environment-competitiveness relationship. *Journal of Economic Perspectives*, *9*(4), 97–118. https://doi.org/10.1257/jep.9.4.97

Powell, W. W., Koput, K. W., & Smith-Doerr, L. (1996). Interorganizational collaboration and the locus of innovation: Networks of learning in biotechnology. *Administrative Science Quarterly*, *41*(1), 116–145.

Teece, D. J., Pisano, G., & Shuen, A. (1997). Dynamic capabilities and strategic management. *Strategic Management Journal*, *18*(7), 509–533.

Contents

Innovation Networks, Entrepreneurship and Regional Growth

Smart Enough to Matter: Rebranding the Rural with Innovation-Driven Growth

Sara Brito Filipe and Paulo Duarte

Abstract The concept of smart cities is evolving beyond its initial technological focus to encompass comprehensive development models that improve the quality of life, drive economic growth, and strengthen government responsiveness. This chapter examines how innovation, as part of the economic dimension of smart city frameworks, contributes to place branding strategies in three Portuguese inland municipalities. This study analyzes how small and medium-sized municipalities, often marginalized in mainstream smart city discussions, use innovation as a tool for place branding within the broader framework of smart territorial development. The importance of this topic is its potential to change how innovation is understood and applied at the regional level. It provides a novel perspective on how municipalities can leverage creativity, technology, and sustainability not only to enhance service delivery but also to create unique, credible, and forward-looking territorial identities. The findings show how the economic aspect, through entrepreneurship, technological progress, and ecosystem development, enhances both the symbolic and tangible attractiveness of places. The study innovates by linking smart city concepts with place branding in smaller territories. It demonstrates that peripheral municipalities can use innovation as both a development tool and a narrative of differentiation, enhancing their economic image and overall competitiveness.

Keywords Smart cities · Place branding · Innovation

S. B. Filipe (✉)
Polytechnic University of Castelo Branco, Castelo Branco, Portugal
e-mail: sarafilipe@ipcb.pt

P. Duarte
University of Beira Interior, Covilhã, Portugal

S. B. Filipe · P. Duarte
NECE-UBI, Covilhã, Portugal

L. Farinha et al. (eds.), *Sustainable Innovations in Mechanical Engineering and Management Sciences*, Lecture Notes on Multidisciplinary Industrial Engineering,
https://doi.org/10.1007/978-3-032-30372-1_1

The significance of this topic lies in its ability to reshape how innovation is understood and implemented at the regional level. This is particularly critical given the increasing recognition that regional competitiveness in peripheral areas depends not only on implementing technological solutions but also on constructing compelling territorial narratives that attract investment, retain talent, and foster civic engagement [11]. By examining how innovation, conceptualized within the economic dimension of smart city frameworks, serves as a development mechanism and a strategic branding element in three Portuguese inland municipalities, this chapter addresses the critical gap in the study of innovation in rural areas. Specifically, it provides a new perspective on how municipalities can use creativity, technology, and sustainability not only to enhance service delivery but also to build unique, credible, and forward-looking territorial identities. Specifically, this study shows how the economic aspect, through entrepreneurship, technological progress, and ecosystem development, boosts both the symbolic and tangible attractiveness of places. Therefore, the main objective of this chapter is to examine how innovation, as part of the economic dimension of smart city frameworks, contributes to place branding strategies in the selected municipalities. Specifically, this chapter contributes to bridging the gap by examining how small and medium-sized municipalities leverage innovation not only as an operational instrument of smart development but also as a strategic narrative for territorial rebranding and competitive positioning. By adopting a qualitative, data-driven approach rooted in media discourse and thematic content analysis, the study offers original empirical insights into the discursive and institutional frameworks of local innovation strategies. It advances knowledge on smart territorial development by expanding theoretical discussions beyond large urban centers and demonstrating the multidimensional nature of innovation as a key driver of local competitiveness and image building. The analysis is guided by an integrated theoretical framework that combines the six dimensions of smart cities [1, 12] with critical perspectives from place branding literature [13, 14], providing a comprehensive understanding of innovation-led place transformation.

2 Conceptual Foundations

2.1 Smart Cities

The concept of the smart city has attracted particular interest and popularity in recent years, becoming a global trend [6, 7, 15]. Smart cities aim to improve development and the quality of life of their citizens by implementing technological innovations in urban lifestyles [2, 16].

Since 2014, more than 17,000 articles addressing smart cities topics have been written, mainly in the fields of computer science, social sciences, and engineering [17]. Exploring the large body of literature, at least three themes emerge. The first refers to the frameworks, concepts, and methods used to study smart cities as socio-technical systems [1, 18, 19]. The second focuses on smart city performance and the application of specific technologies [20–23]. The third important theme addresses work that challenges the traditional definition of smart cities, grounded itechnological assumptions and economic efficiency objectives [24–26]. These works highlight the significance of democratic politics, accountability, citizen engagement, justice, equality, power, knowledge, sustainability, and the transformation of cities by this values-based approach [17].

Despite the ongoing debate, several authors have agreed that the "Smart City" concept is associated with six main dimensions: economy, environment, governance, mobility, life, and people [12, 27–29]. The smart economy is associated with industries that deal with ICT or use it in their production processes. Smart mobility involves the use of ICT in modern transport technologies to improve urban traffic management. Smart people embody qualities such as a commitment to lifelong learning, an appreciation of social and ethnic diversity, adaptability, creativity, a global outlook, open-mindedness, and engagement in public life. Smart governance means that various stakeholders are involved in decision-making and public services. A smart environment relates to the preservation of the natural environment. Smart living relates to safety and quality of life [1, 3, 28].

Despite the abovementioned dimensions, Nam and Pardo [3] argue that the key components of a smart city are technology, its people (who contribute creativity, diversity, and education), and its institutions (which provide governance and policies), integrating research into public management [30]. Therefore, the smart city label should refer to the ability of smart people to develop smart solutions to urban problems [1]. Accordingly, a city can be considered truly smart when investments in human and social capital, alongside ICT infrastructure, drive sustainable growth and improve the quality of life for its citizens. Vitálišová et al. [4] argue that a successful transition to a smart city requires substantial contributions from politicians, city officials, residents, entrepreneurs, and local community groups [1]. From this viewpoint, the idea appears to be linked to large urban areas. However, changes in the urban environment, economic growth, technological progress, ecological movements, and stakeholders' demands on local authorities have led more cities to aspire to be smart and to meet the expectations of their residents, investors, tourists, and other stakeholders.

Although the challenges faced by small and medium-sized cities and their limited resources can present real opportunities for applying "smart cities" principles within the broader context of intelligent territorial development, the ability to promote economic growth, enhance competitiveness and sustainability, and improve the quality of life for citizens through ICTs has mainly been limited to large cities, lacking a wider territorial perspective.

This dominant metropolitan-centric perspective in smart city literature reveals a conceptual limitation that merits critical examination: the undisputed adoption of smart city models designed for large urban centers risks perpetuating what Vanolo [25] terms "smartmentality", normative discourse that assumes technological solutions are universally applicable regardless of territorial specificities. Furthermore, the emphasis on ICT infrastructure and technological innovation often obscures fundamental questions about power dynamics, resource allocation, and the social construction of what constitutes "smartness" [24]. In smaller municipalities and rural territories, this technology-driven paradigm may be particularly problematic, as it overlooks endogenous assets such as social capital, community cohesion, and place-based knowledge systems that could serve as alternative foundations for intelligent territorial development.

The six-dimensional framework adopted in this study, while analytically useful, should not be interpreted as a prescriptive checklist but rather as a flexible heuristic that requires adaptation to local contexts and critical interrogation of whose interests are served by particular smart city interventions. This raises the crucial question of evaluating how the economic dimension of smart cities truly contributes to inclusive territorial development, or if it primarily serves to attract mobile capital while potentially displacing existing communities and traditional economic activities.

2.2 Place Branding

The growing prominence of the Smart City paradigm has increasingly captured the attention of scholars and practitioners in the field of place branding [31, 32]. Although policies for promoting, marketing, and branding places are not a recent phenomenon, their strategic relevance has intensified in recent decades [9]. Place branding has evolved into a comprehensive process that goes beyond promotional activities, encompassing the strategic management of a city's identity to enhance its competitiveness and socio-economic development [33]. Central to this process is the construction of a distinctive, recognizable, and meaningful image to foster positive associations among external audiences [8].

The concepts of identity and image are fundamental to place branding. According to Boisen et al. [13], in place branding, a place's identity is managed to reinforce its brand image by fostering positive associations. Ultimately, these efforts aim to enhance the place's reputation. A place's identity helps distinguish it from others and select the essential tangible and intangible elements that best suit it. This is achieved through a process of identification and a sense of belonging to the place [34]. Once a place's identity is established, it becomes a promise and a guarantee. Therefore, a place's image is defined by how it is experienced and perceived. A strong image exists when most people share similar associations, and an image is positive when these associations are favorably viewed within a specific context. The ideal is a strong correspondence between a place's identity and image.

It is important to acknowledge that place branding is a highly selective political process. This political understanding of places clearly contradicts the need for unity. While branding is indeed about creating 'positive associations' with a place [31], it is also clear that these associations will differ from person to person, or at least from one social group to another. Nevertheless, there are inherent choices that require consensus, and without which branding cannot be achieved. The criteria used for these choices depend on the responsibilities of the organizations tasked with developing a branding system for a place, and these choices can be legitimized through tough practices such as participatory forms of urban governance [14].

This acknowledgment of place branding as a "highly selective political process" demands more critical scrutiny than the literature typically provides. While scholars increasingly recognize that branding involves contested choices and power relations, the field often remains wedded to managerial and promotional logics that privilege economic competitiveness over social equity and democratic participation [35]. The process of selecting which aspects of territorial identity to emphasize inevitably privileges certain stakeholders, narratives, and futures while marginalizing others. In the context of smaller municipalities, this selectivity becomes particularly consequential. Decisions to brand a territory around innovation and technology may inadvertently devalue traditional knowledge, agricultural practices, or cultural heritage that do not align with contemporary "smart" or "innovative" discourses. Moreover, the emphasis on creating "positive associations" raises fundamental questions about authenticity and the commodification of place identity. When innovation becomes a branding strategy, there is a risk of performative smartness-municipalities adopting the language and imagery of innovation without substantive transformation, or worse, implementing changes that serve external audiences (investors, tourists, skilled workers) rather than existing residents. This tension between external positioning and internal development reveals a critical gap in place branding literature: insufficient attention to how branding processes affect power distributions, resource allocation, and the everyday lives of diverse community members, particularly in peripheral territories where vulnerability to external economic pressures is heightened.

Sabilaturrizqi and Susanto [36] identified ten factors influencing the success of a smart place branding strategy. Among them are creativity and innovation. The following section focuses specifically on the significance of innovation.

2.3 Cities as Innovation Hubs

Studies (e.g., Cooke [37]) have examined innovative systems and how knowledge is exploited and capitalized, with the aim of finding new solutions to help territories improve their competitiveness and achieve cultural, social, and economic benefits. More recently, it has been recognized that, while new knowledge is produced on a global scale, many innovative processes occur at a local level [38]. Consequently, models have been developed that link innovation processes to concepts such as proximity, social integration, learning capacity, and economic performance. These models argue that specialization, proximity, and industrial density generate a concentration of resources (e.g., technology, skills, information, and capital) that can provide community participants with competitive advantages through economies of scale and reduced transaction costs.

These studies emphasize the evolving role of cities in the knowledge-based economy and the potential for urban centers and abandoned areas to undergo transformation [39].

However, the literature on innovation systems and urban transformation warrants critical interrogation on several fronts. First, the assumption that specialization, proximity, and industrial density automatically generate competitive advantages overlooks the potential negative externalities of such concentration, including gentrification, social exclusion, and environmental degradation [40]. Second, the focus on knowledge-based economy transitions often implicitly devalues forms of economic activity and knowledge that do not fit within innovation-centric frameworks. This is particularly problematic in rural and peripheral contexts where traditional sectors remain economically and culturally vital. Third, these models frequently exhibit a "success bias," highlighting cases of transformative innovation while neglecting the far more common instances of failed initiatives, widening regional disparities, and communities left behind by innovation-driven development strategies [41]. For smaller municipalities in Portugal's interior, the imperative to position themselves as "innovation hubs" may create unrealistic expectations and divert resources from more contextually appropriate development paths. The rhetoric of innovation, while politically attractive and potentially mobilizing, may mask structural constraints, such as demographic decline, brain drain, limited fiscal capacity, and peripheral location, that cannot be overcome through innovation strategies alone. This raises a fundamental question about the role of innovation in territorial development: is it a genuine pathway to sustainable competitiveness for peripheral regions, or does it represent a new form of territorial competition that ultimately reinforces existing hierarchies between metropolitan centers and peripheries?

Contemporary perspectives on regional innovation systems have evolved to recognize the heterogeneity of territorial contexts and the need for differentiated innovation strategies. Traditional regional innovation system approaches, originally developed to explain industrial clusters and technology-intensive regions, have been reviewed for their limited applicability to peripheral, rural, and less-favored regions [42]. These territories often lack the critical mass of firms, research institutions, and skilled labor that characterize metropolitan innovation systems, leading scholars to propose alternative conceptualizations, such as "organizational thinness" and "peripheral innovation systems," that acknowledge distinct resource configurations and development pathways.

In rural and peripheral contexts, innovation increasingly manifests through non-technological dimensions, such as organizational, social, and business model innovation, rather than exclusively through R&D-intensive activities [5]. Social innovation has gained particular prominence in rural development discourse as a mechanism for addressing demographic decline, service provision challenges, and economic restructuring [43]. This expanded understanding of innovation recognizes that rural territories may innovate precisely through their distinctive characteristics: strong social networks, proximity to natural resources, cultural heritage, and alternative quality-of-life propositions rather than through direct competition with urban innovation models.

In this context, bioeconomy has emerged as a particularly significant innovation trajectory for rural territories, offering potential pathways to leverage local biomass resources, agricultural systems, and ecological assets for sustainable economic development [44]. Bioeconomy strategies encompass diverse activities ranging from organic agriculture and forestry to biotechnology applications and circular economy initiatives, creating opportunities for rural regions to position themselves within contemporary sustainability transitions. However, critical perspectives on bioeconomy development caution against uncritical adoption of biomass-intensive production models that may prioritize industrial applications over local food systems and community well-being [45]. The intersection of bioeconomy with place-based innovation strategies requires careful attention to issues of value capture, environmental sustainability, and social equity to ensure that rural territories benefit substantively rather than merely serve as resource-extraction sites for urban-centered bioeconomy value chains.

Recent scholarship has also emphasized the importance of innovation intermediaries that facilitate knowledge exchange, support collaboration, and reduce transaction costs within innovation systems [46]. In rural and peripheral contexts, such intermediaries, including technology transfer centers, business incubators, and collaborative platforms, play particularly crucial roles given the spatial dispersion of actors and the limited institutional density. The effectiveness of these intermediaries depends not only on their technical support capabilities but also on their embeddedness in local contexts and their ability to bridge external knowledge sources with territorial specificities. Digital technologies have created new possibilities for overcoming spatial constraints through virtual collaboration and remote access to expertise, which may compensate for or reinforce existing territorial inequalities.

As the notion of territorial innovation systems increasingly recognizes that innovation capacity extends beyond narrowly defined economic dimensions to encompass institutional quality, social capital, environmental sustainability, and cultural vitality [47], it aligns with smart city frameworks that integrate multiple development dimensions. This perspective acknowledges that territorial competitiveness ultimately depends on holistic attractiveness encompassing quality of life, environmental quality, and authentic identity. For place branding purposes, this multidimensional understanding of innovation suggests that territories can differentiate not solely through technology but through distinctive combinations of innovation domains (technological, social, ecological, and cultural) that reflect genuine territorial assets and capabilities rather than generic smart city templates. Understanding that innovation is not merely a technical process, but a socially constructed narrative embedded in power relations, is essential for critically assessing its role in place branding and territorial development.

3 Methodology

The main objective of this chapter is to discuss the economic dimension as an element that enhances the image of a smart territory. The unit of analysis consisted of three municipalities in the interior of Portugal classified as Smart Cities in the Smart Cities 2023 Yearbook: Castelo Branco, Fundão, and Idanha-a-Nova. This yearbook is published annually and is the main dissemination channel for knowledge of existing smart cities initiatives in Portugal.

This study adopts a qualitative research approach, selected for its ability to provide a deep understanding of complex social phenomena by exploring the meanings, interpretations, and contextual factors that influence them [48].

The data collection process involved a careful selection of publicly accessible news articles published in newspapers, magazines, and online platforms. The focus was on content related to the municipalities of Castelo Branco, Fundão, and Idanha-a-Nova. Selection criteria were based on the relevance of the material to the research topic and the temporal representativeness of the data, ensuring a comprehensive and contextually grounded analysis [49]. The choice to use news as a data source for this research was based on its accessibility, timeliness, and representativeness [50]. The news reflect the discourses and narratives present in society, providing a thorough and contextual view of the social phenomena under study [51]. Additionally, the availability of online news enables continuous data collection over time, tracking the dynamics and evolution of the phenomenon under study [52]. Data collection began with a keyword-based search using combinations of the terms "Castelo Branco," "Fundão," and "Idanha-a-Nova," along with "Smart Cities," "innovation," and "technology." This initial exploratory search aimed to find relevant publications that addressed the intersection of innovation and local development within the specified municipalities. A second, more targeted search was then conducted within the newspapers and magazines identified in the preliminary results, to refine the focus on materials that matched both the thematic scope and geographical context of the study. In total, 219 news items were identified and analyzed: 113 related to Fundão, 67 to Castelo Branco, and 39 to Idanha-a-Nova, covering the period from 2013 to 2023.

All news items were categorized according to the six dimensions outlined in the theoretical framework. In this chapter, the analysis focused specifically on the economic dimension. A content analysis was conducted to identify the unique features of this dimension across the three municipalities and to highlight areas of agreement among them. The data were systematically categorized and interpreted in accordance with the adopted theoretical framework, following the analytical model shown in Fig. 1.

Data processing was conducted using NVivo 15, a widely used qualitative data analysis software. NVivo helps develop detailed and precise qualitative research by supporting the organized, coded, and interpreted handling of complex textual data. [53]. A conventional content analysis was used to identify recurring themes within the corpus, supported by word clouds that visually show the most frequently mentioned

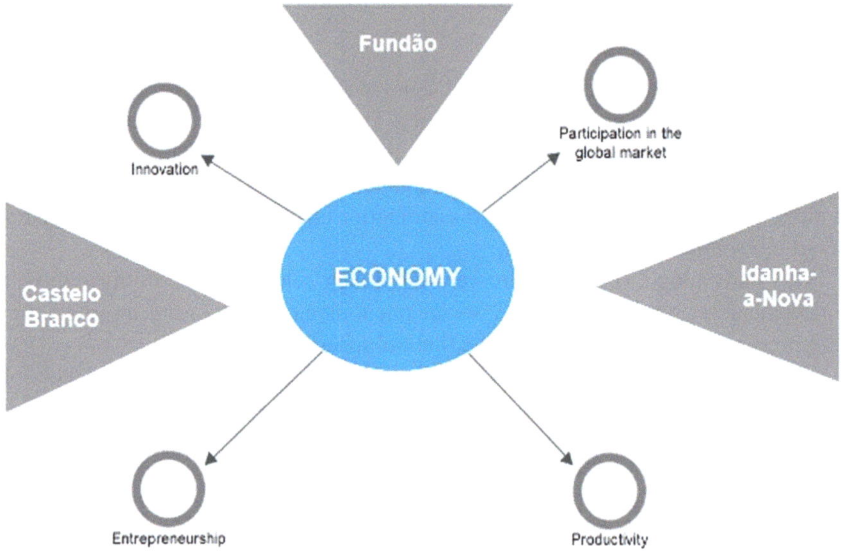

Fig. 1 Conceptual model. *Source* Own elaboration based on Anthopoulos et al. [12], Lai et al. [27], and Nevado Gil et al. [29]

terms across the news items. Theme identification was performed through manual (self-)coding, ensuring it aligned with the research objectives. Themes with fewer than five coded references were combined into other, thematically similar categories to maintain analytical clarity. When themes were merged, their original labels were retained and linked with underscores ("_") to facilitate easy traceability and clarity in the final classification.

4 Results

Within the economic dimension, Castelo Branco, Fundão, and Idanha-a-Nova exhibit distinct approaches to regional development, with a shared emphasis on entrepreneurship and innovation.

The content analysis showed that, in the case of Castelo Branco, the most prominent themes were innovation, entrepreneurship, technology, agri-food, projects, industries, companies, and business (Fig. 2). These keywords reflect the municipality's strategic focus on technological development and sectoral diversification as drivers of local economic growth.

In Castelo Branco, innovation is closely connected to the development of the local business fabric and entrepreneurship, as well as the agri-food sector. Recognized as a European Entrepreneurial Region in 2021, the municipality has actively fostered a diversified and innovation-driven entrepreneurial ecosystem. This ecosystem is

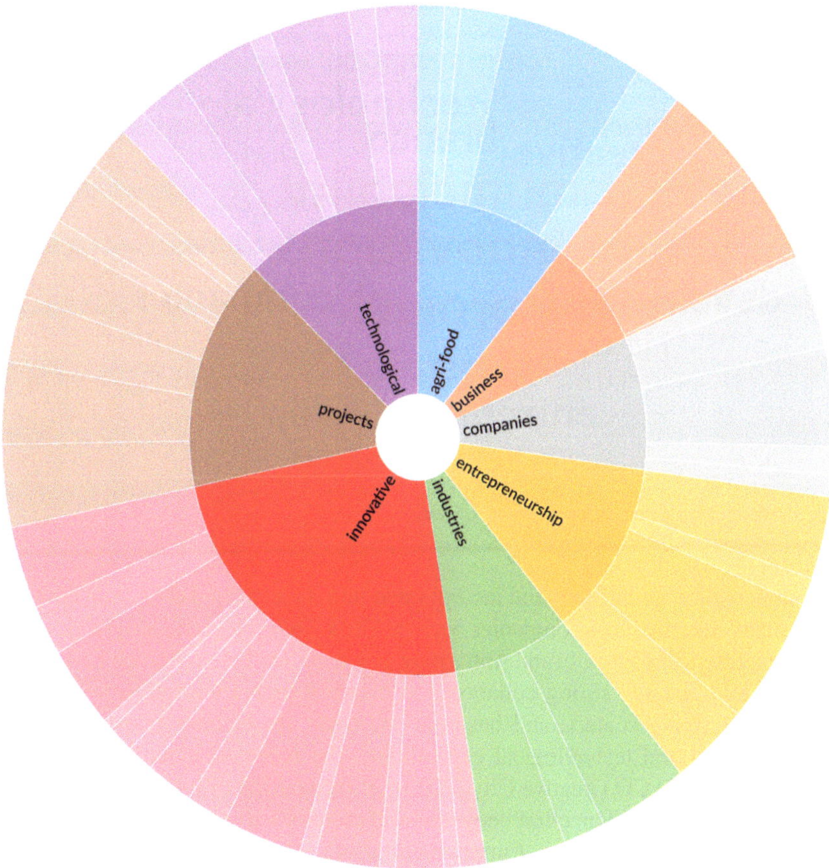

Fig. 2 Key themes identified in the Economy dimension for Castelo Branco

supported by dedicated infrastructure such as the Center for Innovative Companies (CEI), the Industrial Incubator, the Social Incubator (Social IN), and the Creativity Factory. Additionally, the People and Investment Attraction Support Office (GAPI) provides targeted assistance to local businesses. At the same time, institutions like the Agri-Food Technology Support Center (CATAA) and INOVCLUSTER play a central role in advancing the agri-food sector.

Castelo Branco has also increased its visibility through events like the DISRU.PT Summit and participation in European projects, including INOV2B, INOV2AGRO, and FUSILLI. National initiatives, such as the Interior+ program, further strengthen the territory's positioning by promoting agri-food innovation, gastronomic tourism, and business appeal.

Technological advancement, especially through Smart City initiatives, has become a major focus, with support for key sectors such as the automotive and agri-food industries. Several start-ups have created innovative technological solutions that

Fig. 3 Word cloud associated with Castelo Branco

the municipality has adopted and incorporated into its urban and economic development plans. International companies such as Evox Technologies, along with established exporters like Mecalbi and Aptiv, play a vital role in Castelo Branco's economic activity. Additionally, strong collaboration between higher education institutions, vocational training centers, and business and trade associations boosts the city's entrepreneurial and technological capabilities. The recent designation of Castelo Branco as a UNESCO Creative City (2023), in recognition of its embroidery tradition, strengthens the connection between local cultural heritage, creative industries, and sustainable development. Entrepreneurial efforts in the cultural and creative sectors are helping create a more artist-friendly environment while promoting job growth and economic resilience.

The word cloud for Castelo Branco is shown in Fig. 3.

The words *Municipality*, *Innovation*, and *Project* stand out, suggesting that Castelo Branco prioritizes and invests in innovation and the development of technological and economic projects. The presence of words such as *Entrepreneurship*, *Companies*, *Business*, *Technological*, and *Entrepreneurial* points to a supportive environment for entrepreneurial activity, characterized by mechanisms that facilitate the creation and scaling of businesses. Additionally, the repeated use of terms such as Infrastructure and Region suggests the presence of physical and institutional resources that support the municipality's overall development strategy.

The recurrence of terms such as *Promotion* and *Cooperation* indicates that Castelo Branco is committed to promoting its distinctive assets while fostering collaboration with diverse stakeholders. The presence of words like *Network*, *Associated,* and *Ecosystem* underscores the municipality's commitment to building an interconnected development environment, where various actors, public, private, and academic, collaborate to drive innovation and growth. Additionally, the use of

*'National' and 'International' suggests a strategic perspective that extends beyond local and regional boundaries, positioning Castelo Branco within broad*er networks of influence and opportunity.

The presence of the terms *Creative* and *Cultural* underscores the municipality's recognition of creativity and culture as integral pillars of its development strategy. Likewise, the recurrence of *Sustainability* and *Management* suggests a conscious, forward-looking governance approach marked by environmental awareness and long-term planning.

Taken together, these elements portray Castelo Branco as a dynamic municipality, focused on innovation and entrepreneurship, fostering an environment conducive to technological and business development. The emphasis on promotion, cooperation, and sustainability further reflects a commitment to building a diverse and resilient entrepreneurial ecosystem, with creativity and culture playing a central role in its identity and development model.

Data analysis for Fundão revealed eleven prominent thematic clusters that enlighten the *municipality's strategic positioning and development path: innovation_hub, areas_sector, technological, project_network, production_value, inland_cities, development, companies_jobs, business_center, solutions, and systems* (See Fig. 4).

Deliberate institutional measures have shaped Fundão's technological positioning since 2012, when the municipality began a systematic transformation to become a recognized technology hub. This strategic shift is evident through efforts to attract high-tech companies and promote innovation in traditional economic sectors, especially agriculture. The establishment of the Business and Shared Services Center (CNSP) represents a pivotal institutional innovation that has successfully attracted multinational corporations, including IBM and Capgemini, thereby anchoring Fundão's credentials as a viable location for advanced business services and technological operations. The municipality's unique approach to sectoral innovation is exemplified by its focus on agrotechnology, where collaborative project development has become a key feature. Initiatives like the Agrotech Challenge show Fundão's ability to create innovation ecosystems that connect traditional agriculture with advanced technological solutions. This approach reflects a sophisticated understanding of sectoral transformation dynamics, where technological advancement serves not to replace traditional economic activities but to enhance their productivity and competitiveness.

The Experimental Farms are especially notable for their role as integrated agricultural testing and demonstration sites. These facilities serve two main functions: they allow farmers to observe and try out new agricultural technologies in real-world settings, and they also help develop and test solutions for agricultural use. This approach of real-world testing and validation is an innovative way to transfer and adopt new technologies, creating clear links between research and practice that are vital for spreading sustainable innovation in traditional farming sectors.

The municipality's innovation strategy is further reinforced through systematic partnerships with the regional scientific and technological infrastructure, creating

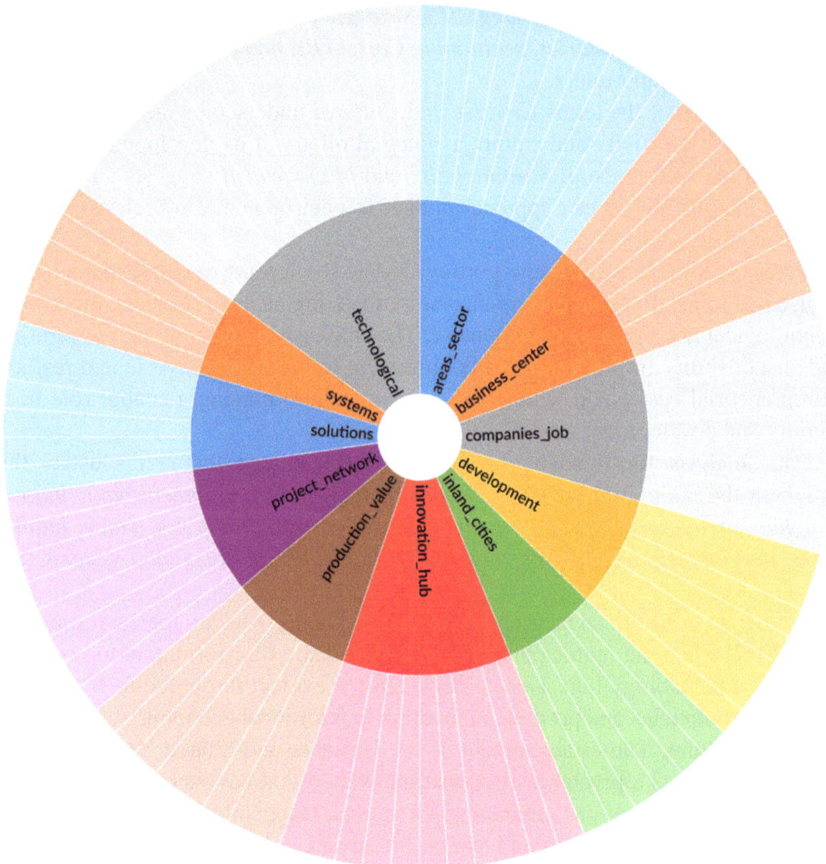

Fig. 4 Main themes identified in the Economy dimension for Fundão

synergistic relationships that amplify agricultural innovation capacity. These collaborative arrangements leverage existing research capabilities and knowledge networks to enhance the municipality's technological development potential, demonstrating a sophisticated understanding of how territorial innovation systems function through interconnected institutional relationships.

The Castelo Branco case illustrates a critical tension within smart city literature regarding the scalability and adaptability of innovation frameworks beyond metropolitan contexts. While Albino et al. [1] argue that smart cities require investments in both human capital and ICT infrastructure, Castelo Branco's experience suggests that smaller municipalities may prioritize institutional infrastructure, such as incubators, support offices, and collaborative networks, over technological infrastructure. This finding challenges the technology-deterministic assumptions prevalent in early smart city conceptualizations [24] and supports more recent arguments that smartness emerges from coordinated ecosystems rather than technological deployment alone [3]. The municipality's strategic emphasis on diversification across creative, cultural, and technological sectors reflects what Camagni and Capello [47] term "territorial innovation models," where innovation capacity derives from distinctive combinations of assets rather than specialization in high-tech sectors. However, this diversification strategy raises questions about resource dispersion and achieving critical mass in any particular domain. From a place branding perspective, Castelo Branco's approach exemplifies Kavaratzis and Kalandides' [14] notion of "participatory place branding" through its emphasis on ecosystem development and stakeholder cooperation. The integration of traditional craft heritage (embroidery) with the discourse of innovation seeks to reconcile authenticity with modernity [8]. Whether this integration appears credible to external audiences or represents merely a symbolic juxtaposition requires further investigation.

Fundão's strategic ambitions extend beyond regional boundaries through the implementation of pilot projects and participation in European-funded initiatives. These transnational engagements provide access to advanced technological knowledge and financial resources while simultaneously positioning the municipality within broader networks of innovation and development. The systematic pursuit of European project opportunities reflects Fundão's recognition that establishing itself as a technological hub requires integration with international innovation ecosystems and adherence to global standards of technological excellence. The implementation of pilot projects and European initiatives helps establish the region as a technological hub for the global market. The visibility of technological projects at fairs and events boosts awareness, and Fundão is now recognized for its agricultural products, such as cherries, as well as for its work and value creation in technological fields. Fundão's current regional identity showcases a sophisticated blend of traditional agricultural excellence with modern technological innovation.

The municipality's strategic positioning has received significant external validation through prestigious awards, including the European Business Promotion Award and the designation as "Most Innovative and Intelligent Municipality." This recognition boosts the municipality's credibility within national and international networks and may also help access additional resources and partnership opportunities.

The technological infrastructure supporting Fundão's innovation ecosystem shows significant depth and sophistication. The extensive deployment of LoRa network coverage is a key investment in Internet of Things (IoT) connectivity, enabling advanced data collection and monitoring in both agricultural and urban areas. Complementing this digital infrastructure, the creation of the Agrotech 4.0 Living Lab is a notable institutional innovation that provides dedicated spaces and resources for developing and testing agricultural technology. This Living Lab model reflects modern approaches to innovation support, in which physical and institutional infrastructures come together to promote collaborative research and development.

Fundão's participation in Digital Innovation Hub consortia further reinforces its commitment to digital transformation and sustainable agriculture. This engagement with broader innovation networks ensures access to cutting-edge knowledge, best practices, and collaborative opportunities that might otherwise be unavailable to individual municipalities. The consortium model also reflects recognition that contemporary innovation challenges require coordinated responses that transcend individual organizational or territorial boundaries.

The economic impact of Fundão's technological strategy is best seen in the transformation of the "Cereja do Fundão" cherry brand, which has increased the product's market value. This traditional agricultural product has been systematically improved through technological innovations that boost both production efficiency and product quality. The enhancement of cherry quality and production illustrates how innovation can strengthen traditional economic sectors without altering their core identity, creating value-added products that achieve premium market positions while preserving cultural and territorial authenticity. Integrating technology with traditional production reflects a sophisticated approach to economic development that leverages historical assets while adapting to modern competitive challenges.

Fundão's commitment to preserving and enhancing traditional agricultural excellence extends beyond cherry production to encompass artisanal cheese manufacturing, a sector that has achieved remarkable international recognition. The municipality's support for local cheese production is a strategic investment in high-quality, region-specific products that leverage territorial specificities and traditional knowledge systems. This approach to agricultural development prioritizes quality over quantity, positioning Fundão's products within premium market segments that value authenticity and craftsmanship.

The international recognition of this strategy was demonstrated when sheep's milk cheese produced by Queijaria Quinta do Pomar in Soalheira received the highest honors at the World Cheese Awards 2024, standing out as the top product among 4786 entries from 47 countries. This achievement signifies more than just individual business success; it serves as regional validation that local production systems, when properly supported and developed, can meet global standards. The international competition victory is strong proof of Fundão's ability to develop world-class productive activities in traditional sectors.

The municipality's strategy for social and economic growth shows greater sophistication by proactively incorporating international migration flows. Fundão has strategically established itself as a welcoming destination for refugees and emigrants, representing 74 different nationalities, turning potential demographic challenges into opportunities for internationalization and economic growth. This multicultural integration serves multiple strategic purposes: it addresses local labor market needs while simultaneously introducing diverse skills, perspectives, and networks that can enhance municipal competitiveness.

The systematic approach to accommodating international residents reflects a progressive strategy for territorial development that values cultural diversity and global connectivity. By actively supporting the integration of different populations, Fundão creates conditions to advance innovation, expand market access, and build

Fig. 5 Word cloud associated with Fundão

social capital. This demographic approach complements the municipality's technological and agricultural initiatives by providing access to diverse human resources and fostering an international perspective, which is crucial to modern competitiveness.

Figure 5 displays the word cloud for Fundão.

The prominence of the words *Technology*, *Innovation*, and *Fundão* in the news corpus reflects a deliberate discursive connection between territorial identity and technological progress. This language pattern suggests that technological innovation has become deeply integrated into the municipality's strategic narrative, going beyond simple policy measures to become a core part of its territorial brand and identity. The consistent association of the municipality's name with innovation-related terms indicates an intentional effort to create mental links between the place and its technological capabilities within relevant stakeholder groups.

The centrality of *Development* within the thematic landscape reflects Fundão's comprehensive approach to territorial transformation, encompassing both economic modernization and social advancement. This omnipresent developmental discourse suggests a holistic understanding of municipal progress that extends beyond sectoral improvements to embrace integrated territorial enhancement. The pervasive use of development-oriented language shows that growth and progress are common themes through which municipal initiatives and goals are understood and communicated.

The convergence of *Agricultural*, *Production*, and *Cherry* within the municipal discourse demonstrates Fundão's strategic commitment to preserving and enhancing traditional agricultural competencies while simultaneously integrating them with contemporary technological paradigms. This word convergence indicates a sophisticated understanding of sectoral transformation, in which traditional agricultural excellence serves as the basis for technology-enhanced value creation rather

1 Introduction

Over the past twenty years, cities have become crucial centers for social and economic change, innovation, and global competition. Traditionally shaped by industrial growth and population concentration, urban areas now face a complex set of issues, including digital change, demographic shifts, environmental sustainability, and regional inequalities. In this changing landscape, the idea of the smart city has gained importance as a strategic approach that combines technological innovation with governance, sustainability, and citizen participation [1, 2]. Smart cities are increasingly seen not just as technology projects, but as comprehensive development models aimed at improving quality of life, economic growth, and government responsiveness [3, 4].

Despite growing academic interest in both smart city development and place branding, a significant scientific gap persists in understanding how these concepts intersect in non-metropolitan contexts, particularly regarding innovation as a strategic territorial asset. Recent scholarship has begun to acknowledge that innovation strategies in rural and peripheral territories require fundamentally different approaches than those applied in metropolitan areas [5]. However, the mechanisms through which innovation functions both as an economic development tool and a branding asset in smaller municipalities remain underexplored. This study investigates how small and medium-sized municipalities, often neglected in mainstream smart city discussions, use innovation as a place-branding tool within the larger framework of smart territorial development. Focusing on three inland municipalities in Portugal (Castelo Branco, Fundão, and Idanha-a-Nova), the chapter examines how innovation helps shape their economic image and competitive edge. Located in a peripheral region, these municipalities offer a valuable case for understanding the dynamics of territorial innovation beyond big metropolitan centers. The study thus bridges several fields, including urban studies, innovation systems, public policy, and regional development.

Previous research has primarily concentrated on smart city development in large urban areas, often highlighting infrastructure, mobility, and ICT applications [6, 7]. Less attention has been paid to how smaller cities and rural areas implement smart strategies, especially those related to branding and innovation, to redefine their roles within national and international networks. While the concept of place branding has shifted from marketing-focused practices to more strategic, participatory, and identity-driven approaches [8, 9], its connection to smart cities remains largely underexplored, particularly outside metropolitan areas. Although studies have examined innovation ecosystems in rural contexts and the adaptation of smart city principles to smaller urban areas [10], few have integrated these perspectives with place branding theory to understand how innovation shapes territorial identity and competitive positioning. This gap is especially significant given the rising policy focus on balanced regional development in Europe and the increasing competition among regions to attract talent, investment, and recognition.

trends. The municipality exhibits proactive governance that integrates ecosystem thinking with organized project execution, regional connectivity with local asset development, and economic attraction strategies with sustainability goals. It actively creates projects and initiatives with a regional and European outlook and values collaboration within its ecosystem, aiming for quality and sustainability in all its actions. This multidimensional strategic profile indicates a municipality that has gained a thorough understanding of current territorial development needs while maintaining a clear identity rooted in unique local assets and skills.

Fundão's development trajectory provides empirical support for recent arguments that peripheral regions can establish innovation systems through organizational economies [42]. The municipality's creation of innovation intermediaries (Business Center, Agrotech Living Lab, and Digital Innovation Hub) addresses the "organizational thinness" that characterizes many rural regions by deliberately constructing institutional density where it does not emerge spontaneously. This strategic institution-building challenges the assumption in classical regional innovation systems literature [37] that innovation capacity primarily develops through organic clustering processes. Instead, Fundão demonstrates that purposeful policy intervention can establish innovation infrastructures, aligning with more recent scholarship on "constructed advantage" in regional development [54]. The municipality's focus on agricultural technology innovation exemplifies what Naldi et al. [5] identify as sector-specific innovation in rural contexts: enhancing traditional economic activities through technological application rather than replacing them with knowledge-intensive industries. This approach resonates with smart specialization principles [55] leveraging existing territorial assets (cherry production, agricultural knowledge) while adding new dimensions (IoT, precision agriculture) that enhance competitiveness without abandoning established comparative advantages. However, Fundão's strategy also reveals potential contradictions. The attraction of multinational technology companies (IBM, Capgemini) through the CNSP represents a classic "branch plant" economic development model that the regional development literature has long critiqued for generating limited local embeddedness, knowledge spillovers, and upgrading opportunities [56]. The coexistence of this exogenous investment attraction with endogenous agricultural innovation raises questions about strategic coherence and whether these parallel tracks genuinely reinforce each other or represent disconnected development pathways serving different audiences. From a place branding perspective, Fundão's dual identity as both "technology hub" and "cherry capital" exemplifies the challenge of creating coherent territorial narratives when pursuing multiple positioning strategies simultaneously. While this multi-positioning may reflect genuine territorial complexity, place branding literature suggests that excessive breadth can undermine brand clarity and memorability [9]. The international recognition of cheese production alongside technological innovation credentials indicates successful quality differentiation across sectors, yet whether these elements cohere into a unified territorial identity or remain parallel narratives targeting distinct audiences remains unclear.

Data analysis for Idanha-a-Nova revealed six distinctive themes driving the municipality's strategic orientation and territorial development approach: innovative_business_collaborative laboratory, green economy_bio-region_organic farming, food_chains, development_wealth_tourism_resources, projects_supporting, policies_program_positioning (See Fig. 6).

Idanha-a-Nova's territorial development strategy has materialized through sophisticated institutional mechanisms designed to integrate well-being enhancement, entrepreneurship assistance, and innovation promotion within a coherent programmatic framework. The Recomeçar program exemplifies this integrated approach, encompassing multiple specialized components that address different dimensions of territorial development. The I-Danha Social Innovation Incubator shows the municipality's dedication to promoting social entrepreneurship and community-driven innovation. Meanwhile, the I-Danha Food Lab is a groundbreaking initiative as the first green accelerator on the Iberian Peninsula, positioning Idanha-a-Nova as a regional

Fig. 6 Main themes identified in the Economy dimension for Idanha-a-Nova

leader in sustainable food system innovation. The Green Valley Food Lab Business Center supports incubation facilities by providing dedicated infrastructure for business growth in the green economy sector.

The municipality's designation as Portugal's first bioregion establishes a foundational strategic position that aligns territorial identity with modern sustainability ideas. This bio-regional framework serves as a guiding principle for attracting companies committed to sustainable practices while providing structured support for organic farming pilot projects. The bioregion concept goes beyond just environmental protection to include broad territorial development strategies that blend ecological care with economic growth and social well-being. Strategic initiatives like CoLAB Food4Sustainability reflect Idanha-a-Nova's commitment to collaborative research and development networks that connect academic knowledge with real-world application. The IdanhaEmprego platform illustrates the municipality's systematic approach to facilitating employment, integrating job-creation objectives with sustainability commitments through coordinated digital infrastructure and targeted support mechanisms.

The municipality's achievements have earned international recognition, most notably as the Best Organic District in Europe at the EU Organic Awards 2023. This prestigious award recognizes Idanha-a-Nova's consistent dedication to organic production excellence and its effective implementation of comprehensive organic development strategies. The award specifically recognizes the municipality's large organic production zone and its consistent support for initiatives that promote short supply chains and organic product development.

As Portugal's pioneering organic municipality, Idanha-a-Nova has developed comprehensive green economy initiatives that involve multiple sectors and stakeholder groups. The implementation of organic school canteens shows a consistent commitment to sustainable consumption patterns while supporting local organic producers. Additionally, the Educational Gardens project aims to spread knowledge and provide practical demonstrations, creating tangible links between educational institutions and sustainable agriculture practices. The organic regional market offers essential infrastructure to connect local producers with consumers and reinforces the municipality's commitment to developing short supply chains.

The annual I-Danha Food Lab event has become a leading gathering in the agrifood sector, serving as a platform for innovation, sustainability, and the growth of the green economy. It has gained recognition as the "Green Web Summit" for its influential role in bringing together key stakeholders and promoting knowledge exchange. Complementary initiatives such as the Rural Innovation Forum and the Raiana Fair offer additional opportunities to showcase innovative business ideas within the broader EUROACE region, highlighting Idanha-a-Nova's dedication to regional collaboration and knowledge sharing.

Idanha-a-Nova's development strategy goes beyond just economic and environmental factors to include cultural and creative industries as ways to stand out and generate value. The municipality's recognition as a UNESCO Creative City of Music shows its ongoing commitment to cultural assets and the growth of its creative economy. This cultural identity has been strategically leveraged to attract major international productions, such as HBO's "House of the Dragon," and the famous Boom Festival, proving the municipality's ability to serve as a destination for important cultural and entertainment events.

While tourism development remains a strategic priority, the municipality has adopted modest expectations for this sector, acknowledging both its potential and limitations within the broader framework of territorial development. The effective attraction of international productions and cultural events demonstrates a refined understanding of how cultural assets can drive economic growth while preserving authenticity and territorial identity.

In line with European policies and global production and consumption trends, Idanha-a-Nova's membership in the International Network of Bio-Regions positions the municipality within global networks of sustainability-focused territories, facilitating knowledge exchange and best-practices dissemination while reinforcing its commitment to environmental and quality enhancement.

The municipality's strategic use of modern technologies supports current work patterns, including remote work options. By enabling remote work, Idanha-a-Nova promotes environmental sustainability through reduced transportation needs and attracts knowledge workers and creative professionals. The agreement with the WINDS platform demonstrates a systematic approach to market access, allowing local producers to connect with global markets while keeping their local roots and sustainable practices.

The word cloud developed for Idanha-a-Nova is presented in Fig. 7.

Fig. 7 Word cloud associated with Idanha-a-Nova

The prominence of *Sustainable*, *Innovative*, and *Idanha* within the municipal news corpus reflects a deliberate strategy that fuses territorial identity with contemporary sustainability and innovation paradigms. This systematic association between the municipality's name and transformative development concepts indicates sophisticated territorial branding that positions ecological practices and technological advancement as fundamental components of local identity. The convergence of *organic*, *green*, and *food4sustainability* within the dominant lexical framework demonstrates a comprehensive commitment to ecological practices that extend beyond mere environmental protection to encompass integrated approaches to sustainable development and innovation, particularly within food and agricultural systems.

The words *Production*, *Farming*, and *Organic* reinforce Idanha-a-Nova's strategic positioning within sustainable agricultural sectors while highlighting the municipality's systematic preference for organic production methods. This convergence suggests territorial development strategies that prioritize qualitative enhancement of traditional agricultural activities rather than quantitative expansion or sectoral diversification. The persistent association of production terminology with organic concepts indicates comprehensive integration of sustainability principles within core economic activities.

The prominence of words such as *Rural*, *Territory*, *Region*, and *Development* within the municipal discourse highlights awareness, importance, and commitment to place-based development approaches that recognize the distinctive characteristics and potential of rural territories. This word's emphasis on territorial development suggests governance philosophies that prioritize endogenous development strategies while acknowledging the importance of regional integration and coordination. The systematic reference to rural contexts indicates recognition that contemporary territorial competitiveness requires specific approaches tailored to rural characteristics rather than generic development models.

The focus on *Entrepreneurship*, *Business*, *Companies*, and *Incubators* within the thematic landscape reveals Idanha-a-Nova's systematic commitment to fostering environments that encourage business creation and entrepreneurial development. This word association suggests institutional recognition that sustainable rural development requires dynamic entrepreneurial ecosystems supported by appropriate infrastructure, facilitation infrastructures, and instruments. The emphasis on incubation processes exposes the understanding that contemporary entrepreneurship requires systematic support and structured development pathways rather than spontaneous market emergence.

The prominence of *International*, *Global*, and *World* expresses Idanha-a-Nova's strategic orientation toward trans-local connectivity and global market engagement, suggesting a municipality that has transcended traditional rural insularity to embrace international networking and global opportunity identification. The systematic reference to global contexts reinforces that contemporary territorial competitiveness requires integration with international networks and alignment with global trends while maintaining distinctive local characteristics.

The presence of *Relevant* and *Award* words within the discourse reflects external validation of territorial development strategies and achievements. This emphasis on recognition and relevance suggests a municipality that actively seeks external validation of its development approaches, using it to enhance territorial credibility and attractiveness. The focus on awards and recognition indicates understanding that contemporary territorial marketing requires verified excellence and third-party validation.

The appearance of *Culture* and *Social* within the thematic framework demonstrates Idanha-a-Nova's acknowledgment that sustainable territorial development requires integration of cultural heritage preservation with social well-being enhancement. The systematic reference to cultural and social dimensions indicates comprehensive territorial development strategies that address multiple aspects of community life rather than focusing exclusively on economic indicators, prioritizing community cohesion and cultural continuity alongside economic advancement.

The terms highlighted in the word frequency analysis depict Idanha-a-Nova as a municipality that has developed a comprehensive model of sustainable rural innovation characterized by organic agricultural excellence, entrepreneurial ecosystem development, and strategic global connectivity. The systematic integration of sustainability principles with innovation capabilities, along with the valorization of cultural heritage and the development of a knowledge economy, suggests a territorial development approach that successfully balances the preservation of tradition with modern competitiveness requirements. This multidimensional profile indicates a municipality that has gone beyond conventional rural development limits to establish itself as a unique model of integrated sustainable territorial progress, combining local roots with global relevance and traditional agricultural skills with modern innovation capabilities.

Broadly, the evaluation suggests that Idanha-a-Nova is a municipality committed to sustainable, innovative rural development and to organic production and food safety. It has an international outlook and actively promotes entrepreneurship and tourism, valuing its culture and knowledge as distinctive assets to attract visitors while simultaneously developing knowledge-based economic activities.

Idanha-a-Nova's development model provides compelling evidence for alternative innovation pathways in rural territories that challenge mainstream smart-city and innovation-system frameworks. The municipality's bioregion designation and organic agriculture focus align with emerging scholarship on bioeconomy as a distinctive innovation trajectory for rural regions [53]. This strategic positioning challenges the technological determinism prevalent in smart city literature [25] by prioritizing environmental quality, social innovation, and community well-being alongside technological advancement. The municipality's emphasis on social innovation through the I-Danha incubator exemplifies what Neumeier [43] identifies as innovation for rural contexts, addressing community needs and enhancing collective capacity rather than pursuing innovation as an abstract goal. However, the approach also reveals inherent tensions in reconciling sustainability-centered development with competitiveness imperatives. The simultaneous pursuit of organic agriculture authenticity and international market integration through platforms like WINDS raises questions about whether genuine commitment to short supply chains and local food systems can coexist with global market ambitions within the municipality's distinctive positioning [45].

From an innovation systems perspective, Idanha-a-Nova challenges the assumption that innovation capacity requires industrial density, the presence of research institutions, or high-tech specialization. The municipality's innovation system operates primarily through network relationships and institutional entrepreneurship rather than traditional innovation infrastructure. This configuration aligns with recent arguments that peripheral regions can develop effective innovation systems through organizational modes distinct from those of metropolitan areas [42]. Though questions remain about long-term sustainability and whether such network-dependent systems can generate sufficient endogenous capacity or remain perpetually reliant on external connections and validation.

The comparative analysis across the three municipalities reveals that while all embrace innovation as a strategic priority, they operationalize innovation in fundamentally different ways that reflect distinctive territorial assets, institutional capacities, and strategic choices. This diversity challenges universalistic smart city frameworks that prescribe standardized dimensions and metrics [1, 12] and supports arguments for context-sensitive, place-based approaches to territorial development [57].

5 Conclusions

This study examined how small and medium-sized municipalities leverage innovation to shape their economic identity and strengthen their positioning as smart territories. Grounded in a qualitative analysis of media narratives from Castelo Branco, Fundão, and Idanha-a-Nova, a comparative analysis reveals differentiated yet complementary approaches to innovation-driven territorial development within Portugal's inland.

The three municipalities share a strong commitment to innovation, entrepreneurship, and territorial differentiation. However, their approaches differ. Castelo Branco uses a diverse model, blending technological infrastructure, creative industries, and international networks to support a multi-sector entrepreneurial ecosystem. Fundão emphasizes integrating advanced technology into traditional sectors, especially agriculture, with a project-based, high-tech strategy to strengthen its identity as a rural innovation center. In contrast, Idanha-a-Nova focuses on sustainability, highlighting organic agriculture, social innovation, and cultural branding, positioning itself as a leader in ecological and inclusive development.

This comparative perspective enhances theoretical understanding by demonstrating that smart city principles, especially the economic aspect, are not limited to large urban areas but can be adapted to various territorial scales and identities. The study strengthens the link between smart cities and place branding, showing that innovation serves as both a catalyst for economic growth and a narrative tool for building a competitive, authentic, and resilient place image. Indeed, innovation serves not only as a development tool but also as a symbolic asset in creating distinctive, competitive territorial identities. Overall, from a theoretical perspective, the findings support the view that smart city principles can be effectively tailored to non-metropolitan contexts through locally rooted, strategically coordinated, and coherent collaborative innovation models.

From a practical perspective, the findings provide actionable insights for local governments and policymakers. They emphasize the importance of context-sensitive innovation strategies, cross-sector collaboration, and narrative coherence in developing smart territories. The territorial branding of innovation, whether technological, cultural, or ecological, proves vital not only for attracting investment but also for fostering civic pride, retaining talent, and promoting sustainable development.

Looking ahead, these cases suggest that the future of smart territorial management will involve adopting diverse models of innovation that are not limited by urban size but are characterized by strategic clarity, ecosystem collaboration, and built-in identity. In a rapidly changing socio-technical environment, small and medium-sized cities can serve as testing laboratories for agile, sustainable, and inclusive growth, provided they align their innovation approaches with broader territorial strategic narratives and stakeholder involvement.

This study's originality lies in its focused analysis of how innovation functions as both a symbolic and practical driver in the economic identity of non-metropolitan smart territories. By combining place branding theory with the smart city framework and applying it empirically to understudied rural municipalities, this chapter helps address a gap in both theory and practice in the existing literature.

Building on the comparative analysis of Castelo Branco, Fundão, and Idanha-a-Nova, several governance lessons emerge for sustaining innovation in low-density regions. Local governments should shift from project-based experimentation to the institutionalization of innovation support. Specifically, municipalities and regional agencies should: (1) establish intermediary structures such as innovation brokers, living labs, and shared service centres that act as connectors between firms, academia, and civic actors; (2) formalize intermunicipal consortia with clear governance rules, shared funding mechanisms, and measurable outcomes to pool scarce resources and build critical mass; and (3) secure multi-year, place-based funding combined with technical capacity-building programs that ensure continuity of human resources and knowledge transfer [58, 59].

Addressing institutional barriers is equally critical. Limited financial autonomy, shortage of skilled personnel, and weak coordination across municipalities remain major obstacles to replicating successful models. These barriers can be mitigated through fiscal incentives for intermunicipal cooperation, joint hiring of innovation managers, and secondment programs with universities to enhance local absorptive capacity [60, 61]. Finally, embedding innovation within bioeconomy and social innovation frameworks ensures that economic modernization also reinforces social cohesion and environmental sustainability. In doing so, municipalities can translate "smartness" from a technological paradigm into a durable governance capability and a distinct territorial narrative that strengthens regional competitiveness and identity.

Limitations and future research While the analysis provides deep insight into three municipalities, it relies mainly on media discourse, which may better reflect institutional communication priorities than public opinion. Future research could include stakeholder interviews, citizen perceptions, or policy impact evaluations to deepen understanding. Broadening the geographic scope to cover other Portuguese and European regions could also offer comparative insights and enhance the typology of smart territorial trajectories.

Acknowledgement NECE, Research Centre for Business Sciences, Research Centre and this work are funded by FCT—Fundação para a Ciência e a Tecnologia, IP, project UID/04630/2025 and DOI identifier 10.54499/UID/04630/2025\.

Annex 1: Main Topics Identified in the Economy Dimension for Castelo Branco

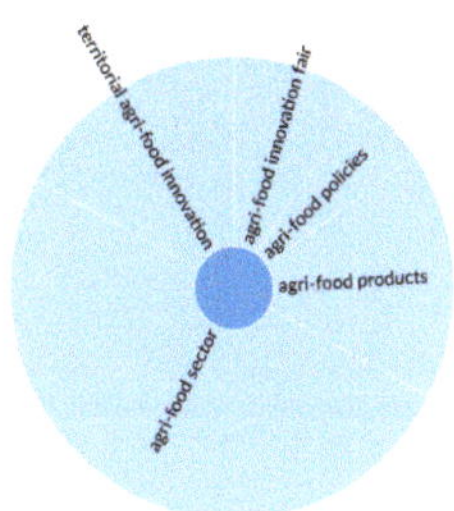

Topics identified in the agri-food theme

Topics identified in the business theme

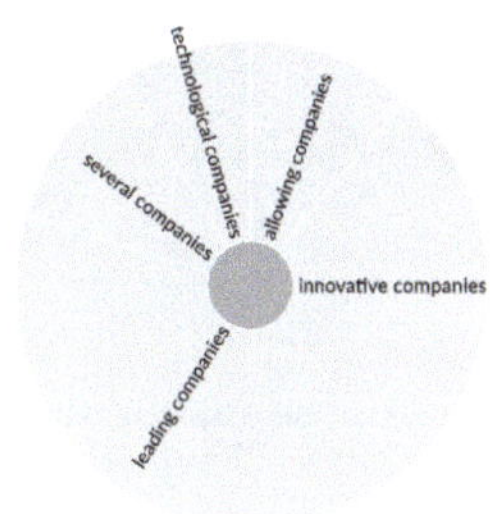

Topics identified in the companies theme

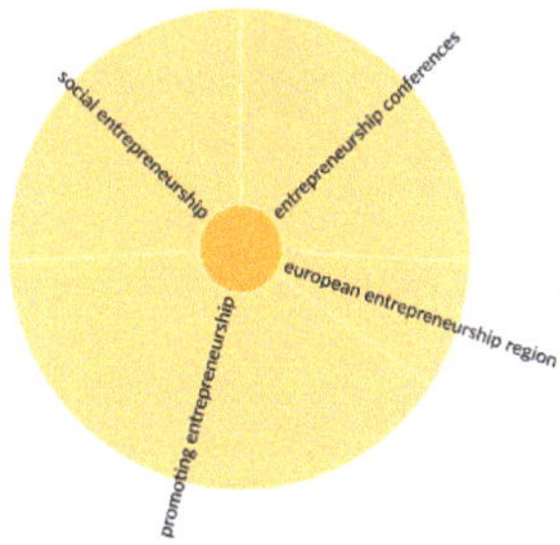

Topics identified in the entrepreneurship theme

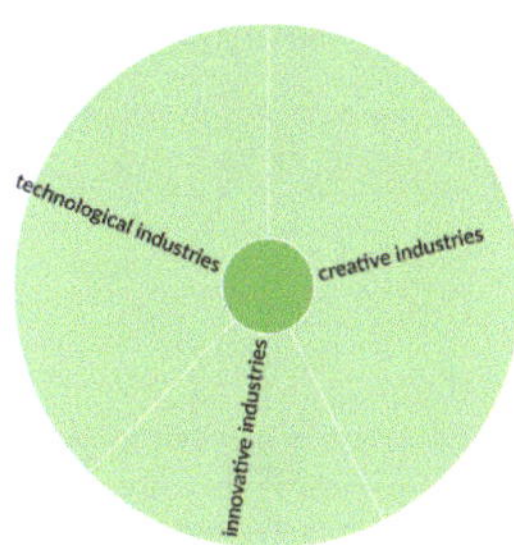

Topics identified in the industries theme

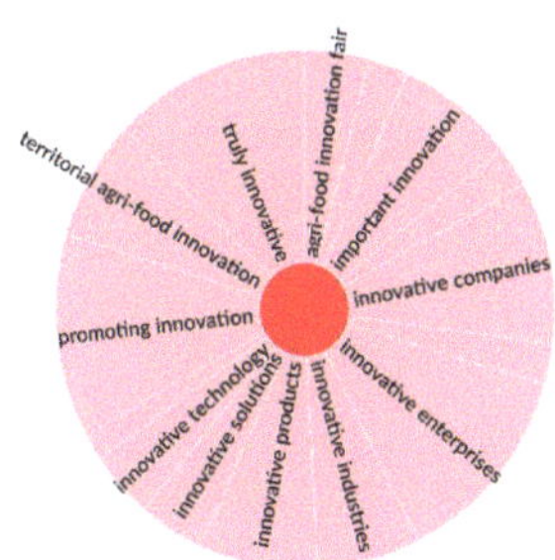

Topics identified in the innovative theme

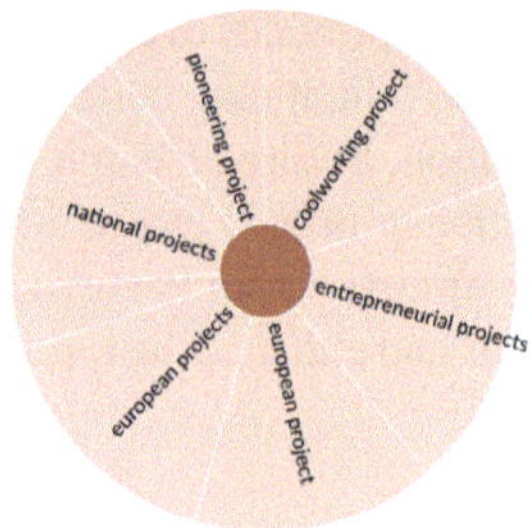

Topics identified in the projects theme

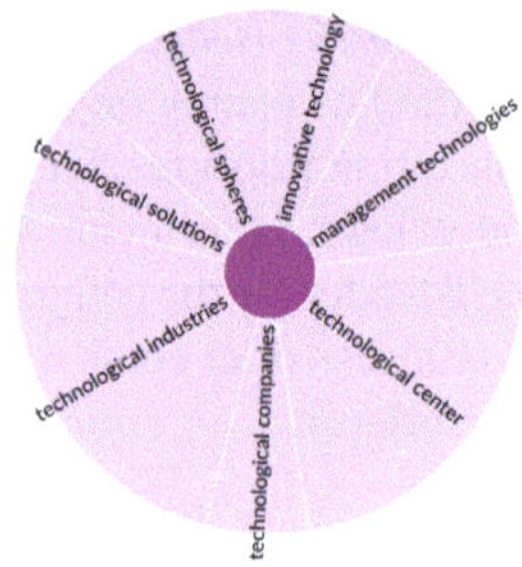

Topics identified in the technological theme

Annex 2: Main Topics Identified in the Economy Dimension Themes for Fundão

Topics identified in the sector_areas theme

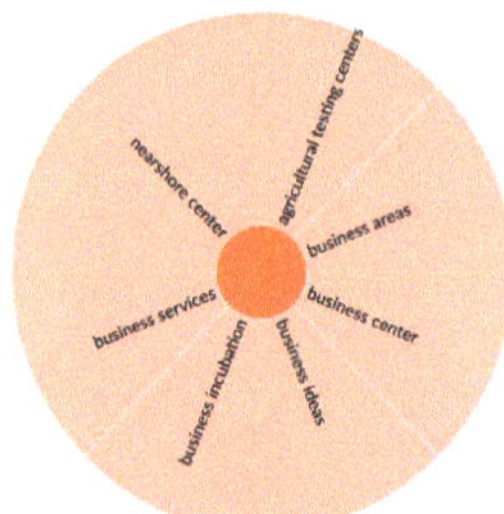

Topics identified in the business_center theme

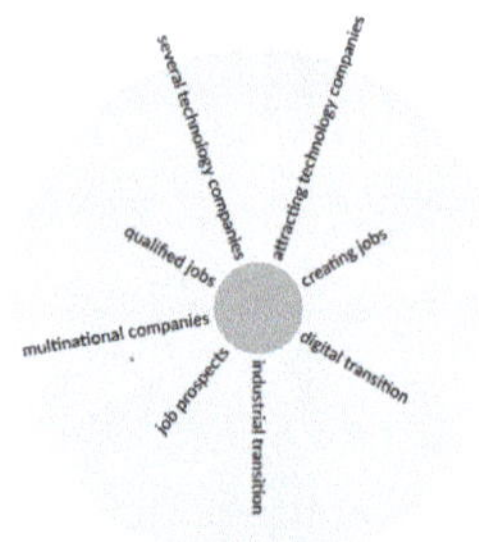

Topics identified in the company_jobs theme

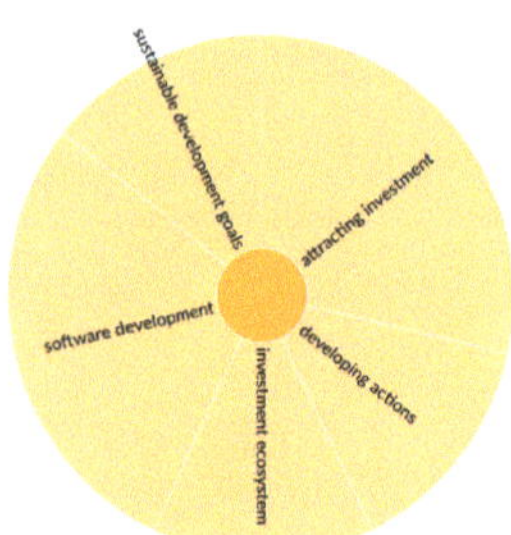

Topics identified in the development theme

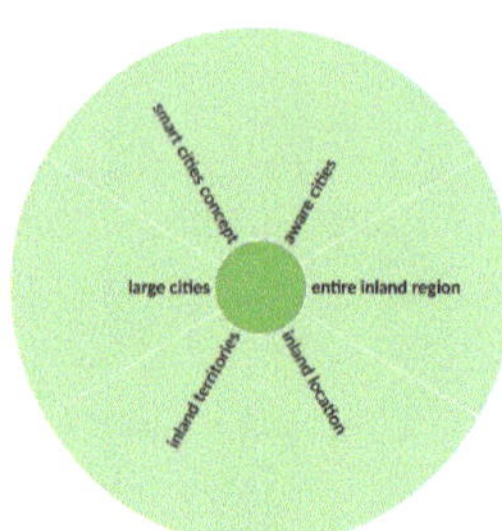

Topics identified in the inland_cities theme

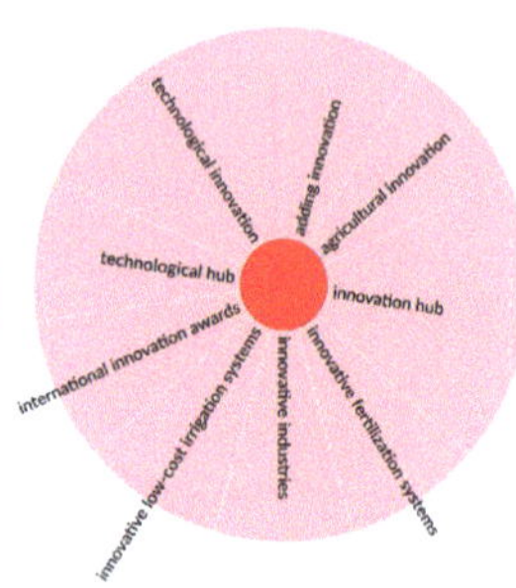

Topics identified in the innovation_hub theme

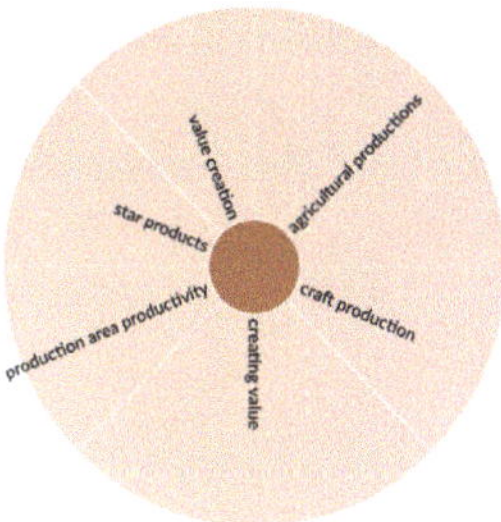

Topics identified in the production_value theme

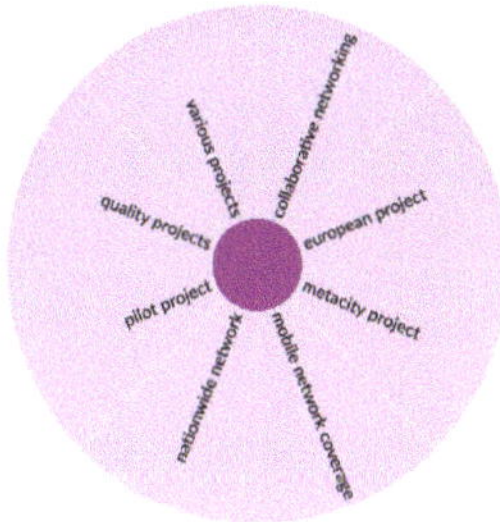

Topics identified in the project-network theme

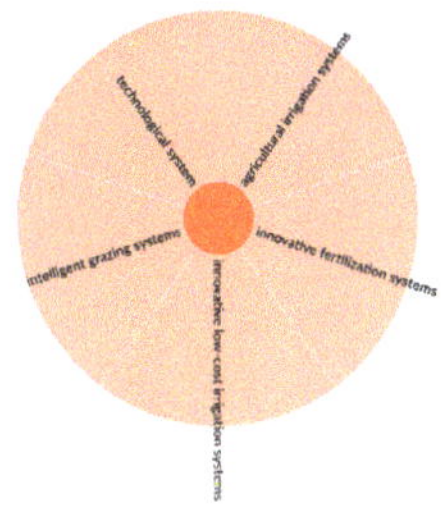

Topics identified in the systems theme

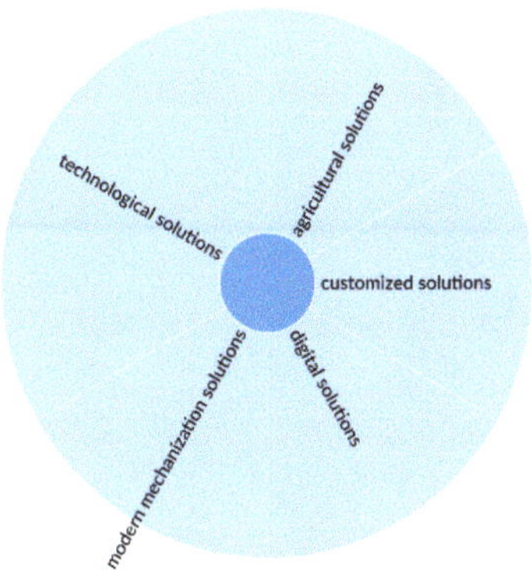

Topics identified in the solutions theme

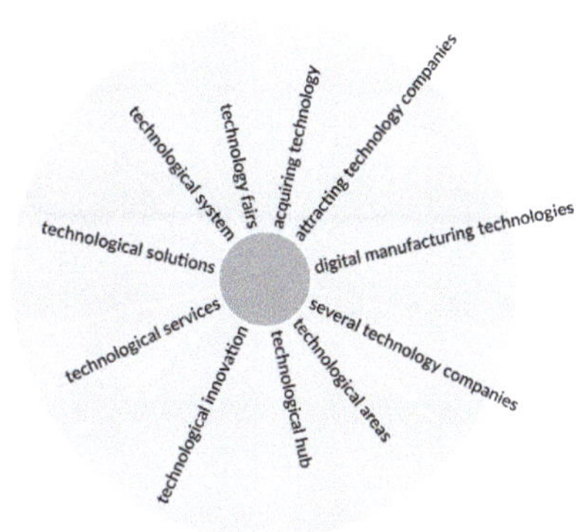

Topics identified in the technological theme

Annex 3: Main Topics Identified in the Economy Dimension Themes for Idanha-a-Nova

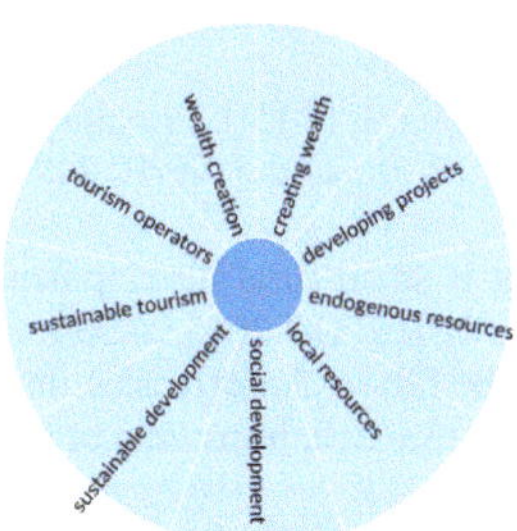

Topics identified in the development_wealth_tourism_resourc

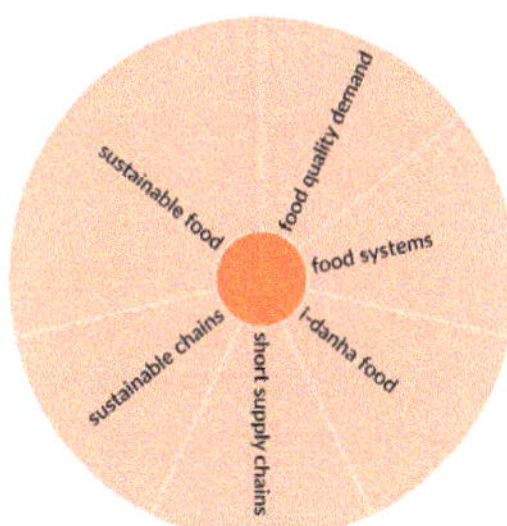

Topics identified in the food_chains theme

Topics identified in the green economy_bio-region_organic farming

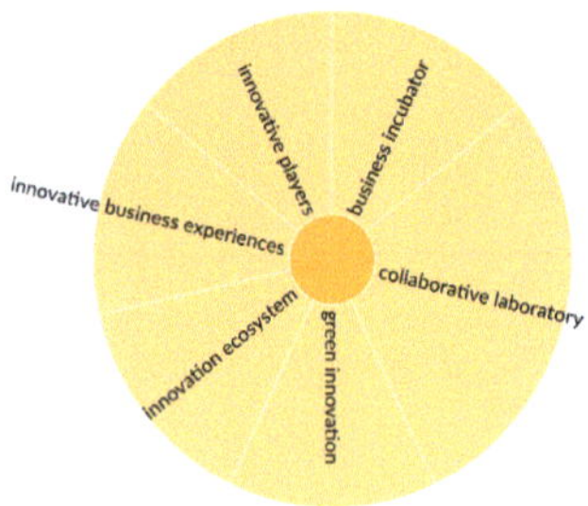

Topics identified in the innovative_business_collaborative laboratory

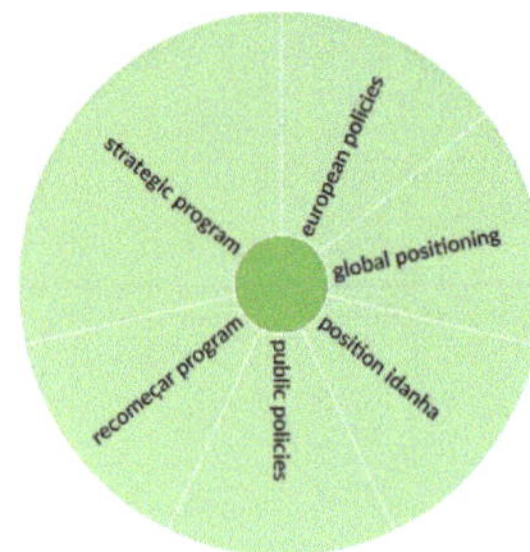

Topics identified in the policies_program_positioning theme

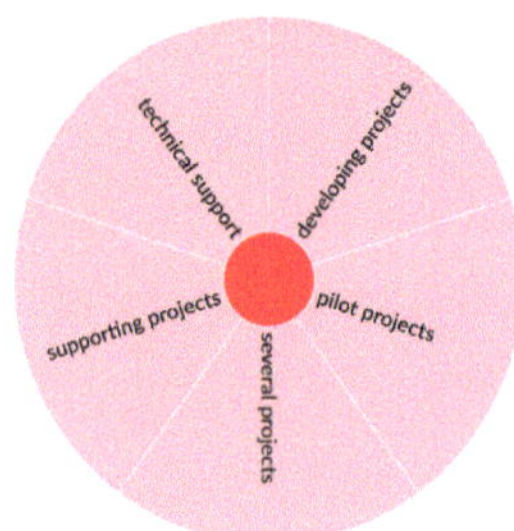

Topics identified in the projects_supporting theme

References

1. Albino, V., Berardi, U., & Dangelico, R. M. (2015). Smart cities: Definitions, dimensions, performance, and initiatives. *Journal of Urban Technology, 22*(1), 3–21. https://doi.org/10.1080/10630732.2014.942092
2. Castelnovo, W., Misuraca, G., & Savoldelli, A. (2015). Smart cities governance: The need for a holistic approach to assessing urban participatory policy making. *Social Science Computer Review, 34*(6), 724–739. https://doi.org/10.1177/0894439315611103
3. Nam, T., & Pardo, T. A. (2011). Conceptualizing smart city with dimensions of technology, people, and institutions. In *ACM international conference proceeding series* (Vol. 282–291). https://doi.org/10.1145/2037556.2037602
4. Vitálišová, K., Vaňová, A., Vavrúšová, M., Turečková, K., & Nevima, J. (2024). Principles of smart governance in cities. *Administratie Si Management Public, 42*. https://doi.org/10.24818/amp/2024.42-11
5. Naldi, L., Nilsson, P., Westlund, H., & Wixe, S. (2015). What is smart rural development? *Journal of Rural Studies, 40*, 90–101. https://doi.org/10.1016/J.JRURSTUD.2015.06.006
6. Bibri, S. E., & Krogstie, J. (2017). Smart sustainable cities of the future: An extensive interdisciplinary literature review. *Sustainable Cities and Society, 31*, 183–212. https://doi.org/10.1016/J.SCS.2017.02.016
7. Gracias, J. S., Parnell, G. S., Specking, E., Pohl, E. A., & Buchanan, R. (2023). Smart cities—A structured literature review. In *Smart cities* (Vol. 6, Issue 4, 1719–1743). Multidisciplinary Digital Publishing Institute (MDPI). doi:https://doi.org/10.3390/smartcities6040080.
8. Govers, R. (2013). Why place branding is not about logos and slogans. *Place Branding and Public Diplomacy, 9*, 71–75. https://doi.org/10.1057/pb.2013.11
9. Kavaratzis, M., & Ashworth, G. (2010). Place branding: Where do we stand? In *Towards effective place Brand Management: Branding European cities and regions*

(pp. 1–14) https://www.scopus.com/inward/record.uri?eid=2-s2.0-80052359965&partnerID=40&md5=80b04d7f470f9a76aac2b8f10d746b41
10. Bravaglieri, S., Åberg, H. E., Bertuca, A., & de Luca, C. (2025). Multi-actor rural innovation ecosystems: Definition, dynamics, and spatial relations. *Journal of Rural Studies, 114*, 103492. https://doi.org/10.1016/j.jrurstud.2024.103492
11. Kavaratzis, M. (2024). Place branding 'in colours bold.' *Place Branding and Public Diplomacy, 20*(1), 25–28. https://doi.org/10.1057/s41254-024-00375-0
12. Anthopoulos, L., Janssen, M., & Weerakkody, V. (2016). A unified Smart City model (USCM) for smart city conceptualization and benchmarking. *International Journal of Electronic Government Research, 12*(2), 77–93. https://doi.org/10.4018/IJEGR.2016040105
13. Boisen, M., Terlouw, K., & Van Gorp, B. (2011). The selective nature of place branding and the layering of spatial identities. *Journal of Place Management and Development, 4*(2), 135–147. https://doi.org/10.1108/17538331111153151
14. Kavaratzis, M., & Kalandides, A. (2015). Rethinking the place brand: The interactive formation of place brands and the role of participatory place branding. *Environment and Planning A, 47*(6), 1368–1382. https://doi.org/10.1177/0308518X15594918
15. Aldegheishem, A. (2019). Success factors of smart cities a systematic review of literature from 2000-2018. *TEMA-Journal of Land Use Mobility and Environment, 12*(1), 53–64. https://doi.org/10.6092/1970-9870/5893
16. Cortés-Cediel, M. E., Cantador, I., Bolívar, M. P., & Rodríguez. (2019). Analyzing citizen participation and engagement in European smart cities. *Social Science Computer Review, 39*(4), 592–626. https://doi.org/10.1177/0894439319877478
17. Araral, E. (2020). Why do cities adopt smart technologies? Contingency theory and evidence from the United States. *Cities, 106*, 102873. https://doi.org/10.1016/J.CITIES.2020.102873
18. Neirotti, P., De Marco, A., Cagliano, A. C., Mangano, G., & Scorrano, F. (2014). Current trends in Smart City initiatives: Some stylised facts. *Cities, 38*, 25–36. https://doi.org/10.1016/J.CITIES.2013.12.010
19. Yigitcanlar, T., Kamruzzaman, M., Buys, L., Ioppolo, G., Sabatini-Marques, J., da Costa, E. M., & Yun, J. H. J. (2018). Understanding 'smart cities': Intertwining development drivers with desired outcomes in a multidimensional framework. *Cities, 81*, 145–160. https://doi.org/10.1016/J.CITIES.2018.04.003
20. Botta, A., De Donato, W., Persico, V., & Pescapé, A. (2016). Integration of cloud computing and internet of things: A survey. *Future Generation Computer Systems, 56*, 684–700. https://doi.org/10.1016/J.FUTURE.2015.09.021
21. Haarstad, H., & Wathne, M. W. (2019). Are smart city projects catalyzing urban energy sustainability? *Energy Policy, 129*, 918–925. https://doi.org/10.1016/J.ENPOL.2019.03.001
22. Yigitcanlar, T., Kamruzzaman, M., Foth, M., Sabatini-Marques, J., da Costa, E., & Ioppolo, G. (2019). Can cities become smart without being sustainable? A systematic review of the literature. *Sustainable Cities and Society, 45*, 348–365. https://doi.org/10.1016/J.SCS.2018.11.033
23. Zanella, A., Bui, N., Castellani, A., Vangelista, L., & Zorzi, M. (2014). Internet of things for smart cities. *IEEE Internet of Things Journal, 1*(1), 22–32. https://doi.org/10.1109/JIOT.2014.2306328
24. Hollands, R. G. (2015). Critical interventions into the corporate smart city. *Cambridge Journal of Regions, Economy and Society, 8*(1), 61–77. https://doi.org/10.1093/cjres/rsu011
25. Vanolo, A. (2013). Smartmentality: The Smart City as disciplinary strategy. *Urban Studies, 51*(5), 883–898. https://doi.org/10.1177/0042098013494427
26. Wiig, A., & and Wyly, E. (2016). Introduction: Thinking through the politics of the smart city. *Urban Geography, 37*(4), 485–493. doi:https://doi.org/10.1080/02723638.2016.1178479.
27. Lai, C. S., Jia, Y., Dong, Z., Wang, D., Tao, Y., Lai, Q. H., Wong, R. T. K., Zobaa, A. F., Wu, R., & Lai, L. L. (2020). A review of technical standards for smart cities. *Clean Technologies, 2*(3), 290–310. https://doi.org/10.3390/cleantechnol2030019
28. Lombardi, P., Giordano, S., Farouh, H., & Yousef, W. (2012). Modelling the smart city performance. In *Innovation: The European journal of social science research* (Vol. 25, Issue 2, pp. 137–149) doi:https://doi.org/10.1080/13511610.2012.660325.

29. Nevado Gil, M. T., Carvalho, L., & Paiva, I. (2020). Determining factors in becoming a sustainable Smart City: An empirical study in Europe. *Economics & Sociology, 13*(1), 24–39. https://doi.org/10.14254/2071-789X.2020/13-1/2
30. Zhu, H., Shen, L., & Ren, Y. (2022). How can Smart City shape a happier life? The mechanism for developing a happiness driven Smart City. *Sustainable Cities and Society, 80*, 103791. https://doi.org/10.1016/J.SCS.2022.103791
31. Filipe, S. B., Duarte, P., & Henriques, D. (2025). From smart cities to smart territories: Cooperation as a place branding strategy|De smart cities a território inteligente: A cooperação Como estratégia de place branding. *Janus Net, 15*(2), 48–72. https://doi.org/10.26619/1647-7251.DT0125.3
32. Garcia Sousa, R., Galina, V. R., & S., & Lourenção, M. (2025). Unpacking the role of place branding and smart cities in attracting foreign direct investment: Theoretical perspectives and future research paths. *Journal of Place Management and Development, 18*(2), 279–299. https://doi.org/10.1108/JPMD-06-2024-0066
33. Kavaratzis, M., & Ashworth, G. (2008). Place marketing: How did we get here and where are we going? *Journal of Place Management and Development, 1*(2), 150–165. https://doi.org/10.1108/17538330810889989
34. Kalandides, A. (2011). The problem with spatial identity: revisiting the "sense of place." *Journal of Place Management and Development, 4*(1), 28–39. https://doi.org/10.1108/17538331111117142
35. Eshuis, J., & Edwards, A. (2013). Branding the City: The democratic legitimacy of a new mode of governance. *Urban Studies, 50*(5), 1066–1082. https://doi.org/10.1177/0042098012459581
36. Sabilaturrizqi, M., & Susanto, T. D. (2024). The Main success factors of smart branding: A systematic literature review. *International Journal of Computer and Information System (IJCIS) Peer Reviewed-International Journal, 05*. https://ijcis.net/index.php/ijcis/index
37. Cooke, P. (2002). Regional innovation systems: General findings and some new evidence from biotechnology clusters.
38. Marceau, J. (2008). Introduction: Introduction: Innovation in the city and innovative cities. *The Innovation, 10*(2–3), 136–145. https://doi.org/10.5172/impp.453.10.2-3.136
39. Engel, J. S., Berbegal-Mirabent, J., & Piqué, J. M. (2018). The renaissance of the city as a cluster of innovation. *Cogent Business and Management, 5*(1), 1–20. https://doi.org/10.1080/23311975.2018.1532777
40. Carvalho, L. C. (2018). Entrepreneurial ecosystems. In *E-planning and collaboration* (pp. 1120–1138). IGI Global. https://doi.org/10.4018/978-1-5225-5646-6.ch053
41. Rodríguez-Pose, A. (2018). The revenge of the places that don't matter (and what to do about it). *Cambridge Journal of Regions, Economy and Society, 11*(1), 189–209. https://doi.org/10.1093/cjres/rsx024
42. Isaksen, A., & Trippl, M. (2016). Path development in different regional innovation systems: A conceptual analysis. In *Innovation drivers and regional innovation strategies*. https://doi.org/10.4324/9781315671475
43. Neumeier, S. (2012). Why do social innovations in rural development matter and should they be considered more seriously in rural development research?—Proposal for a stronger focus on social innovations in rural development research. *Sociologia Ruralis, 52*(1), 48–69. https://doi.org/10.1111/j.1467-9523.2011.00553.x
44. Schmidt, O., Padel, S., & Levidow, L. (2012). The bio-economy concept and Knowledge Base in a public goods and farmer perspective. *Bio-Based and Applied Economics, 1*(1), 47–63. https://doi.org/10.13128/BAE-10770
45. Birch, K., Levidow, L., & Papaioannou, T. (2010). Sustainable capital? The Neoliberalization of nature and knowledge in the European knowledge-based bio-economy. *Sustainability, 2*(9), 2898–2918. https://doi.org/10.3390/su2092898
46. Howells, J. (2006). Intermediation and the role of intermediaries in innovation. *Research Policy, 35*(5), 715–728. https://doi.org/10.1016/J.RESPOL.2006.03.005
47. Camagni, R., & Capello, R. (2013). Regional innovation patterns and the EU regional policy reform: Toward smart innovation policies. *Growth and Change, 44*(2), 355–389. https://doi.org/10.1111/grow.12012

48. Creswell, J. W., & Poth, C. N. (2018). *Qualitative inquiry and research design: Choosing among five approaches* (4th edition). SAGE Publications.
49. Bryman, A. (2016). *Social research methods* (5th edition). Oxford University Press.
50. Flick, U. (2018). *The SAGE handbook of qualitative data collection.* SAGE Publications.
51. Altheide, D. L., & Schneider, C. J. (2013). *Qualitative media analysis* (2nd edition). SAGE Publications.
52. Hsieh, H.-F., & Shannon, S. E. (2005). Three approaches to qualitative content analysis. *Qualitative Health Research, 15*(9), 1277–1288. https://doi.org/10.1177/1049732305276687
53. Silver, C., & Lewins, A. (2014). *Using software in qualitative research: A step-by-step guide.* SAGE Publications Ltd. https://doi.org/10.4135/9781473906907
54. Asheim, B. T., Boschma, R., & Cooke, P. (2011). Constructing regional advantage: Platform policies based on related variety and differentiated knowledge bases. *Regional Studies, 45*(7), 893–904. https://doi.org/10.1080/00343404.2010.543126
55. Foray, D., David, P. A., & Hall, B. H. (2009). Smart specialization. The concept. In *Knowledge economists policy brief* (issue June). European Commission.
56. Isaksen, A. (2015). Industrial development in thin regions: Trapped in path extension? *Journal of Economic Geography, 15*(3), 585–600. https://doi.org/10.1093/jeg/lbu026
57. Barca, F., McCann, P., & Rodríguez Pose, A. (2012). The case for regional development intervention: Place-based versus place-neutral approaches. *Journal of Regional Science, 52*(1), 134–152. https://doi.org/10.1111/j.1467-9787.2011.00756.x
58. Ferreira, H., Marques, C. S., & Farinha, L. (2025). Regional smart specialisation strategies: A systematic literature review. *Journal of the Knowledge Economy.* https://doi.org/10.1007/s13132-025-02736-3
59. Rietmann, C. (2024). Hidden champions and their integration in rural regional innovation systems: Insights from Germany. *68*(1), 63–78. doi:https://doi.org/10.1515/zfw-2021-0024.
60. Huggins, R., & Thompson, P. (2025). Behavioural theory and regional development: Nurturing cultures of possibility. *Spatial Economic Analysis, 20*(3), 368–395. https://doi.org/10.1080/17421772.2025.2474769
61. OECD. (2024). *Assessing the framework conditions for social innovation in rural areas.* www.oecd.org.

Driving Agricultural Innovation: The Role of Collaborative Networks in the INTERIOR+ Project under the Triple Helix Framework

Teresa Paiva, Daniel Raposo, Luís Pinto de Andrade, Fernanda Delgado, Fátima Verissímo, Paula Coutinho, Ana Rodrigues, and Nuno Brito

Abstract Fostering and developing agricultural innovation is critical to addressing global challenges such as food security, climate change, and sustainable development. This study adopted a descriptive case study design combining document analysis and semi-structured interviews with representatives from academia, industry, and government involved in the INTERIOR+ project. It explores the potential of

T. Paiva (✉) · P. Coutinho
BRIDGES—Biotechnology Research, Innovation, and Design of Health Products, Polytechnic of Guarda, Guarda, Portugal
e-mail: tpaiva@ipg.pt

T. Paiva
Research Centre in Business Sciences, University of Beira Interior, Covilhã, Portugal

D. Raposo · F. Verissímo
ReThink—Centro de Investigação em Design para o Território, Polytechnic Institute of Castelo Branco, Castelo Branco, Portugal

CIAUD, Lisbon School of Architecture, Lisbon, Portugal

L. P. de Andrade · F. Delgado
Polytechnic Institute of Castelo Branco, Castelo Branco, Portugal

CERNAS-Research Centre for Natural Resources, Environment and Society, Coimbra, Portugal

F. Delgado
Lab. de Fitoquímica do CBPBI do Centro de Biotecnologia de Plantas da Beira Interior, Castelo Branco, Portugal

A. Rodrigues
Centro de Apoio Tecnológico Agroalimentar de Castelo Branco, Zona Industrial de Castelo Branco, Castelo Branco, Portugal

N. Brito
CISAS, Polytechnic Institute of Viana do Castelo, Viana do Castelo, Portugal

1H–TOXRUN—One Health Toxicology Research Unit, University Institute of Health Sciences, Paredes, Portugal

L. Farinha et al. (eds.), *Sustainable Innovations in Mechanical Engineering and Management Sciences*, Lecture Notes on Multidisciplinary Industrial Engineering,
https://doi.org/10.1007/978-3-032-30372-1_2

cross-sectoral collaboration within a triple helix context to accelerate the adoption of innovative practices and technologies in the agriculture sector. The analysis identified key collaborative mechanisms, knowledge sharing, digital connectivity, and stakeholder intermediation, that enhance innovation capacity in low-density agricultural territories. Findings show that integrating Collaborative Innovation Networks (COINs) within a Triple Helix framework fosters resilience, accelerates technology adoption, and supports sustainable rural development. Moreover, the study provides practical insights on how to establish and sustain these networks, highlighting the critical role of stakeholders in leveraging the potential of COINs and cross-sectoral collaboration to propel agricultural innovation and development, and, therefore, in designing an intersectoral collaboration strategy.

Keywords Collaborative innovation network · Triple helix · Agriculture innovation systems · Rural development · Low-density territories

1 Introduction

The concept of Collaborative Innovation Networks (COINs) has emerged as a significant area of study within the broader field of innovation ecosystems. In particular, COINs have been identified as a promising approach for fostering knowledge-sharing and joint problem-solving among diverse stakeholders [1, 2]. The fundamental premise of COINs is to convene participants from disparate fields and sectors with the objective of propelling innovation in a manner that is more efficacious and inclusive than the isolated pursuit of such endeavours [3–5].

The open innovation models that underpin COINs present distinctive prospects for small and medium-sized enterprises (SMEs) to engage in collaborative endeavours and capitalise on external expertise [6]. Such collaboration fosters a more dynamic and resilient ecosystem, which is of particular importance in industries such as agriculture, where innovation can directly impact productivity and sustainability [7–10]. The existing literature on COINs emphasises the significance of inter-organisational knowledge sharing and collaborative R&D as pivotal drivers of innovation, disseminating success stories, and the benefits of digital tools that facilitate collaboration represent a key strategy for fostering innovation within COINs [11–13]. Such mechanisms facilitate accelerated innovation cycles and enhanced accessibility to state-of-the-art solution.

The role of collaborative spaces within COINs demonstrates how organisations can unite to address shared concerns [13]. Such spaces are of critical importance for the creation of networks in which innovation can flourish, enabling the pooling of resources and expertise that are often fragmented across sectors. Furthermore, the literature emphasises the inclusivity of these networks, which are perceived as a means of encouraging the participation of smaller, frequently underrepresented actors in the innovation process [14–16].

The COINs have been the subject of extensive investigation within sectors characterised by a constant imperative for adaptation and innovation. One such sector is agriculture, where the collaboration between farmers, researchers, agribusinesses, and policymakers is paramount in addressing challenges such as climate change and food security [7, 8, 10, 17]. In a rapidly evolving world facing challenges such as climate change, food security and sustainability, agricultural innovation is no longer an option but a necessity [9]. COINs, which facilitate partnerships between different stakeholders [4, 5, 18], such as farmers, researchers, agribusinesses and policymakers, provide the collaborative platforms needed to tackle complex problems that no single entity can solve alone. They have become essential for driving innovation in various sectors, and their importance in agriculture is particularly profound [17] since they are able to address urgent global challenges and significantly enhance the performance of agricultural value chains. The integration of different stakeholders ensures that innovation is inclusive and benefits all members of the chain, including smallholder farmers and agribusinesses. The formation of COINs permits cross-sectoral collaboration, which is essential for responding to challenges such as climate change and resource depletion [17]. By fostering knowledge sharing, resource pooling and collaborative problem solving, these networks can accelerate the development and adoption of new technologies and practices, ultimately leading to more productive and resilient agricultural systems [19]. Even the changing landscape of consumer behaviour as Generation Z and Alpha begin to influence markets, particularly in the agri-food sector [20]. These generations place a strong emphasis on sustainability, ethical sourcing, and technological integration. COINs are increasingly valuable in this context, as they allow agrifood businesses to collaborate with researchers and policymakers to develop sustainable and tech-driven innovations that meet the demands of younger consumers.

However, the effectiveness of COINs depends heavily on the structure and dynamics of the collaboration itself [17, 18]. This is where the triple helix model - an approach that emphasises the interplay between academia, industry and government - proves indispensable.

The triple helix framework brings together three critical sectors [21] academia, which generates research and new knowledge; industry, which applies this knowledge to create practical, scalable solutions; and government, which provides the regulatory and policy support needed to implement these innovations. In the context of agriculture, this collaboration is particularly valuable in bridging the gap between theoretical research and practical application [22]. For example, academia could develop new sustainable farming practices or resilient crop varieties, but without industry partners to commercialise these innovations and government support to create a supportive policy environment, these breakthroughs could remain in the laboratory. Without this multi-sectoral collaboration, breakthroughs could remain confined to the laboratory rather than being effectively scaled and implemented [8].

One of the key contributions of the Triple Helix model to COINs is its ability to facilitate knowledge sharing across sectors [19]. This promotion of cross-sector knowledge exchange is particularly valuable in agriculture since it is a field that requires continuous adaptation to changing environmental conditions, market demands, and technological advancements, particularly in low-density territories (LDTs), where access to cutting-edge knowledge and technology may be limited, the Triple Helix model becomes a key mechanism for bridging the gap between local needs and global expertise [8, 23–27]. The Triple Helix framework encourages open communication and collaboration between all relevant actors, allowing them to share insights, data and strategies [21]. For example, universities researching climate-smart agriculture can share their findings with agribusinesses developing precision farming tools while governments provide the data and regulatory support needed to implement these technologies on a wider scale. Through its emphasis on collaboration across sectors, it helps build networks of connectivity that can break down these barriers. In this model, government agencies often play a crucial role by facilitating infrastructure development, offering subsidies for innovation, and creating policies that encourage public-private partnerships [24, 28]. For example, governments can prioritise the development of digital infrastructure, which is essential for COINs to function effectively in low-density territories, like in the case of the Interior+ project. Digital platforms can enable local actors to participate in virtual collaboration spaces, linking them to national and global markets, research hubs, and industrial innovation clusters [29]. This increased connectivity allows LDTs to be part of a broader innovation ecosystem, despite their geographical limitations. These regions often have untapped potentials—such as unique agricultural products, renewable energy resources, or cultural heritage—that can be harnessed through targeted innovation. By integrating the expertise of academia, the practical know-how of local industries, and the policy support of governments, COINs in low-density areas can focus on place-based innovation strategies [27].

The government's role in this process is essential, as it often provides the necessary incentives—such as innovation grants, subsidies, or tax incentives—to attract businesses and research institutions to collaborate in these regions [22]. Meanwhile, academia ensures that local populations have access to education and training programs that prepare them to participate in new industries, while local businesses can leverage academic expertise to diversify their operations and tap into new markets [30]. LDTs often face constraints when it comes to attracting investment and maintaining financial resources necessary for innovation. In the Triple Helix model, government involvement is critical to ensure that these regions receive the support they need to become innovation hubs [22]. Policies that provide incentives for research and development (R&D), particularly those targeted at sectors like agriculture, renewable energy, or rural tourism, can help align local industry interests with national and international development goals.

Another advantage of the triple helix approach in agricultural COINs is its ability to align research and development (R&D) efforts with real-world needs [21, 30]. Collaborative R&D projects between universities, industry and government can accelerate the creation and refinement of new technologies, such as drought-resistant seeds, smart irrigation systems and sustainable pest control methods. By involving all three sectors early in the innovation process, the Triple Helix model ensures that innovations are not only scientifically sound but also practical and commercially viable.

This open innovation perspective provides a valuable opportunity for businesses in agriculture to collaborate, share knowledge and leverage external expertise, ultimately fostering a more resilient and dynamic agricultural ecosystem [6]. This collaborative approach not only addresses pressing issues but also resonates with the values of the newer generation of consumers, which is very important in the agricultural sector [14]. These generations emphasise sustainability, ethical sourcing, and the integration of technology, pushing agrifood companies to innovate continually in response to these priorities. By aligning with these expectations, agrifood companies can help pave the way for a more sustainable and prosperous future for all stakeholders in the agricultural ecosystem.

The aim of this research is to demonstrate how the triple helix approach can be used to develop a collaborative innovation network in the agricultural sector. This will entail an examination of the role played by digital platforms, such as the one associated with the Interior+ project, in facilitating such collaboration.

2 Materials and Methods

The purpose of this research is to describe the strategies and practices used to facilitate and advance collaborative innovation networks involving academia, industry, and government, with a particular focus on the Triple Helix model. To this end, the study documents the implementation of these strategies, the roles of the various stakeholders, and the outcomes of collaborative efforts.

2.1 Research Method

This research followed a descriptive case study methodology. Data were collected through document analysis (policy reports, project documentation) and semi-structured interviews with 15 stakeholders from academia, industry, and government. Interviews explored collaboration practices, innovation barriers, and perceived impacts. Data were thematically analysed to identify recurring patterns in collaborative behaviour and innovation outcomes. Methodological triangulation between documentation and interview data enhanced validity and reliability.

Table 1 summarises the descriptive methodological approach used to identify and detail the practices and strategies of collaborative innovation networks operating within the Interior+ consortium and under the scope of the Triple Helix model [31].

The descriptive study offers a comprehensive overview of how collaborative innovation networks are promoted within a Triple Helix framework, providing insights into effective strategies, roles and contributions of different stakeholders (academia, industry, government), collaboration dynamics, and the impacts in the promotion of collaborative innovation networks. In addition, the description of the dynamics of collaboration within the networks, including how stakeholders interact, share

Table 1 Main research steps of the case study

Defining the scope:	Define the triple helix model and its relevance to the study, focusing on the interactions between academia, industry, and government, and identify specific collaborative networks
Specific population:	Triple helix stakeholders of a low-density territory
Data collection and analysis:	Documentation review understand: How the networks are promoted and managed, what strategies are employed to encourage participation and collaboration, the specific roles and contributions of different stakeholders within the networks

information, and work together on projects, is detailed. These data support the design of the Interior+ project's intersectoral strategy outlining potential collaborative networks, such as successful innovations or improvements in research and development processes.

2.2 *Interior+ Project*

The Interior+ project presents an opportunity to rethink and define a sustainable development model for rural and mountainous regions in Portugal's interior. In particular, it will address the challenges and opportunities facing these areas, with a particular focus on the Beiras and Serra da Estrela (BSE), Beira Baixa (BB), and Alto Minho (AM).

The organisation and human occupation of Portugal are the result of a complex interplay between multiple development policies and the evolution of agricultural and industrial activities over the past centuries. These territories are typified by low population density, which is indicative of the typical social and economic characteristics observed in such areas. The vicious circle that is difficult to counteract in these territories can be addressed by the creation of rural and mountain communities that bring together economic and social agents with the goal of building a more cohesive, innovative, and sustainable community in the medium and long term. It is therefore proposed that a digital platform be established, comprising pertinent content tailored to the most urgent needs of economic actors within the agricultural sector from a "Farm to Fork" perspective. This will facilitate enhanced connectivity between all stakeholders. Consequently, this will enhance the communication and training capabilities of these actors while also drawing attention to the social aspect. This is the primary objective of the Interior+ project, and it is within this context that the developed platform will be presented.

3 Results

Following a descriptive analysis, the Interior+ territory target and its agriculture sector were characterised. At the same time, a reflection was made on how COINs may fit in these low-density agriculture territories. It is possible to say that COINs can play a pivotal role in revitalising rural areas through the promotion of agricultural innovation and resilience.

3.1 Characteristics of the Portuguese Agriculture Sector

The agricultural sector in Portugal, currently eighth in turnover growth and one of the most stable in the market, reflects a trend of financial autonomy with lower economic pressures, even in a scenario of global oscillation. In 2022, the sector was composed of around 40,483 companies, which generated approximately 298,583 jobs and a turnover of around 68,994 million euros (INE-Instituto Nacional de Estatística[1]).

The companies in the sector have a predominant structure of micro and small companies, with 85.54% classified as micro companies and 11.75% as small. Medium and large companies represent, respectively, 2.28% and 0.43% of the total (INE[1]). Despite being a minority, large companies concentrate 40.35% of the turnover, suggesting high productivity and the use of advanced technology, which compensates for the smaller number of workers compared to smaller companies.

This disparity between the number of enterprises and the turnover suggests some underlying factors, such as the size of the enterprises, the type of activity carried out, and the proximity to densely populated areas and logistics centres, which directly impact regional economic performance. In 2022, the agricultural sector recorded one of the lowest rates of new business creation and a high rate of closures (84%), highlighting the difficulty of new businesses in establishing themselves in the sector and the need for supportive policies to promote sustainability and innovation (INE[1]).

Comparing the regions, the North of Portugal stood out in the creation of new companies, but the agricultural sector remained at a relatively low growth rate. In addition, individual companies predominate in the agricultural sector, which points to an entrepreneurial profile focused on family support and small operations.

In the agricultural segment, the highest concentration of companies is in Agriculture, which represents 49.47% of the companies in the sector, although it accounts for only 10.78% of the turnover and employs 27.63% of the total number of workers (INE[1]). The Agricultural Products Industry is the most economically relevant segment, corresponding to 48.79% of the turnover and concentrating 50.9% of the jobs. The Trade of Agricultural Products accounts for 40.43% of the turnover and 24.32% of the companies in the sector.

In this scenario, the Interior+ project is in a sector with a majority of micro and small businesses, of farmers and entrepreneurs mainly seeking to produce and

[1] www.ine.pt

commercialise agri-food products for family support, but also some export in collaboration with clusters and associations [32]. It has a smaller number of medium and big companies that produce nationally and internationally.

3.2 Characteristics of the Low-Density Territories of the Interior+ Project

The problem identified, for which Interior+ presents an innovative response with a view to mitigating it, is the demographic desertification of Rural Mountain Regions (RMR) of Beira Serra da Estrela (BSE), Beira Baixa (BB) and Alto Minho (AM) associated with the weak capacity to attract new talent for agricultural and agri-food and related activities. The RMR fall within the LDTs and has suffered a significant reduction in population and abandonment of activities, which are fundamental for the dynamism of these territories [33]. The disconnect in socio-economic structure is a strong constraint on rural revitalisation: the negative impact of the lack of employment opportunities in breaking the cycle of desertification. Counteracting this cycle requires transversal, concerted measures to create conditions to revitalise the territories, attracting and retaining people and retaining their value [33]. It is crucial to develop bottom-up processes appropriate to local specificities advocating rurality (EU Action Plan for Rural Regions[2]) that sustain strong, connected, resilient and prosperous rural regions. It is necessary to mobilise resources and identify the potential of territories that value human capital and that are based on knowledge transfer and recognition of their products and identity. The RMR must be constituted as an empowered and attractive local community, with citizens participating in decision-making.

All Mountain LDTs have common characteristics [34, 35]: (a) weakened agroecosystems based on small-scale and low-income family farming; (b) ageing and low-skilled workforce; (c) traditional production system without added value; (d) practices that do not facilitate environmental, economic and social sustainability contributing to their vulnerability to climate change. In addition, the primary sector has little attractiveness for investors and/or highly qualified workers who could bring the necessary innovation for the competitiveness and sustainability of the sectors associated with the agricultural and agri-food sectors [33].

The competitiveness of the agro-industrial sector depends not only on its specific performance but also on the character and degree of innovation performance, which is vital for the development of added value and differentiation in the value chain [36]. The agri-food sector, based on EU and national policies and regulations, should focus on improving competitiveness by promoting technology transfer, modernisation, innovation and quality throughout the food chain. The agri-food industry has begun to renew the concept of the role of agriculture, assuming a new paradigm

---

[2] https://ec.europa.eu/regional_policy/pt/newsroom/news/2021/06/30-06-2021-long-term-vision-for-rural-areas-for-stronger-connected-resilient-prosperous-eu-rural-areas

focused on the sustainable use of natural resources, the creation of public goods, equity in access to quality food, and increasingly the delivery of attractive value to society [37]. Traditions and cultural values, for example, play a decisive role in the potential of resources for the generation of value if properly used by entrepreneurs. According to this model, knowledge and development in agriculture also become a more complex and systemic concept that breaks with the model of linear relationships between stakeholders [34]. In this sense, the business has become increasingly sensitive to innovation opportunities, not only related to technological changes but also related to strategy, marketing, organisation and management.

3.3 Analysis and Discussion of the Results

Low-density regions are confronted with a distinctive set of challenges, including an ageing population, scarce resources, and geographical isolation, which collectively impede the capacity for local innovation [34, 35, 37]. In such contexts, COINs can provide indispensable platforms for the exchange of knowledge, the sharing of resources, and collaborative problem-solving, leveraging the Triple Helix model of academia, industry, and government [21, 27]. As illustrated in Fig. 1, an understanding of the theoretical background and the Interior+ proposal allows for the identification of potential roles that may contribute to the collaborative innovation network (Fig. 2).

Universities or research institutions often play a role in network promotion and management in low-density agricultural areas centred on innovation, technology advisement, and research to create accessible, centralised networks for small farmers

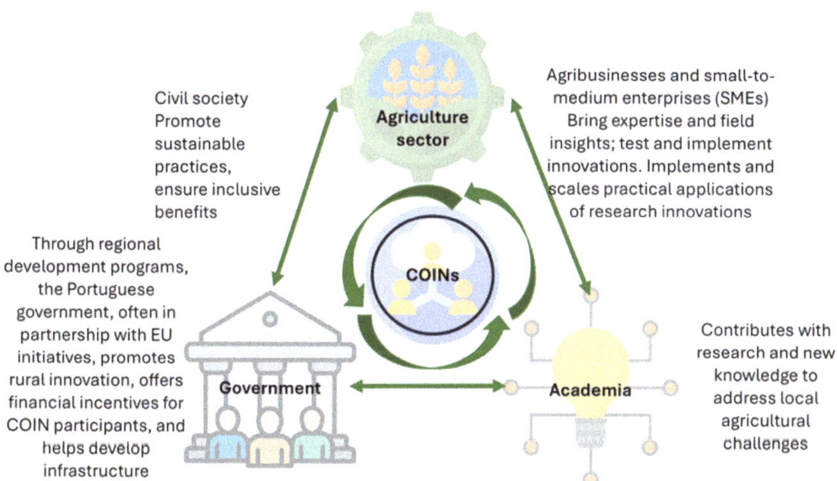

Fig. 1 Collaborative innovation networks and the main actors of the triple helix model

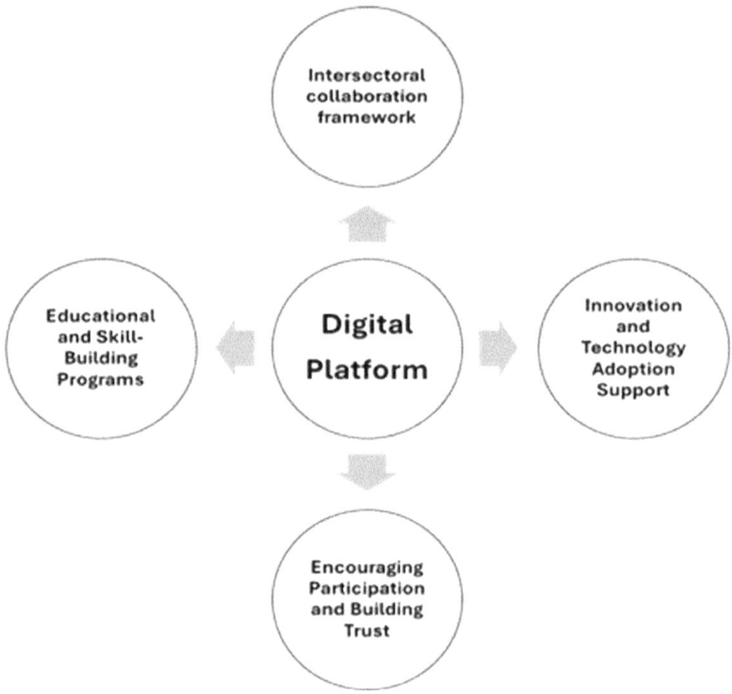

Fig. 2 Intersectoral strategy of interior + project

and entrepreneurs in remote regions, and as previously addressed in our own research [36, 38–40]. For example, the Polytechnic Institutions across Portugal often work directly with local communities to provide research-based solutions tailored to the unique needs of each area. They also lead training that allows stakeholders to develop skills to engage effectively in COINs. These agricultural innovation hubs can offer workshops, training, and demonstrations of new technologies (e.g., precision agriculture or sustainable irrigation) that are critical for smallholders who may not have easy access to such resources. There are some organisations that can play the role of "innovation brokers," helping research centres and Higher Education Institutions (HEI), such as regional development agencies, agricultural associations, and extension services, which are critical for initiating and managing networks [41]. They may serve as coordinators, connecting stakeholders from different sectors and facilitating information flow, financing, and collaboration. In low-density areas, pooled resources like shared machinery, storage facilities, or research stations enable smaller farms to benefit from advanced technology without the high individual costs [15, 16, 41]. This strategy encourages collaboration since farmers rely on shared assets and, in turn, share their knowledge and experiences with one another. Given the close-knit nature of these territories, trust-building activities like regular meetings, field visits, and joint projects help strengthen the social fabric within COINs. Trust is essential in rural areas where personal relationships are often key to successful collaboration.

In the agriculture sector, the "industry" side of the Triple Helix model is characterised by agribusinesses, most of them micro or small-to-medium enterprises (SMEs), that have a role in implementing and scaling practical applications of research innovations [32, 35]. These entities can introduce technologies like data-driven farming techniques or environmentally friendly pest control solutions, which enhance productivity and sustainability in agriculture [9, 40]. Farmers and cooperatives bring practical, on-the-ground expertise and insight into the real-world applicability of proposed innovations. Cooperatives often play a dual role, both as beneficiaries of COIN initiatives and as promoters who disseminate knowledge and technology among their members. They serve as the operational backbone of COINs, directly testing, providing feedback, and implementing innovations.

Stakeholders frequently emphasised that collaborative platforms reduce geographical isolation, enabling continuous exchange of knowledge and resources. This aligns with findings from Cronin et al. [8], and Klerkx and Leeuwis [15], who argue that structured interaction mechanisms within Triple Helix networks enhance innovation diffusion in peripheral regions. Empirical evidence from the INTERIOR+ case demonstrates that COINs facilitate the integration of local actors into broader innovation ecosystems by providing structured opportunities for collaboration and learning.

The government's role is central in low-density areas, as it provides both funding and policy support for agricultural innovation [7, 8]. Through regional development programs, the Portuguese government, often in partnership with European Union (EU) initiatives, promotes rural innovation, offers financial incentives for COIN participants, and helps develop infrastructure. Subsidies and grants from the government or EU (e.g., from the EU's Common Agricultural Policy) provide direct incentives for farmers, cooperatives, and agribusinesses to join collaborative networks. These funds encourage participation by reducing the financial risk of innovation adoption for smallholders in these regions.

Therefore, an intersectoral strategy to promote COINs is key to creating agricultural communities, fostering collaboration, and responding to challenges that are difficult to overcome as a single stakeholder. This perspective will allow to combine science and policy definitions of building resilience. The collaboration of the local government, industry, and academic resources accelerates the adoption of new technologies, making them more accessible and better suited to local needs. This will stimulate socioeconomic development through employment and business opportunities. As rural communities engage in collaborative networks, they can tap into new markets, diversify their products, and improve their livelihoods. They enable local farmers to adapt to modern challenges, contribute to environmental sustainability, and play an active role in transforming the agricultural landscape.

3.4 Intersectoral Strategy

The intersectoral strategy must emphasise the creation of a collaborative, adaptable, and resourceful environment capable of fostering ongoing innovation to encourage innovation and the implementation of technology within a digital platform aligned with the Triple Helix concept (which involves the participation of academia, industry, and government).

The Interior+ project proposes a digital platform built for collaborative innovation network promotion and an intersectoral strategy within the Triple Helix model to achieve the implementation goal of an effective network of collaboration to promote innovation and technology adoption in the agriculture sector.

Digital platforms help manage and sustain interactions in remote regions by facilitating virtual meetings, online training, and collaboration forums. These digital tools allow stakeholders in low-density areas to stay connected with broader national or international agricultural networks. The digital platform design has to have a user-centred approach, with direct input from stakeholders in an intuitive manner, to address users' specific needs and facilitate a high adoption rate. It has also ensured high standards for data management, privacy, and security to enable trust among stakeholders for data sharing and transparency in collaborative projects. This platform must be developed with different access levels and dashboards tailored to the unique needs of each Triple Helix role: academia - tools for publishing research, accessing data sets, and connecting with industry partners for collaborative R&D projects; industry - access applied research results and innovation trends and partner with startups or universities for technology development; Government - monitor innovation impacts, facilitate regulatory compliance, and fund specific projects or collaborations.

The strategy is based on a framework that establishes clear channels for cross-sector communication, including forums and workshops, where all sectors can engage in open discussions. To facilitate this communication, some "innovation brokers" have to be selected, within or outside the Interior+ partnership, to help match research or innovation needs with resources or expertise available across sectors. The platform should feature a resource hub where government funding opportunities, industry R&D tools, and academic resources (like research labs or test fields) are accessible, allowing stakeholders to leverage shared assets efficiently.

In order to facilitate the adoption and implementation of innovative technologies, academic institutions and industry partners can engage in collaborative pilot projects on the platform, which will serve to demonstrate and evaluate the efficacy of emerging technologies or innovations. It is also within the remit of government partners to ensure that these projects are in accordance with the relevant regulations and to facilitate their wider implementation. It would be beneficial to present case studies and success stories that illustrate the positive impact of new technologies on productivity, sustainability, and social outcomes. Such an approach may foster confidence among stakeholders and motivate participation. The platform can provide funding information and technical assistance to encourage industry or academic partners to pilot or adopt new innovations that address real-world challenges.

Encouraging participation and building trust are key to creating COINs, so they enable a transparent tracking system on the platform that shows how contributions from different sectors lead to real-world outcomes in innovation, policy improvement, or economic growth. A transparent impact record encourages continuous engagement by demonstrating each sector's importance. Structured feedback mechanisms, allowing users to provide input on the platform's functionalities, collaboration experiences, and resource availability. Regularly update the platform in response to this feedback is needed to build trust in the platform's responsiveness and effectiveness.

To engage the stakeholders in the platform within an intersectoral strategy, it is necessary to offer programmes in a continuous learning environment in different areas, from innovation and technology implementation to more technical competences development in partnership with vocational education institutions. Also, hosting workshops and bringing together students, researchers, industry experts, and policymakers to work on pressing challenges collaboratively is a way to engage the stakeholders.

An intersectoral strategy focused in a digital platform within the Triple Helix framework has the potential to accelerate technology adoption by providing a centralised, adaptable, and collaborative space where academia, industry, and government each play distinct yet interconnected roles. By facilitating collaboration through structured resources, incentives, and knowledge-sharing opportunities, this platform can support dynamic, responsive innovation that meets both local needs and global challenges.

4 Conclusions

The Triple Helix model strengthens COINs in agriculture by fostering interdisciplinary collaboration and resource sharing. By integrating the expertise and resources of academia, industry, and government, this framework accelerates the pace of innovation, ensures that new technologies are both scientifically sound and commercially viable and creates a policy environment conducive to sustainable agricultural development. In a sector as critical as agriculture, where the stakes include global food security and environmental sustainability, the Triple Helix approach provides a robust foundation for driving transformative innovation.

The descriptive analysis provides insight into how a collaborative innovation network within a triple helix approach can improve technology adoption and innovation research to meet agriculture's needs, as the Interior+ project wants to do.

The findings support the Interior+ project's intersectoral strategy, which enables collaboration across sectors, not only through a triple model approach but also within industry sectors linked to agriculture that are part of its value chain.

The study demonstrates that COINs within the Triple Helix framework can strengthen innovation ecosystems in low-density agricultural territories by promoting knowledge exchange, digital collaboration, and intersectoral trust. However, as a single case study, the findings are context-specific and require further comparative analysis across regions.

Although the scientific and managerial contributions are positive, as they offer a case of structuring collaborative innovation networks within a triple model concept in the agriculture sector of RMR and LDTs, this research has the limitations of any case study and a qualitative approach.

Future research should explore quantitative indicators of network performance and the long-term impact of digital platforms on rural innovation sustainability.

Acknowledgments We thank the financial support for developing this work within the project Interior+, Comunidades Rurais do Interior e de Montanha, PRR-C05-i03-I-000233-LA7.2; LA7.5; LA7.6, co-financed by Programa de financiamento -PRR—Plano de Recuperação e Resiliência, Medida—Agenda de investigação e inovação para a sustentabilidade da agricultura, alimentação e agroindústria—componente Investimento e inovação.

References

1. Li, X., & Liu, X. (2023). The impact of the collaborative innovation network embeddedness on enterprise green innovation performance. In *Frontiers in environmental science 11*. Frontiers. https://doi.org/10.3389/fenvs.2023.1190697
2. Song, W., Ming, X., & Wang, P. (2013). Collaborative product innovation network: Status review, framework, and technology solutions. *Concurrent Engineering Research and Applications, 21*(1), 55–64. https://doi.org/10.1177/1063293X12468457
3. Xue, X., Zhang, R., Liang, W., Fan, H., Yang, R. J., & Dai, J. (2018). Collaborative innovation in construction project: A social network perspective. *KSCE Journal of Civil Engineering, 22*, 417–427. https://doi.org/10.1007/s12205-017-1342-y
4. Xie, X., Fang, L., & Zeng, S. (2016). Collaborative innovation network and knowledge transfer performance: A fsQCA approach. *Journal of Business Research, 69*, 5210–5215. https://doi.org/10.1016/j.jbusres.2016.04.114
5. Cinelli, M., Ferraro, G., & Iovanella, A. (2019). Network processes for collaborative innovation. *International Journal of Entrepreneurship and Small Business, 36*(4), 430–452.
6. West, J., Salter, A., Vanhaverbeke, W., & Chesbrough, H. (2014). Open innovation: The next decade. *Research Policy, 43*(5), 805–811. https://doi.org/10.1016/j.respol.2014.03.001
7. Berthet, E. T., Hickey, G. M., & Klerkx, L. (2018). Opening design and innovation processes in agriculture: Insights from design and management sciences and future directions. *Agricultural Systems, 165*, 111–115. https://doi.org/10.1016/j.agsy.2018.06.004
8. Cronin, E., Fosselle, S., Rogge, E., & Home, R. (2021). An analytical framework to study multi-actor partnerships engaged in interactive innovation processes in the agriculture, forestry, and rural development sector. *Sustainability (Switzerland), 13*(11), 6428. https://doi.org/10.3390/su13116428
9. Gumbi, Nametshego, Lucas Gumbi, and Hossana Twinomurinzi. 2023. Towards sustainable digital agriculture for smallholder farmers: A systematic literature review. Sustainability (Switzerland) 15(16):12530. doi:https://doi.org/10.3390/su151612530.
10. Bitzer, V., Glasbergen, P., & Arts, B. (2013). Exploring the potential of intersectoral partnerships to improve the position of farmers in global agrifood chains: Findings from the coffee sector in Peru. *Agriculture and Human Values, 30*(1), 5–20. https://doi.org/10.1007/s10460-012-9372-z

11. Jamali, H., Dascalu, S. M., & Harris, F. C. (2024). Fostering joint innovation: A global online platform for ideas sharing and collaboration. In S. Latifi (Ed.), *ITNG 2024: 21st International Conference on Information Technology-New generations* (pp. 305–312). Springer. https://doi.org/10.1007/978-3-031-56599-1_40
12. De Falco, E., Salvatore, A. R., Orlando, B., & Cucari, N. (2017). Open collaborative innovation and digital platforms. *Production Planning & Control, 28*, 1344–1353. https://doi.org/10.1080/09537287.2017.1375143
13. Swaans, K., Boogaard, B., Bendapudi, R., Taye, H., Hendrickx, S., & Klerkx, L. (2014). Operationalizing inclusive innovation: Lessons from innovation platforms in livestock value chains in India and Mozambique. *Innovation and Development, 4*, 239–257. https://doi.org/10.1080/2157930X.2014.925246
14. Klerkx, L., Van Mierlo, B., & Leeuwis, C. (2012). Evolution of systems approaches to agricultural innovation: Concepts, analysis and interventions. In *Farming Systems Research into the 21st Century: The New Dynamic* (pp. 457–483). Springer. https://doi.org/10.1007/978-94-007-4503-2_20
15. Klerkx, L., & Leeuwis, C. (2009). Establishment and embedding of innovation brokers at different innovation system levels: Insights from the Dutch agricultural sector. *Technological Forecasting and Social Change, 76*, 849–860. https://doi.org/10.1016/J.TECHFORE.2008.10.001
16. Jialu, S., Ma, Z., Zhu, B., Xie, H., Oteng, A. F., & Hu, W. (2021). Collaborative innovation network, knowledge base, and technological innovation performance-thinking in response to COVID-19. *Frontiers in Psychology, 12*, 648276. https://doi.org/10.3389/fpsyg.2021.648276
17. Ma, H. (2023). The dynamics of China's collaborative innovation network in agricultural biotechnology: A spatial-topological perspective. *Systems, 11*(2), 73. https://doi.org/10.3390/systems11020073
18. Wang, F., Qing, S., & Zhang, Z. (2024). The influence of collaborative innovation network characteristics on firm innovation performance from the perspective of innovation ecosystem. *Kybernetes, 53*, 1281–1305. https://doi.org/10.1108/K-04-2022-0553
19. Dong, X. (2023). Research on the construction model of "innovation-network-space" industrial system of agricultural product processing in Hebei Province. *Frontiers in Business, Economics and Management, 10*, 72–75.
20. Sjahruddin, H., & Adif, R. M. (2024). Gen Z consumer trends: Understanding the next wave of buying behavior. *Management Studies and Entrepreneurship Journal, 5*, 480–485.
21. Leydesdorff, L., & Etzkowitz, H. (1998). Triple helix of innovation model. *Science and Public Policy, 25*, 358–364.
22. Li, E., Yao, F., Xi, J., & Guo, C. (2018). Evolution characteristics of government-industry-university-research cooperative innovation network for China's agriculture and influencing factors: Illustrated according to agricultural patent case. *Chinese Geographical Science, 28*, 137–152. https://doi.org/10.1007/s11769-017-0924-4
23. Etzkowitz, H., & Leydesdorff, L. (2000). The dynamics of innovation: From national systems and "mode 2" to a triple helix of university–industry–government relations. *Research Policy, 29*, 109–123. https://doi.org/10.1016/S0048-7333(99)00055-4
24. Hermans, F., Geerling-Eiff, F., Potters, J., & Klerkx, L. (2019). Public-private partnerships as systemic agricultural innovation policy instruments—Assessing their contribution to innovation system function dynamics. *NJAS-Wageningen Journal of Life Sciences, 88*, 76–95. https://doi.org/10.1016/j.njas.2018.10.001
25. Pant, L. P. (2010). Creative commons: Non-proprietary innovation triangles in international agricultural and rural development partnerships. *The Innovation Journal: The Public Sector Innovation Journal, 15*(2), 1–22.
26. Klerkx, L., Aarts, N., & Leeuwis, C. (2010). Adaptive management in agricultural innovation systems: The interactions between innovation networks and their environment. *Agricultural Systems, 103*, 390–400. https://doi.org/10.1016/J.AGSY.2010.03.012
27. Janahi, N. A., Durugbo, C. M., & Al-Jayyousi, O. R. (2022). Exploring network strategies for eco-innovation in manufacturing from a triple helix perspective. *Cleaner Logistics and Supply Chain, 4*, 100035. https://doi.org/10.1016/j.clscn.2022.100035

28. Wanzenböck, I., Wesseling, J. H., Frenken, K., Hekkert, M. P., Matthias, K., & Weber. (2020). A framework for mission-oriented innovation policy: Alternative pathways through the problem–solution space. *Science and Public Policy, 47*(4), 474–489. https://doi.org/10.1093/scipol/scaa027
29. Veile, J. W., Schmidt, M.-C., & Voigt, K.-I. (2022). Toward a new era of cooperation: How industrial digital platforms transform business models in industry 4.0. *Journal of Business Research, 143*, 387–405. https://doi.org/10.1016/j.jbusres.2021.11.062
30. Razak, A. A., & Saad, M. (2007). The role of universities in the evolution of the triple helix culture of innovation network: The case of Malaysia. *International Journal of Technology Management and Sustainable Development, 6*(3), 211–225. https://doi.org/10.1386/ijtm6.3.211/1
31. Yin, R. K. (2018). *Case study research and applications* (6th ed.). Sage Publications.
32. Paiva, T., Domingues, C., & de Andrade, L. P. (2016). Innovation and knowledge transference in a cluster user-driven innovation perspective—the Inovcluster experience. *International Journal of Food Studies, 5*, 1. https://doi.org/10.7455/ijfs/5.1.2016.a5
33. Monteiro, A. (2019). Territórios do interior, coesão territorial e modelos de governança: a propósito do programa nacional para a coesão territorial. *Sociologia on Line, 19*, 127–151. https://doi.org/10.30553/SOCIOLOGIAONLINE.2019.19.6
34. Knickel, K., Brunori, G., Rand, S., & Proost, J. (2009). Towards a better conceptual framework for innovation processes in agriculture and rural development: From linear models to systemic approaches. *The Journal of Agricultural Education and Extension, 15*, 131–146. https://doi.org/10.1080/13892240902909064
35. Coutinho, P., Simões, M., Pereira, C., & Paiva, T. (2021). Sustainable local exploitation and innovation on meat products based on the autochthonous bovine breed Jarmelista. *Sustainability, 13*, 2515. https://doi.org/10.3390/su13052515
36. Paiva, T., Ribeiro, M., & Coutinho, P. (2020). R&D Collaboration, Competitiveness Development, and Open Innovation in R&D. *Journal of Open Innovation: Technology, Market, and Complexity, 6*, 116. https://doi.org/10.3390/joitmc6040116
37. Brunori, G., Barjolle, D., Dockes, A.-C., Helmle, S., Ingram, J., Klerkx, L., Moschitz, H., Nemes, G., & Tisenkopfs, T. (2013). Reform and innovation: The role of learning and innovation networks. *EuroChoices, 12*, 27–33. https://doi.org/10.1111/1746-692X.12025
38. Gaspar, P. D., Pinheiro, R., Domingues, C., Almeida, C., Paiva, T., Pereira, C. D., & Velho, M. V. (2015). Training requirements for the agro-food industry in Portugal. *International Journal of Food Studies, 4*, 12–28. https://doi.org/10.7455/ijfs/4.1.2015.a2
39. Paiva, T., Ribeiro, M. P., & Coutinho, P. (2022). Capacity-building model to promote innovation and sustainability in the Portuguese agro-industrial sector. *Sustainability, 14*, 15873. https://doi.org/10.3390/su142315873
40. Gaspar, P. D., Soares, V. N. G. J., Caldeira, J. M. L. P., Andrade, L. P., & Soares, C. D. (2022). Technological modernization and innovation of traditional Agri-food companies based on ICT solutions—The Portuguese case study. *Journal of Food Processing and Preservation, 46*. https://doi.org/10.1111/jfpp.14271
41. Klerkx, L., van Mierlo, B., & Leeuwis, C. (2012). Evolution of systems approaches to agricultural innovation: Concepts, analysis and interventions. In I. Darnhofer, D. Gibbon, & B. Dedieu (Eds.), *Farming Systems Research into the 21st Century: The New Dynamic* (pp. 1–488). Springer.

Incubators as Catalysts of Dynamic Capabilities: Strategic Contributions to the Sustainability of Entrepreneurial Ecosystems in Low-Density Territories

Anderson R. Galvão, Carla S. Marques, and Margarida Cabral

Abstract This study aims to analyze the role of incubators in entrepreneurial ecosystems in low-density territories, based on the theory of dynamic capacity. To develop this study, a qualitative approach was chosen, using semi-structured interviews applied to three incubators located in the interior regions of North and Center Portugal, as well as to incubated entrepreneurs. Data analysis was performed using NVIVO software. The results show that incubators are central elements in the dynamization of entrepreneurial ecosystems, promoting entrepreneurship, innovation, attracting investment and the creation of competitive and sustainable businesses. Its role in coordinating key local stakeholders (municipalities, higher education institutions, and business associations) and mitigating structural challenges is also noteworthy. However, significant limitations persist, such as a shortage of skilled labor and limited access to financing. By using dynamic capability theory as a lens of analysis, the study offers an innovative perspective that highlights how incubators act as catalysts for adaptation, knowledge absorption, and innovation in low-density regions, thus contributing to enriching scientific debate and expanding understanding of entrepreneurship in challenging territorial contexts.

Keywords Incubators · Entrepreneurial ecosystems · Low-density territories · Entrepreneurship · Sustainable

A. R. Galvão (✉) · C. S. Marques
CETRAD & University of Tras-os-Montes and Alto Douro, Vila Real, Portugal
e-mail: anderson@utad.pt

A. R. Galvão
NECE—Research Centre for Business Sciences, Covilhã, Portugal

M. Cabral
University of Tras-os-Montes and Alto Douro, Vila Real, Portugal
e-mail: smarques@utad.pt

L. Farinha et al. (eds.), *Sustainable Innovations in Mechanical Engineering and Management Sciences*, Lecture Notes on Multidisciplinary Industrial Engineering,
https://doi.org/10.1007/978-3-032-30372-1_3

1 Introduction

Low-density territories have less capacity to face market evolution, having difficulties in implementing social and environmental measures, reducing their continuous path of growth and development [23]. Over the years, there has been an increase in the movement of the population from low-density territories in the interior of Portugal to the coast, increasing cross-border migratory flows, composed mainly of adults and young people [55]. In low-density territories, incubators are seen as important innovation facilitators that provide spaces, equipment, monitoring services, as well as financial, legal and administrative support for technology entrepreneurs and start-ups [72]. Hackett and Dilts [29] state that business incubators are an economic development tool designed to accelerate the growth and success of companies through a wide range of resources and support services for new businesses. Business incubators support the establishment of new ventures by providing financing to entrepreneurs, creating an entrepreneurial ecosystem, characterized by being a vital element for economic growth, facilitating the establishment of new ventures and stimulating business activity in a given region [2].

Business incubators are distinguished by their role in promoting regional development and fostering economic growth by increasing entrepreneurship, innovation, technology, and job creation. Incubators are part of the "ecosystem" that influences how entrepreneurs understand the need to plan and control [7]. The entrepreneurial ecosystem resembles a complex network of interconnected elements, where several factors act together to influence and strengthen the dynamics of entrepreneurship in business incubators [48]. According to Zaidi et al. [77], the entrepreneurial ecosystem has a relevant impact not only on the decision to start a new business, but also on its continuity, expansion and sustainable success over time, through dynamic interactions between local actors.

However, it is important to highlight that the growing attention devoted to entrepreneurial ecosystems in low-density territories reflects a cross-cutting concern for territorial cohesion and sustainable economic development. These territories, often marked by phenomena such as depopulation, aging populations, lack of infrastructure, and a poorly diversified economic base, face significant challenges when it comes to promoting innovation and stimulating business activity. In this context, business incubators emerge as strategic actors in strengthening local entrepreneurial ecosystems. Samuelsson and Jutterström [59] state that the support offered by incubators to companies affects their performance, just as the resources of the incubators affect the performance of the incubated companies.

Although there is research on incubators in urban contexts or rural areas, studies that simultaneously address incubators and low-density territories remain rare. For example, Virkkala and Saarela [71] develop a framework of good practices for rural incubators, but argue that understanding of how territorial factors influence the performance of these incubators remains weak. At the same time, Almeida and Daniel [5] explore the governance of entrepreneurial ecosystems in low-density territories, highlighting the role of local government in peripheral municipalities, without, however, incorporating the role of incubators as mediators of innovation and development. These limitations point to a clear gap: research has not yet comprehensively addressed how incubators contribute to strengthening entrepreneurial ecosystems in low-density contexts, particularly with regard to the creation of dynamic capabilities and regional resilience.

This study aims to analyze the role of incubators in entrepreneurial ecosystems in low-density territories, based on the theory of dynamic capacity.

The research was applied in person to three incubators located in low-density territories in the northern interior and in the Beira Baixa region. The interviews were conducted with technicians and managers from the three incubators, and to triangulate the information, 10 entrepreneurs hosted at the incubator were interviewed through semi-structured interviews. In the data processing process, content analysis was performed using NVIVO software.

Following this introductory section, we present a literature review on entrepreneurship in low-density areas and the role of incubators in fostering entrepreneurial ecosystems. Section three presents the methodology used to develop this study. Section four describes the data analysis and discusses the results. Finally, in the last point, the main conclusions, implications, limitations and future lines of research will be presented.

2 Literature Review

2.1 Entrepreneurship in Low-Density Territories

Entrepreneurship is a key factor for the economy of a country or region, being responsible for the creation of new companies and increased employability [18]. According to Alerasoul et al. [3], entrepreneurship is defined as a process of identifying and taking advantage of new business opportunities, through the creation of new products or services in the market.

This goes beyond any attempt to create a new business or a new venture, relating to demand, supply, value creation, wealth innovation, resource management, organizational capacity, covering all types of entrepreneurship, economic, social, public, organizational and collaborative, covering micro, small, medium, start-ups and new ventures [37]. Many governments around the world promote entrepreneurship itself and entrepreneurial education as a means of fostering entrepreneurship [51].

In entrepreneurship, small and medium-sized enterprises play an important role in economic development, especially in developing countries, through the creation of new technologies, for the development of incubators and accelerators [66]. According to Gómez-Araujo and Bayon [28], entrepreneurship has a wide range of benefits for rural areas, through innovation, the adequate use of rural resources, in order to reduce rural poverty, modernizing LDT.

Low-density territories are territories with characteristics that are limited compared to urban areas. Depopulation, a lack of human and cultural capital, and government support lead to limited business activity in relation to entrepreneurship [34]. These territories face major challenges in developed countries, associated with factors such as population aging, emigration, lower labor productivity, lower educational levels, and a lower level of public services. Therefore, entrepreneurship is a fundamental mechanism to counter the problem of development in low-density territories [38]. The loss of young population has indispensable consequences for the economic, social and environmental sustainability of low-density territories [47, 58].

LDTs present economic advantages, such as cooperation networks to start a business, due to the low cost and high potential for job creation [21]. The creation of new companies helps to improve regional competitiveness, causing an impact on innovation and sustainable development, reducing poverty and promoting the well-being of society [13, 56].

The main function of creating new companies is to face economic and social challenges, through the implementation of sustainable business models, for future entrepreneurs [24, 40]. Entrepreneurs who choose to settle in low-density territories tend to differentiate themselves from other types of entrepreneurs, as they take advantage of local resources, causing the local economy to develop [27]. Entrepreneurship based on innovation strengthens the socioeconomic development of territories through job creation, improvement of regional resources, cultural heritage and the quality of life of populations [17]. Therefore, it is crucial to implement entrepreneurship strategies and programs that meet the needs of LDT populations [11].

2.2 The Role of Incubators in Promoting Entrepreneurial Ecosystems

Incubators can play an important role in promoting entrepreneurship [66]. Business incubators can be used for knowledge transfer, contributing to the performance of newly created companies [19]. According to Grimaldi and Grandi [25], incubators can be classified into four categories: business innovation centers, university business incubators, independent private incubators and corporate private incubators.

Business incubators support small businesses through monetary support, internal and external networking activities, provision of office space, and support in area knowledge, advice, allowing the start-up to implement its business model in operational processes, products or services, in order to be successful in the future [29].

An incubator acts institutionally, connecting incubated entrepreneurs with different actors and institutions in the business ecosystem, promoting the flow of information [16]. In startups, the environment among employees within incubators can generate innovation, fostering networking opportunities and training to solve common problems [60]. Coworking spaces allow the sharing of collaborative tools and equipment, in order to benefit from affordable access to essential resources, eliminating the need for an initial investment [70]. Sharing a space contributes to sustainable entrepreneurship [53].

The proximity of startups, mentors and experts from different domains leads to unplanned interactions, generating new ideas and innovative solutions for business development [61]. Through discussions and mentoring in incubators, skills are trained and developed in incubated companies in order to drive innovation, encouraging entrepreneurs to perform entrepreneurial roles [31]. Incubators help entrepreneurs expand in the market by providing business support services such as coaching, training, and mentoring through workshops and seminars [78].

The network generated within incubators extends only to presentations to potential investors, allowing the startup to gain visibility through building relationships with investors, proposing a favorable value for both parties for the formation of strategic partnerships [49].

In order to modernize rural areas, policymakers promote regional development through business incubators as mechanisms to create value for society [46]. Support for incubators goes through a competitive selection process, where a prior assessment is made, from the beginning of the project to final implementation [10].

Incubators play a fundamental role in fostering entrepreneurial ecosystems, providing support, guidance, and access to resources, and strongly contributing to the creation of a network of interconnected companies, institutions, and organizations that drive economic development [35]. Li et al. [42] state that Entrepreneurial Ecosystems are networks of companies, organizations, investors, mentors, accelerators, incubators, universities and governments that work in networks to create and sustain a favorable environment for the development of new businesses and entrepreneurship.

Stam and Van de Ven [63] define an entrepreneurial ecosystem as a set of coordinated interdependent actors and factors, allowing entrepreneurship to be productive within a given territory. An entrepreneurial ecosystem interconnects with entrepreneurship and regional development, allowing researchers to open new research questions and research avenues on policy-related issues on how to support economic growth [75].

Some studies have shown that an entrepreneurial ecosystem supports entrepreneurship and value creation at a regional level, developing a life cycle focused on the internal commercialization of knowledge [12]. One of the main characteristics of entrepreneurial ecosystems is to identify which companies have high growth, comparing them with companies created in other locations, with dynamic ecosystems, in order to create new opportunities for growth and employability [74]. According to Asmit et al. [6], the entrepreneurial ecosystem supports entrepreneurial activities in specific geographic regions.

Portugal has some characteristics that make the entrepreneurial ecosystem distinct, such as specialization in specific sectors, for example, technology, energy, biotechnology and tourism, or its location in strategic regions for certain sectors, which can benefit incubated companies operating in these sectors Startup [64].

The entrepreneurial ecosystem model includes cultural and social attributes, network creation, investment capital, relationships between universities, infrastructure, government, local agents and business programs, in order to reduce operational costs, allowing the redefinition of sustainable business models [36]. Some business incubators also offer acceleration programs that aim to accelerate the growth and development of incubated companies Startup [64]. Entrepreneurial ecosystems encourage and support entrepreneurs through political institutions [62].

2.3 Dynamic Capacity Theory

Dynamic capability theory is defined as the ability to use organizational and strategic resources in order to ensure better performance [22]. This allows companies to coordinate their performance in the short and long term [20]. Dynamic capabilities result from the change in companies' commercial activities towards new market positioning opportunities and from the transaction to the commercialization of new products and services, boosting companies' competitiveness [1]. The theory of dynamic capabilities argues that an organization's varied resources and competencies are capable of being adjusted over time, taking into account their transformation and adaptation to changes in the external environment [68].

According to Jantunen et al. [33], three fundamental dimensions can be distinguished. The first refers to adaptive capacity, that is, the company's ability to identify and seize opportunities in emerging markets. The second is absorptive capacity. In this dimension, the organization must recognize the value of new external information, assimilate it, and apply it effectively in the business context. Finally, innovation capacity. This reflects the company's ability to utilize and articulate its employees' knowledge to generate new skills for creating products or processes.

For Wei et al. [73], the fundamental principle of the dynamic capabilities theory holds that a company is capable of adapting, integrating, renewing and reorganizing its competitive advantages, in order to more effectively take advantage of and exploit the opportunities generated by constantly changing contexts. This aspect is particularly relevant for early-stage or emerging businesses, which generally face greater challenges and have limited resources. The development of dynamic capabilities in incubated companies becomes essential for these companies to be able to use and readjust the resources available in industrial environments characterized by high diversity, ensuring their survival and growth [45].

3 Methodology

3.1 Type of Study and Case Study

With the aim of analyzing the role of incubators in entrepreneurial ecosystems in low-density territories, this study adopted a qualitative approach, using semi-structured interviews as its main technique. This methodology proved suitable for exploring in depth the perceptions, experiences, and practices of the different actors involved, allowing us to understand how incubators contribute to local economic and social development. Qualitative research, as argued by Maxwell [50], makes it possible to capture the meaning and context of interactions, as well as identify nuances relevant to the design of public policies. Semi-structured interviews, characterized by the combination of previously defined topics and flexibility in conducting the dialogue [9, 65], proved particularly useful for collecting detailed information about the functioning of the incubators and their articulation with the territory.

Following Yin's [76] methodological principles, a multiple case study design was adopted, as this approach allows for an in-depth investigation of a contemporary phenomenon within its real-life context and facilitates analytical generalisation. The cases correspond to three business incubators located in low-density territories in Portugal. Two are situated in the northern interior and one in the central interior. These cases were deliberately chosen through theoretical and purposive sampling based on the following criteria: (i) geographical location in officially classified low-density territories; (ii) active integration within a local entrepreneurial ecosystem involving higher education institutions, municipalities, and business associations; (iii) representativeness of different governance and operational models, including university-linked, municipal, and mixed management structures; and (iv) accessibility and feasibility for data collection.

The unit of analysis in this study is twofold. The first level corresponds to the incubator as an organisational entity, focusing on its structure, governance, and strategic orientation. The second level corresponds to the network of relationships established with incubated entrepreneurs and local stakeholders, which are considered embedded sub-units of analysis. This design enables the comparison of organisational practices and interaction mechanisms across different territorial contexts, providing a richer and more nuanced understanding of how incubators foster regional entrepreneurial dynamics.

The three selected incubators display distinctive profiles that enhance the analytical depth of the study. Incubator 1 is associated with science, technology, and eco-construction sectors. Incubator 2 is municipally managed and supports start-ups across a variety of emerging industries. Incubator 3 is closely linked to a regional university and focuses on the agri-food, agro-industrial, and environmental sectors. This diversity of cases strengthens the internal validity of the research and contributes to a more comprehensive interpretation of the interrelations between incubators, local communities, and regional development processes.

3.2 Characterization of the Territory

The territory under study is the North and Center of Portugal, more specifically 3 business incubators located in the Trás-os-Montes and Alto Douro region and in the Beira Interior region.

Figure 1 presents the territory under study, specifically two regions in the interior of northern Portugal and one in the central region. These territories were selected for this study due to the scarcity of research on these types of territories.

The first district is located in the northern interior of Portugal, in the Trás-os-Montes and Alto Douro region. Comprising 12 municipalities, it is characterized by being one of the districts with the lowest population density, depopulation, and aging. The business sector is mainly made up of micro and small businesses, often family-owned.

The second district is also located in the northern interior of Portugal, also in the Trás-os-Montes and Alto Douro region. Comprising 14 municipalities, this district is characterized by marked depopulation, a loss of young people and qualified professionals, an aging population, and low industrial diversification.

The last district is located in the municipality of Beira Baixa. This district stands out for its national example of innovation in the countryside, investing in technology, entrepreneurship, and revitalization of the rural environment, despite being a municipality with an aging population, depopulation, and a need to retain young people.

Fig. 1 Characterization of the territories under study

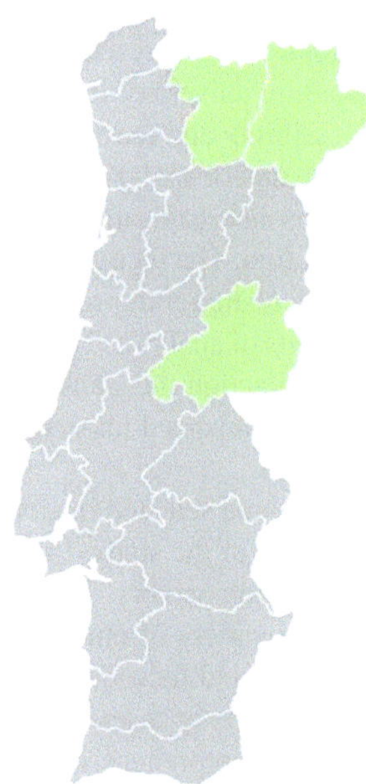

Table 1 Data on the districts under study. *Source* INE

	District 1		District 2		District 3	
	2013	2023	2013	2023	2013	2023
Resident population (no.)	133,121	122,739	198,964	184,525	187,179	177,719
Population density*	20.2	18.5	46.2	42.8	28.0	26.6
Total number of companies (no.)	6570	6399	7212	6399	6131	6399

* Habitants per km^2

According to Table 1, in district 1 over the last 10 years, the resident population has decreased by approximately 7.8%, reflecting emigration and low birth rates. Population density fell from around 20.2 to 18.5 in habitants per km^2, indicating a lower concentration of people in the territory. Regarding the number of companies and societies, this is possibly decreasing due to factors such as population aging and the migration of young people to other regions. Individual businesses have increased, with a trend towards individual entrepreneurship and micro-enterprises.

In District 2, the resident population decreased by about 7.3%. The population density reduced from 46.2 to 42.8 inhabitants per km^2, indicating a decrease in the number of people in the territory. The total number of companies and partnerships has declined, possibly due to factors such as population aging and migration, reducing sole proprietorship to 30.4% of sole proprietorships, suggesting challenges in individual entrepreneurship.

In the last and third district, the population decreased by about 5.1%, and the population density fell from 28.0 to 26.6 inhabitants per km^2, reducing the number of people in the territory. The total number of companies and partnerships increased, possibly due to economic development and entrepreneurship initiatives, however there was a 31.9% decrease in sole proprietorships.

3.3 Data Collection and Analysis

Data collection was carried out between February and April 2025, through 13 interviews with managers, technicians and entrepreneurs of three incubators. The interviews took place in person and via videoconference (via Google Meet), subject to the participants' availability, and lasted an average of approximately one hour. For its implementation, a semi-structured script was prepared, consisting of six main sections: (i) characterization of the interviewee; (ii) main local actors in the entrepreneurial ecosystem; (iii) role of incubators; (iv) challenges and limitations; (v) opportunities; and (vi) importance of the entrepreneurial ecosystem in low-density territories. In total, the guide included 13 open-ended questions, allowing for an in-depth exploration of the different dimensions of the phenomenon under study. Before the beginning of each interview, permission was requested to record and use the data for academic purposes, ensuring the confidentiality of the responses. Table 2 presents the characteristics of the interviewees.

After the interviews were conducted, the recordings were fully transcribed and the collected material was organized. Data analysis followed the content analysis technique, supported by NVIVO 15 software, which allowed the systematic coding of information and the identification of emerging patterns, categories and relationships. This process made it possible not only to answer the research questions, but also to understand in an integrated manner the role of incubators in boosting entrepreneurial ecosystems in low-density territories, as well as the factors that enhance or limit their impact.

Table 2 Data on contact persons for start-ups and the incubator

Region	Interviewee	Gender	Age	Qualifications	Type of organization	Position
1	1.1	Masculine	34	Master's in design	Start-up	CEO
	1.2	Masculine	29	Master's degree in mechanical engineering	Start-up	Director of operations
	1.3	Masculine	35	Postgraduate degree in marketing	Start-up	CEO
	1.4	Masculine	30	Master's degree in computer engineering	Start-up	Managing partner
	1.5	Feminino	47	Master's in management	Incubator	Reception technique
2	2.1	Masculine	33	PhD in computer science	Start-up	Managing partner
	2.2	Masculine	38	PhD in pharmaceutical sciences	Start-up	CEO
	2.3	Masculine	37	Bachelor's degree in computer engineering	Start-up	Managing partner
	2.4	Masculine	47	PhD in development, society and territories	Incubator	Director
3	3.1	Masculine	58	Bachelor's degree in systems analyst	Start-up	Director of operations
	3.2	Masculine	32	Master in innovation	Start-up	CEO
	3.3	Masculine	47	Bachelor's degree in sociology	Start-up	Technical director
	3.4	Masculine	47	Bachelor's degree in architecture	Incubator	Head of innovation

4 Results

4.1 Main Local Actors in the Entrepreneurial Ecosystem in Low-Density Territories

In low-density territories, the entrepreneurial ecosystem can be composed of several actors, such as SMEs, HEIs, incubators and local authorities, among others e.g., [4, 71]. Cooperation between these actors strengthens economic resilience and promotes the sustainability of the local entrepreneurial ecosystem [43]. Figure 2 shows the main local actors in the entrepreneurial ecosystem of the three low-density territories under study.

The results show that the entrepreneurial ecosystem of the three regions under study is composed of a set of local actors who play strategic roles in economic dynamism and the creation of new businesses. Each region presents a unique configuration of collaboration between these actors, reflecting different levels of coordination and effectiveness.

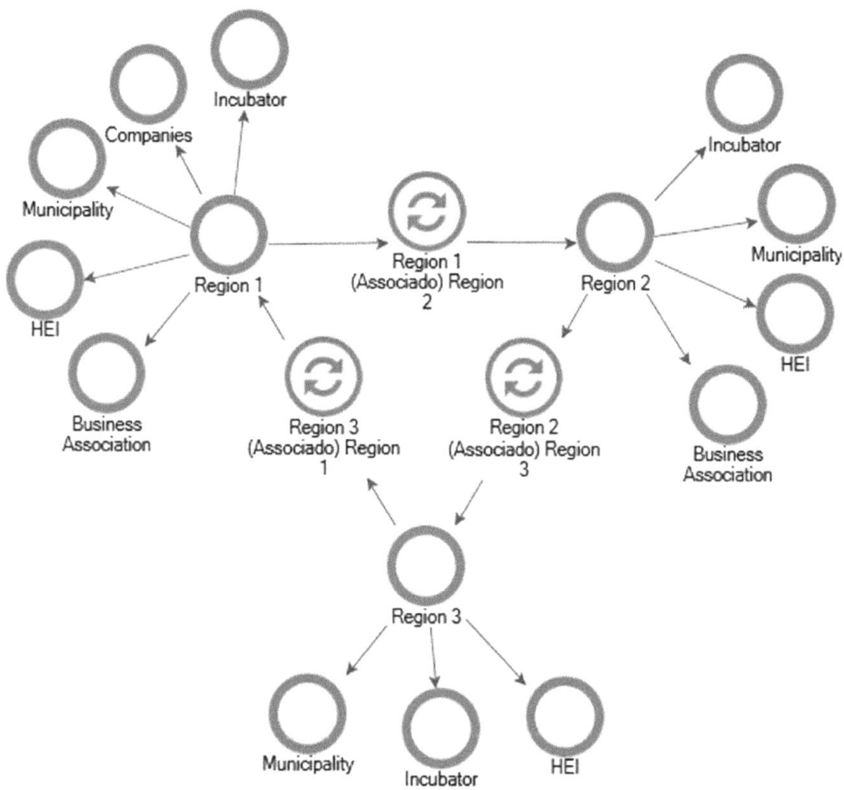

Fig. 2 Main local actors participating in the entrepreneurial ecosystem

In Region 1, the main actors mentioned in the interviews are the local incubator, the business association, higher education institutions (HEIs), the municipality, and private companies. The incubator acts as a central hub of the ecosystem, organizing networking events and creating coworking spaces that foster interaction between companies and universities. The university collaborates through internships and training programs aimed at entrepreneurs. In Region 2, the main actors are the incubator, the business association, the HEIs, and the municipality. Collaboration between these actors occurs primarily through joint projects aimed at regional development and the promotion of entrepreneurship. In Region 3, the main local actors are the incubator, two HEIs, and the municipality. In this region, collaboration is described as quite efficient, with the municipality playing a central role in supporting entrepreneurs and working closely with the universities and the incubator. Integrated innovation and talent attraction plans are in place, including digital observatories, coworking spaces, and ongoing mentoring programs.

In general, in low-density areas, the main actors in the entrepreneurial ecosystem are incubators, business associations, higher education institutions, and municipalities. Collaboration between these actors tends to occur through the sharing of resources, organization of joint events, training and mentoring programs, as well as the development of projects focused on innovation and strengthening the local business community.

Interviewee 1.5: "We are an example of this collaboration, because the incubator was born precisely from a partnership between the municipality and the region's higher education institution. Regarding the HEI and the municipality, there is very active participation, because the members are part of the board. But beyond that, there is always collaboration on projects that share in the economic dynamism. […]".

Interviewee 2.4: "Our incubator emerged from a partnership between the HEI and the municipality. […] We collaborated through a joint project to develop activities, whether promoting entrepreneurship or providing training."

Interviewee 3.3: "They collaborate very well; we have incubation programs. We also have a permanent support office for incubated companies, providing all the services to help us. […] I think there are synergies between everyone."

4.2 The Role of Incubators in the Entrepreneurial Ecosystem

Incubators can play a fundamental role in the entrepreneurial ecosystem, especially in low-density territories, by providing an essential support environment for the development of new businesses [71]. They offer a combination of resources, such as physical space, access to mentors, specialized training, and networks, that facilitate the incubation process and help entrepreneurs overcome initial barriers, such as lack of financing and business experience [41]. Furthermore, incubators act as catalysts for local transformation, driving economic development and sustainability in the communities in which they operate [52]. Figure 3 illustrates some of the roles of incubators in the entrepreneurial ecosystem.

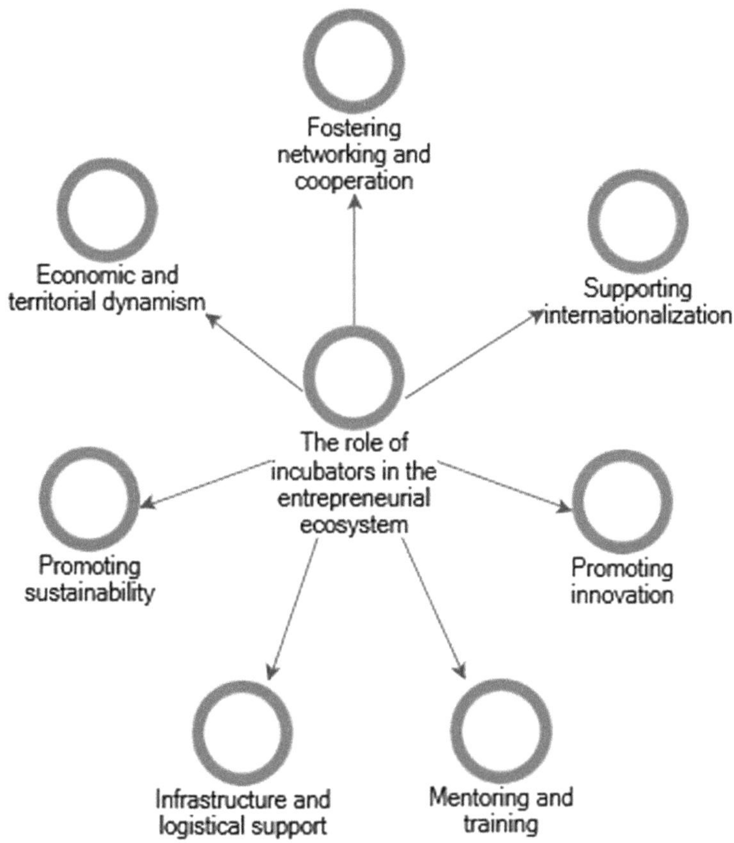

Fig. 3 Role of incubators in the entrepreneurial ecosystem

According to the interviewees, incubators can play a crucial role in the entrepreneurial ecosystem of low-density areas, acting as drivers of economic and social dynamism. They promote networking and cooperation between entrepreneurs, universities, business associations, and municipalities, organizing events, workshops, and mentoring programs that strengthen the local support network.

They also offer infrastructure support, providing accessible and equipped spaces that facilitate the establishment and growth of new businesses, something vital in regions with few private alternatives. In terms of training, incubators support entrepreneurs from the initial idea to the consolidation of their business, fostering skills in management, financing, and internationalization. Furthermore, they encourage innovation through partnerships with higher education institutions, facilitating the transfer of knowledge to the market.

In general, incubators are essential for creating jobs, retaining talent, and economically revitalizing territories, consolidating their position as strategic pillars for sustainable regional development.

Interviewee 1.4: "The incubator plays an important role because it helps you take the first steps by providing physical space [...]. It also [...] acts as an accelerator, providing access to information about support and applications, helping you develop your business plan, and hosting networking events, all of which help to streamline this aspect."

Interviewee 2.2: "I think it's important because perhaps if I wanted to develop my business any other way, without the incubator and without the incubator's support, things would take longer and delay everything. And it helped me leverage my business. The incubator attracts people, attracts them to the city. And the existence of these infrastructures will promote and foster entrepreneurship and innovation."

Interviewee 3.4: "Through networks, talent acquisition, and an innovation plan that we are currently implementing to create an ecosystem, attract talent, and attract people. Over the past 10 years, we have been building an ecosystem that will deliver results. Through these collaborations, we created a digital manufacturing observatory and a coworking space, and we currently have 11 coworking spaces. Through the creation of online platforms and tools, [...] it generates wealth for the region through the creation of companies and population growth, attracting investment for the business."

4.2.1 Promoting Entrepreneurship and Innovation

Promoting entrepreneurship and innovation is essential for economic and social development, especially in regions with structural challenges, such as those with low population density [57]. Fostering entrepreneurship and innovation not only drives the creation of new businesses, but also contributes to economic revitalization, generating employment and promoting the diversification of economic sectors in local communities [8], as illustrated in Fig. 4.

Based on the analysis of the interviews, the role of incubators in fostering entrepreneurship and innovation is multifaceted and essential to boosting local development. Incubators contribute primarily through organizing networking events, workshops, and training sessions, which encourage the exchange of experiences and ideas between entrepreneurs, universities, and companies. They facilitate the development of business plans, offer ongoing mentoring and support in structuring financing applications, strengthening entrepreneurs' skills from the initial phase through to business expansion.

Innovation is promoted primarily through partnerships with higher education institutions, which allows academic research to connect with the market, transforming scientific knowledge into innovative products and services. Some incubators also stand out for implementing specific programs to stimulate new business ideas and encourage sustainability and circular economy practices, preparing companies to meet the demands of current and future markets.

Interviewee 1.4: "The incubator has projects that we develop in partnership with universities [...]. What we try to do is encourage the generation of ideas, to bring together the research work developed by universities and bring it to new business

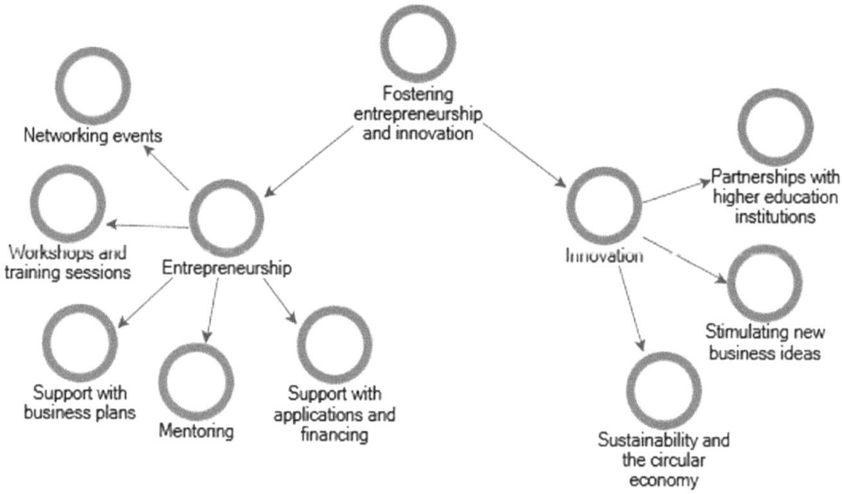

Fig. 4 Fostering entrepreneurship and innovation

ideas [...]. And then there's another aspect, which is the training we provide to current and future entrepreneurs [...] trying to raise awareness, even providing some knowledge of some skills that are important in these areas."

Interviewee 2.2: "Regarding access to a range of infrastructure at a reduced price. Here at the incubator, they're always sending out information about all kinds of training, not only online but also in-person, on business creation, job creation, and funding sources, not only at the national level but also at the European level. [...] By having these training courses, we can promote our products and the services we provide. [...] There's a mutually supportive relationship between companies."

Interviewee 3.1: "In addition to organizing numerous events, attracting people from abroad, it participates in very important events, both in Portugal and abroad. This way, entrepreneurship becomes more visible."

4.2.2 Investment Attraction

Incubators play a crucial role in attracting investment by offering a structured environment that reduces the risks associated with investing in startups and new businesses. According to Lichtenstein and Lyons [44], incubators act as intermediaries between entrepreneurs and investors, creating an ecosystem conducive to capital raising. Furthermore, by fostering strategic partnerships with angel investors, venture capital funds, and other financial institutions, incubators help startups secure the financing they need to expand, directly contributing to the growth of the local entrepreneurial ecosystem [26]. Figure 5 shows the interviewees' opinion on the role of incubators in attracting investment.

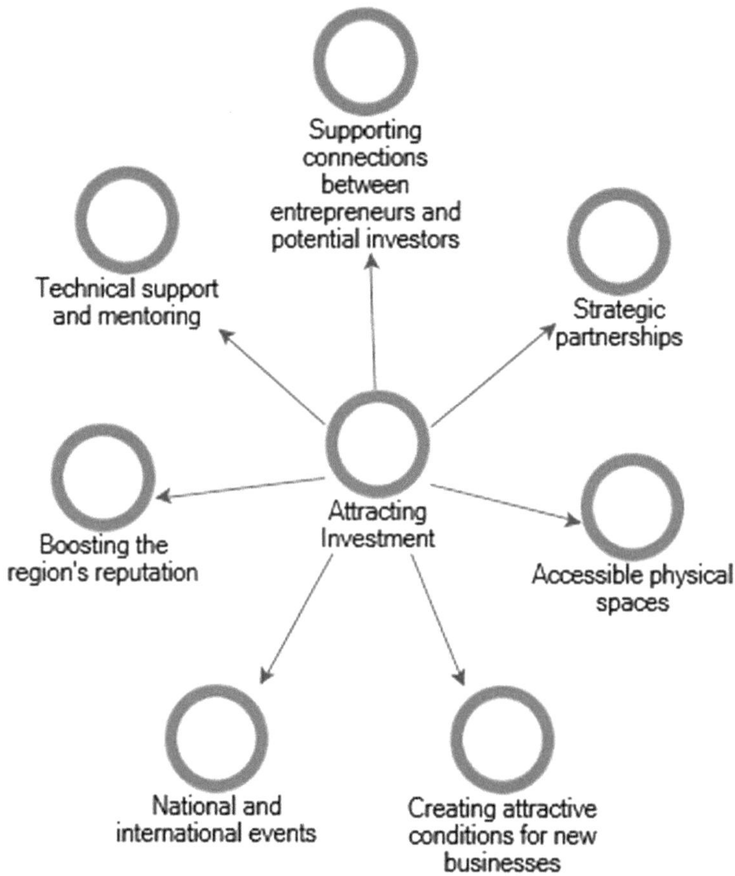

Fig. 5 Role of incubators in attracting investment

Interviewees report that in these low-density territories, incubators are fundamental in attracting investment to their regions, despite the various challenges that exist (e.g., population loss, aging population, etc.). Incubators have contributed primarily by creating attractive conditions for new companies, offering accessible physical spaces, technical support, and mentoring that make business establishment and growth more viable. Furthermore, they participate in national and international events, promoting regions and their ecosystems to external investors and establishing strategic partnerships with other incubators and international organizations.

Another important aspect is supporting connections between entrepreneurs and potential investors, whether through idea contests, business fairs, or local initiatives to attract capital. Incubators also play an indirect role by enhancing the region's reputation as a place conducive to business development, which is essential for attracting new investment.

Some of those interviewed reported that sustainability is integrated into daily practices and training offered to entrepreneurs, through workshops, awareness raising for ecological practices and encouraging the adoption of circular economy models. This effort aims to align incubated businesses with European and global environmental goals, highlighting, for example, actions related to carbon footprint and energy efficiency.

Some incubators also implement internal sustainable measures, such as installing water reuse and waste separation systems, in addition to encouraging businesses to adopt similar practices.

Interviewee 1.5: "We hold sustainability awareness events [...]. We try to talk about the importance of these issues, and especially highlight the need to align them with European targets for these topics."

Interviewee 2.4: "We encourage recycling from the outset with collection points and sensors. In our day-to-day operations, we are an eco-park; the entire park itself is based on sustainability, based on an energy-efficient construction approach. We try to raise awareness about efficient management of building costs."

Interviewee 3.2: "[...] I think the incubator practices sustainability a lot in the training they provide. At my company, we analyze the carbon footprint of products, and I see the incubator already thinking about this, helping companies create sustainable products and fostering a circular economy. Sustainability is part of our daily lives as incubated companies."

However, some interviewees stated that sustainability is not yet a clear priority, pointing to the lack of concrete initiatives in this area or the need to strengthen sustainable policies among incubated companies.

Interviewee 2.1: "As far as I know, the incubator in our region does not promote sustainability practices [...]".

Interviewee 3.3: "In my opinion, I don't remember any initiatives for sustainability practices. [...]".

4.3 Challenges and Limitations

According to Almeida and Daniel [5], incubators in low-density territories face specific challenges, such as a shortage of financial resources, a lack of a robust business network, and difficulty in attracting qualified talent. Furthermore, low population density can result in a restricted local market, which reduces partnership opportunities and the demand for new products and services. Figure 7 illustrates the main challenges or limitations that incubators face in low-density territories.

Based on the interviews conducted, the challenges faced by incubators are quite similar in the three regions under study. One of the main challenges identified is the lack of qualified labor, which limits both the growth of incubated companies and the ability to attract new companies to these regions. Furthermore, talent retention is highlighted as a recurring problem, as many young people and professionals tend to migrate to larger urban centers in search of better opportunities. Another

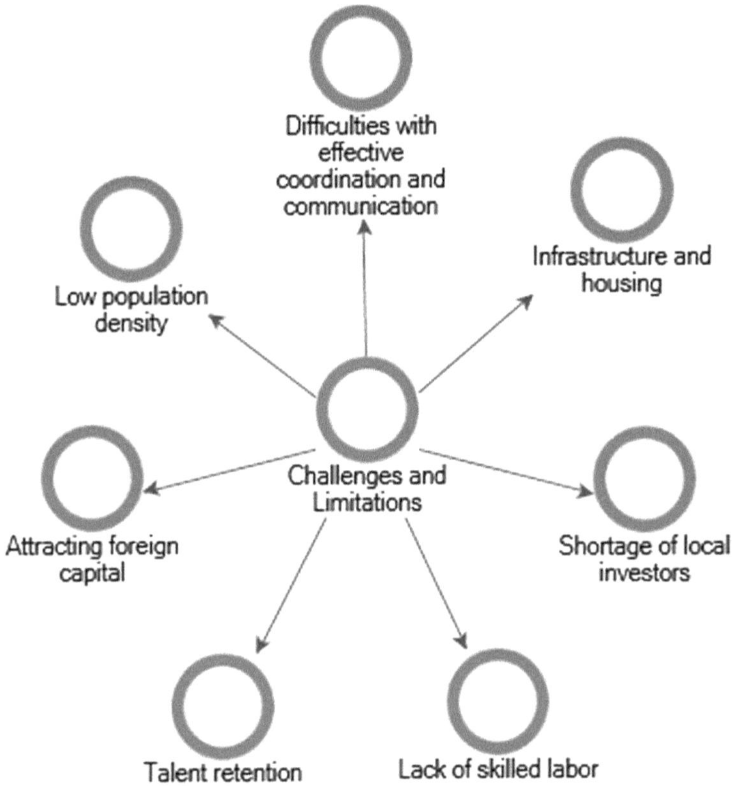

Fig. 7 Main challenges or limitations that incubators face in low-density territories

obstacle highlighted is the low population density, which hinders the creation of strong networks and limits the local market available for the products and services of incubated companies.

The scarcity of local investors and the need to attract external capital also emerge as significant limitations, requiring greater efforts from incubators to present their regions as attractive for investment. Additionally, several interviewees mentioned difficulties in coordination and effective communication between incubators, universities, and other ecosystem actors, which can weaken cooperation and reduce the impact of joint initiatives.

Finally, issues related to infrastructure and housing are also mentioned, especially with regard to providing adequate spaces to accommodate new companies and professionals.

Interviewee 1.5: "The main challenge is the fact that we don't always have specialized labor here, the lack of funding for the development of certain activities […]."

Interviewee 2.4: "[…] The main challenge is to promote the four corners of the region and foster a network that allows us to support entrepreneurs with all the necessary conditions. To achieve this, there must be [...] infrastructure, such as a coworking space in each municipality [...]. Then we also need to consider training, because we often know that municipalities lack the technical teams prepared to efficiently deliver entrepreneurial support in terms of training and concepts. [...] access to information for entrepreneurs can often be ineffective [...]."

Interviewee 3.1: "The external feeling that we're far from everything. People don't want to come to the countryside. Sometimes these are misconceptions. That's the big challenge."

4.4 Opportunities

To strengthen the entrepreneurial ecosystem in low-density regions, there are several specific opportunities, such as creating strategic partnerships with local universities and research centers, which can boost innovation and knowledge transfer [5]. Figure 8 shows the main opportunities for strengthening the entrepreneurial ecosystem in the region.

Based on the 13 interviews analyzed, several opportunities can be identified to strengthen the entrepreneurial ecosystem in low-density regions. One of the main opportunities identified is strengthening connections between universities and entrepreneurs, leveraging the innovation and scientific research potential of higher education institutions to generate new businesses and technological solutions. Creating more incubation and acceleration programs, involving young people and recent graduates, is seen as an effective way to stimulate local entrepreneurship and prevent talent flight.

Another highlighted opportunity is the development of stronger synergies between incubated companies, creating collaborative networks that facilitate the sharing of resources, knowledge, and markets. Exploring endogenous resources, such as local products and regional traditions, also emerges as a competitive advantage, especially when combined with innovation and sustainability. Participation in national and international events and strengthening partnerships with incubators and external investors are mentioned as effective strategies for attracting investment and increasing the visibility of incubated companies.

Finally, interviewees suggest the need to improve local infrastructure, including more affordable coworking spaces and housing programs that attract professionals to the regions, creating favorable conditions for companies and talent to choose to remain.

Interviewee 1.5: "I think the opportunities lie a little in fostering a change in mentality towards greater cooperation [...] companies together can have increased growth potential [...], there is a great opportunity in all business ideas that can combine innovation with tradition, that which is the identity of the territory that is different. [...]."

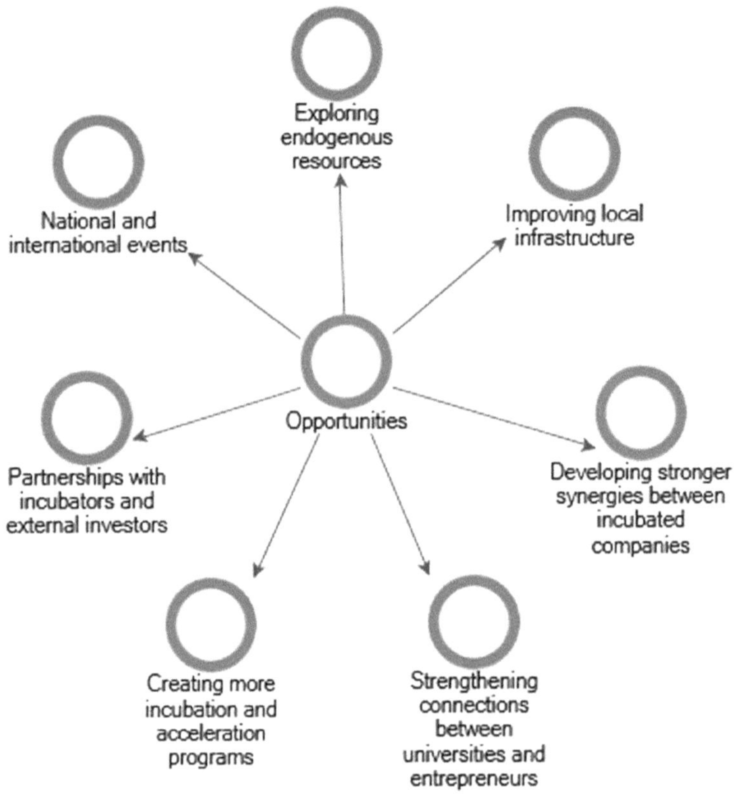

Fig. 8 Main opportunities to strengthen the entrepreneurial ecosystem in the region

Interviewee 2.1: "In my opinion, I think it's the partnerships and synergies that an incubator has. For example, I think that entities can create more and more synergies among themselves. […]".

Interviewee 3.4: "[…] The interior has immense opportunities. We have financing that the big cities don't. We leverage the innovation ecosystem, identifying and developing tools to help businesses grow, in the way and where we position ourselves. Not all interior regions are aware of these programs."

4.5 Importance of the Entrepreneurial Ecosystem in Low-Density Territories

The entrepreneurial ecosystem in low-density territories is crucial for the economic revitalization and sustainability of these regions, as it promotes the creation of new businesses, generates employment, and attracts investment [5]. In a context of scarce resources and limited opportunities, entrepreneurship can be an essential tool to combat depopulation and stimulate local development [15]. A well-structured entrepreneurial ecosystem not only strengthens the regional economy but also improves the quality of life of communities by offering solutions tailored to their specific needs and challenges [67].

The results show that all interviewees recognize the importance of the entrepreneurial ecosystem in the region, as it enables the economic and social development of these regions.

The entrepreneurial ecosystem is seen as an essential driver for job creation, population retention, and economic diversification, critical factors for combating desertification and revitalizing the local social and economic fabric. Furthermore, it provides a supportive and innovative environment that facilitates the creation of new businesses and increases the region's attractiveness to talent and investors. The interaction between companies, universities, and public entities contributes to regional dynamism and creates opportunities for sustainable growth.

Regarding the future of incubators in these regions, interviews indicate that these structures will continue to play a vital role. Incubators are seen as key players in fostering innovation, attracting investment, and supporting business internationalization, in addition to playing a growing role in promoting sustainability and adapting to new economic dynamics. There is a perception that their future success depends on the continued evolution of their offerings, such as more mentoring, acceleration, and strategic partnership programs, as well as strengthening ties with universities and other research centers.

Interviewee 1.5: "I think the entrepreneurial ecosystem is a tool that helps capture ideas based on the territory, which can have greater growth potential here. Specifically in these low-density territories, it's important to have an entrepreneurial ecosystem, so that people also have this as a safe haven to seek any kind of help they need [...]."

Interviewee 2.4: "It's vital because it has a life of its own. The hard part is creating it because then it grows, is destroyed, and regenerates again, and that's very important. [...] Because it's fueled by the very organic nature of the entrepreneurs, but also by the stakeholders who nurture it every day."

Interviewee 3.1: "It's very important because it's what drives the creation of new businesses, new opportunities, and attracts people. I think what the interior needs is people, but to achieve that, we need to provide opportunities and conditions for people to come and want to stay."

5 Discussion of Results

The study results show that incubators play a central role in entrepreneurial ecosystems in LDT, promoting innovation, the creation of competitive businesses, and networking among local stakeholders. In light of the dynamic capability theory [33, 69], it is possible to interpret this evidence considering three fundamental dimensions: adaptive capacity, absorptive capacity and innovation capacity.

First, regarding adaptive capacity, incubators have proven crucial in helping entrepreneurs identify and exploit opportunities in contexts characterized by structural constraints, such as talent shortages and fragile local economic structures. As suggested by Abrate et al. [1], adapting to external changes requires organizational flexibility, and the data reveal that incubators facilitate this flexibility by creating conditions for attracting investment and reconfiguring business strategies. Close collaboration with municipalities and higher education institutions reinforces this adaptive capacity, allowing incubated companies to align their strategies with regional development policies.

Absorptive capacity, defined as the ability to recognize, assimilate, and apply external knowledge [14, 33], also clearly emerges. Incubators function as intermediation mechanisms, connecting entrepreneurs to knowledge and innovation networks, primarily through partnerships with universities and business associations. Supporting funding applications, promoting workshops, and connecting with external investors are practical examples of how external information and resources are internalized and transformed into strategic assets for entrepreneurs. These findings corroborate Wei et al. [73], who argue that incubators strengthen startups' absorptive capacity by fostering the integration of new knowledge in contexts of high uncertainty.

Finally, in terms of innovation capacity, the results reveal that incubators play a decisive role in stimulating the creation of innovative products, services, and business models, often based on the valorization of endogenous resources and sustainability practices. The connection between academic research and practical application in local markets translates into incremental innovation processes and, in some cases, more radical innovation. As Lin & Wu [45] argue, innovation results from the articulation of internal resources and external opportunities, something that is evident in mentoring programs, digital observatories, and circular economy practices supported by the incubators analyzed.

However, the results also confirm some limitations in building these dynamic capabilities. The shortage of skilled labor, low population density, and dependence on external capital restrict the full development of adaptive, absorption, and innovation mechanisms. These barriers align with the conclusions of Almeida and Daniel [5], who highlight the structural constraints that affect entrepreneurial ecosystems in LDT, reinforcing the need for public policies that complement the actions of incubators.

Cross-sectionally, the analysis suggests that incubators not only offer physical and technical resources but, above all, serve as platforms for building dynamic capabilities, essential for incubated companies to respond to the pressures of constantly changing markets. In this sense, the results corroborate previous studies [71, 72], which emphasize that the impact of incubators goes beyond logistical support, translating into the creation of strategic capabilities that increase regional resilience and competitiveness.

6 Conclusion

This study allowed us to analyze the role of incubators in strengthening entrepreneurial ecosystems in low-density territories in Portugal, mobilizing the theory of dynamic capacity as an interpretative framework. The results demonstrate that incubators play a strategic role not only as providers of infrastructure and support services, but above all as catalysts for processes of adaptation, knowledge absorption and innovation, essential to the competitiveness and sustainability of incubated companies.

In territories marked by structural constraints (such as a shortage of skilled labor, low population density, and limited access to financing), incubators emerge as critical platforms for creating collaborative networks between municipalities, higher education institutions, business associations, and entrepreneurs. These networks foster synergies, facilitate knowledge transfer, and contribute to the valorization of endogenous resources, strengthening the economic and social resilience of regions.

The analysis also shows that incubators not only foster entrepreneurship and innovation but also play a key role in attracting investment, creating competitive businesses, and promoting sustainable practices. By combining local resources with external opportunities, these structures help transform adverse contexts into spaces of opportunity, mitigating the risks associated with economic and social desertification.

However, the study also confirms that the effectiveness of incubators remains conditioned by structural factors that limit the full activation of the dynamic capabilities of incubated companies. Therefore, it is imperative to complement the work of incubators with consistent public policies to support entrepreneurship in low-density regions, capable of strengthening talent acquisition and retention, access to capital, and integration into national and international innovation networks.

Based on the results obtained, this study presents theoretical and practical implications. As theoretical implications, this study contributes to the existing literature by deepening the understanding of the role of incubators in low-density territories, a context still underexplored in academic research on entrepreneurial ecosystems. By leveraging dynamic capability theory, it is clear that incubators do not simply provide support services, but also act as platforms for building adaptive, absorptive, and innovative capabilities, essential for the survival and competitiveness of startups. Thus, this work expands the application of dynamic capability theory to the field of entrepreneurship research in peripheral regions, providing a conceptual perspective that articulates resources, networks, and territorial dynamics. Furthermore, the study reinforces the literature on entrepreneurial ecosystems by highlighting the strategic role of incubators as mediators between local and external actors, contributing to regional resilience and sustainability.

In terms of practical implications, the results offer relevant insights for both incubator managers and policymakers. For incubators, the study demonstrates the importance of strengthening cooperation networks with universities, municipalities, and business associations, as well as promoting talent retention and investment strategies, which are crucial to overcoming the structural constraints of low-density regions. For policymakers, this study highlights the need for more consistent public policies to support entrepreneurship in peripheral territories, including tax incentives, tailored financing programs, and measures to promote the retention of a qualified population. By providing evidence on the strategic impact of incubators, this work contributes to the formulation of more effective policies and the construction of entrepreneurial ecosystems capable of supporting sustainable regional development in low-density territories.

Despite its contributions, this study has some limitations that are important to acknowledge. First, the analysis was based on a limited number of incubators and entrepreneurs located in low-density areas in the interior of Portugal, which may limit the generalizability of the results to other national or international contexts. Second, the qualitative approach, while suitable for exploring local perceptions and dynamics in depth, does not allow for systematically quantifying the impact of incubators on economic performance or job creation. Finally, data collection through interviews may be subject to biases in participants' perceptions, reflecting individual experiences that do not always reflect the full reality of the ecosystem.

Given these limitations, opportunities arise for future research. First, it would be appropriate to expand the sample to a larger number of incubators and territories, including comparisons between low- and high-density regions, to identify structural and contextual differences. Second, we recommend using quantitative or mixed methodologies to more objectively assess the impact of incubators on indicators such as business growth, job creation, or investment attraction. Finally, it would be relevant to explore the longitudinal perspective, analyzing the evolution of incubated companies over time, as well as the role of incubators in building more resilient and sustainable ecosystems in scenarios of economic and demographic transformation.

Funding This work is supported by national funds, through the FCT–Portuguese Foundation for Science and Technology under the projects UID/04011/2025 (https://doi.org/10.54499/UID/04011/2025), UID/04630/2025 (https://doi.org/10.54499/UID/04630/2025) and Project EECreLowD, no. 14886 (operation no. COMPETE2030-FEDER-00845000), co-financed by ERDF through COMPETE2030.

References

1. Abrate, G., Bruno, C., Erbetta, F., & Fraquelli, G. (2020). Which future for traditional travel agencies? A dynamic capabilities approach. *Journal of Travel Research, 59*(5), 777–791.
2. Acs, Z. J., Stam, E., Audretsch, D. B., & O'Connor, A. (2017). The lineages of the entrepreneurial ecosystem approach. *Small Business Economics, 49*(1), 1–10.
3. Alerasoul, S. A., Tiberius, V., & Bouncken, R. B. (2022). Entrepreneurship and innovation: The coevolution of two fields. *Journal of Small Business Strategy, 32*(2).
4. Almeida, J., Daniel, A. D., & Botelho, A. (2023). Local development through entrepreneurship and innovation ecosystems. In *Territorial innovation in less developed regions: Governance, technologies, and sustainability* (pp. 245–266). Springer International Publishing.
5. Almeida, J., & Daniel, A. D. (2025). Addressing the distinctive features of entrepreneurial ecosystems in low-density territories. Journal of entrepreneurship and public. *Policy*.

6. Asmit, B., Simatupang, T. M., Rudito, B., & Novani, S. (2024). Uncovering the building blocks of rural entrepreneurship: A comprehensive framework for mapping the components of rural entrepreneurial ecosystems. *Heliyon, 10*(1).
7. Becker, S. D., & Endenich, C. (2023). Entrepreneurial ecosystems as amplifiers of the lean Startup philosophy: Management control practices in earliest-stage Startups. *Contemporary Accounting Research, 40*(1), 624–667.
8. Bernhard, I., Gråsjö, U., & Karlsson, C. (2021). Introduction to unlocking regional innovation and entrepreneurship: The potential for increasing capacities. In *Unlocking regional innovation and entrepreneurship* (pp. 1–13). Edward Elgar Publishing.
9. Bryman, A., Foster, L., Sloan, L., & Clark, T. (2021). *Bryman's social research methods* (6th ed.). Oxford University Press.
10. Butz, H., & Mrożewski, M. J. (2021). The selection process and criteria of impact accelerators. *An Exploratory Study. Sustainability, 13*(12), 6617.
11. Campos, F., Frese, M., Goldstein, M., Iacovone, L., Johnson, H. C., McKenzie, D., & Mensmann, M. (2017). Teaching personal initiative beats traditional training in boosting small business in West Africa. *Science, 357*(6357), 1287–1290.
12. Cantner, U., Cunningham, J. A., Lehmann, E. E., et al. (2021). Entrepreneurial ecosystems: A dynamic lifecycle model. *Small Business Economics, 57*, 407–423.
13. Coduras, A., Urbano, D., Rojas, Á., & Martínez, S. (2008). The relationship between university support to entrepreneurship with entrepreneurial activity in Spain: A gem data based analysis. *International Advances in Economic Research, 14*(4), 395–406.
14. Cohen, W. M., & Levinthal, D. A. (1990). Absorptive capacity: A new perspective on learning and innovation. *Administrative Science Quarterly, 35*(1), 128–152.
15. Del Olmo-García, F., Domínguez-Fabián, I., Crecente-Romero, F. J., & del Val- Núñez, M. T. (2023). Determinant factors for the development of rural entrepreneurship. *Technological Forecasting and Social Change, 191*, 122487.
16. Dutt, N., Hawn, O., Vidal, E., Chatterji, A., McGahan, A., & Mitchell, W. (2016). How open system intermediaries address institutional failures: The case of business incubators in emerging-market countries. *Academy of Management Journal, 59*(3), 818–840.
17. Elisabete, S., Casais, B., & Silva, J. R. (2019). Local development through rural entrepreneurship, from the triple helix perspective. *International Journal of Entrepreneurial Behaviour & Research, 25*(4), 698–716.
18. Faggio, G., & Silva, O. (2014). Self-employment and entrepreneurship in urban and rural labour markets. *Journal of Urban Economics, 84*, 67–85.
19. Ferreira, J. J. M., Fernandes, C. I., & Kraus, S. (2019). Entrepreneurship research: Mapping intellectual structures and research trends. *Review of Managerial Science, 13*, 181–205.
20. Forliano, C., Ferraris, A., Bivona, E., & Couturier, J. (2022). Pouring new wine into old bottles: A dynamic perspective of the interplay among environmental dynamism, capabilities development, and performance. *Journal of Business Research, 142*, 448–463.
21. Fortunato, M. W. P. (2014). Supporting rural entrepreneurship: A review of conceptual developments from research to practice. *Community Development, 45*(4), 387–408.
22. Furnival, J., Boaden, R., & Walshe, K. (2019). A dynamic capabilities view of improvement capability. *Journal of Health Organization and Management, 33*(7/8), 821–834.
23. Giannakis, E., & Bruggeman, A. (2019). Regional disparities in economic resilience in the European Union across the urban–rural divide. *Regional Studies, 54*(9), 1–14.
24. Grijalvo Martín, M., Pacios Álvarez, A., Ordieres-Meré, J., Villalba-Díez, J., & Morales-Alonso, G. (2020). New business models from prescriptive maintenance strategies aligned with sustainable development goals. *Sustainability, 13*(1), 216.
25. Grimaldi, R., & Grandi, A. (2005). Business incubators and new venture creation: An assessment of incubating models. *Technovation, 25*(2), 111–121.
26. Griol-Barres, I., & Morant-Martinez, O. (2023). The role of incubators and accelerators in entrepreneurial fundraising. *Contributions to Finance and Accounting*, 11–22.
27. Gyımah, P., & Lussier, R. N. (2021). Rural entrepreneurship success factors: An empirical investigation in an emerging market. *Journal of Small Business Strategy, 31*(4), 5–19.

28. Gómez-Araujo, E., & Bayon, M. C. (2017). Socio-cultural factors and youth entrepreneurship in rural regions. *Revista brasileira de gestão de negócios, 19*, 200–218.
29. Hackett, S. M., & Dilts, D. M. (2004). A systematic review of business incubation research. *The Journal of Technology Transfer, 29*(1), 55–82.
30. Hakim, A. I., Sukimi, M. F., & Hafiz, A. (2024). Exploring the role of business incubators to sustainable Startups: A systematic literature review. *Paper Asia, 40*(5b), 307–320.
31. Haugh, H. (2019). Call the midwife! Business incubators as entrepreneurial enablers in developing economies. *Entrepreneurship & Regional Development, 32*(1–2), 156–175.
32. Hernández, R., & Carrà, G. (2016). A conceptual approach for business incubator interdependence and sustainable development. *Agriculture and Agricultural Science Procedia, 8*, 718–724.
33. Jantunen, A., Ellonen, H.-K., & Johansson, A. (2012). Beyond appearances—Do dynamic capabilities of innovative firms actually differ? *European Management Journal, 30*(2), 141–155.
34. Kalantaridis, C., & Bika, Z. (2011). Entrepreneurial origin and the configuration of innovation in rural areas: The case of Cumbria, north West England. *Environment and Planning A, 43*(4), 866–884.
35. Karambakuwa, J. K. (2021). Influence of incubation hubs on start-ups in Harare, Zimbabwe (doctoral dissertation).
36. Klofsten, M., Bienkowska, D., Laur, I., & Sölvell, I. (2015). Success factors in cluster initiative management: Mapping out the 'big five'. *Industry and Higher Education, 29*(1), 65–77.
37. Kojo Oseifuah, E. (2010). Financial literacy and youth entrepreneurship in South Africa. *African Journal of Economic and Management Studies, 1*(2), 164–182.
38. Labrianidis, L. (2006). Fostering entrepreneurship as a means to overcome barriers to development of rural peripheral areas in Europe. *European Planning Studies, 14*(1), 3–8.
39. Lalkaka, R. (1996). Technology business incubators: Critical determinants of success. *Annals of the New York Academy of Sciences, 798*(1), 270–290.
40. Lemus-Aguilar, I., Morales-Alonso, G., Ramirez-Portilla, A., & Hidalgo, A. (2019). Sustainable business models through the lens of organizational design: A systematic literature review. *Sustainability, 11*(19), 5379.
41. Lesakova, L. (2012). The role of business incubators in supporting the SME start-up. *Acta Polytechnica Hungarica, 9*(3), 85–95.
42. Li, Y., Kenney, M., Patton, D., & Song, A. (2023). Entrepreneurial ecosystems and industry knowledge: Does the winning region take all? *Small Business Economics, 61*(1), 153–172.
43. Li, Q., & Mostafavi, A. (2021). Local interactions and homophily effects in actor collaboration networks for urban resilience governance. *Applied Network Science, 6*(1), 89.
44. Lichtenstein, G. A., & Lyons, T. S. (2001). The entrepreneurial development system: Transforming business talent and community economies. *Economic Development Quarterly, 15*(1), 3–20.
45. Lin, Y., & Wu, L.-Y. (2014). Exploring the role of dynamic capabilities in firm performance under the resource-based view framework. *Journal of Business Research, 67*(3), 407–413.
46. Lin-Lian, C., De-Pablos-Heredero, C., & Montes-Botella, J. L. (2022). Exploring the relationship between the entrepreneurship motive and value creation in business incubators. *Sustainability, 14*(13), 7758.
47. Llorent-Bedmar, V., Cobano-Delgado Palma, V. C., & Navarro-Granados, M. (2021). The rural exodus of young people from empty Spain. *Socio-educational aspects. Journal of Rural Studies, 82*, 303–314.
48. Lucarelli, C., Marinelli, N., Micozzi, A., Palazzo, M., & Sica, G. (2025). Analysing the influence of gender and gender diversity in network-building behaviour and entrepreneurial ecosystem. *International Entrepreneurship and Management Journal, 21*(1).
49. Madaleno, M., Nathan, M., Overman, H., & Waights, S. (2022). Incubators, accelerators and urban economic development. *Urban Studies, 59*(2), 281–300.
50. Maxwell, J. A. (2020). The value of qualitative inquiry for public policy. *Qualitative Inquiry, 26*(2), 177–186.

51. Mei, W., & Symaco, L. (2022). University-wide entrepreneurship education in China's higher education institutions: Issues and challenges. *Studies in Higher Education, 47*(1), 177–193. https://doi.org/10.1080/03075079.2020.1735330
52. Olkiewicz, M., Wolniak, R., Eva-Grebski, M., & Olkiewicz, A. (2018). Comparative analysis of the impact of the business incubator center on the economic sustainable development of regions in USA and Poland. *Sustainability, 11*(1), 173.
53. Oswald, K., & Zhao, X. (2020). What is a sustainable Coworking space? *Sustainability, 12*(24), 10547.
54. Panda, L. P., Rath, K. C., & Rao, N. J. (2025). Entrepreneurial empowerment: Maximizing potential through business incubators. In *Promoting entrepreneurship and innovation through business incubation* (pp. 1–28). IGI Global.
55. Pato, M., & Teixeira, A. (2020). Determinants of performance of new ventures located in Portuguese incubators and science parks with a focus on institutional factors: Do rural and urban new ventures differ? *European Planning Studies, 28*(8), 1–20.
56. Poblete, C., & Amorós, J. E. (2013). Determinantes en la Estrategia de Internacionalización para las Pymes: el Caso de Chile. *Journal of Technology Management & Innovation, 8*(1), 97–106.
57. Ren, M., & Wu, H. (2023). How Innovation and Entrepreneurship affect economic growth? Government support and financial development as moderators. *International Journal of Scholarly Papers, 22*(3A (60A)), 869–887.
58. Rodríguez-Rodríguez, D., & Larrubia Vargas, R. (2022). Protected areas and rural depopulation in Spain: A multi-stakeholder perceptual study. *Land, 11*(3), 384.
59. Samuelsson, M., & Jutterström, M. (2023). An empirical test of sponsorship theory on a population of business incubators between 2005 and 2017–is it working or not? *The Innovation*, 1–27.
60. Sharma, G. (2024). *'Making'in India: Understanding makerspaces and Fablabs in the Indian informal innovation context* (Vol. 2) for Innovation Research (CIRCLE).
61. Sperindé, S., & Nguyen-Duc, A. (2020). Fostering open innovation in coworking spaces: A study of norwegian startups. In *Fundamentals of software startups: Essential engineering and business aspects* (pp. 161–178). Springer International Publishing.
62. Stam, E. (2015). Entrepreneurial ecosystems and regional policy: A sympathetic critique. *European Planning Studies, 23*(9), 1759–1769.
63. Stam, E., & Van de Ven, A. (2021). *Entrepreneurial ecosystem elements. Small business economics, 56*(2), 809–832.
64. Startup Portugal. (2022). Portugal Startup ecosystem report. Available online: https://startupportugal.com/wp-content/uploads/2022/03/GG-countryreport-Portugal.Pdf. Accessed 5 July 2025.
65. Stuckey, H. (2013). Three types of interviews: Qualitative research methods in social health. *Journal of Social Health and Diabetes, 01*(02), 056–059.
66. Surana, K., Singh, A., & Sagar, A. D. (2020). Strengthening science, technology, and innovation-based incubators to help achieve sustainable development goals: Lessons from India. *Technological Forecasting and Social Change, 157*, 120057.
67. Szerb, L., Lafuente, E., Horváth, K., & Páger, B. (2019). The relevance of quantity and quality entrepreneurship for regional performance: The moderating role of the entrepreneurial ecosystem. *Regional Studies, 53*(9), 1308–1320.
68. Tabaklar, T., Sorkun, M. F., Yurt, O., & Yu, W. (2021). Exploring the microfoundations of dynamic capabilities for social innovation in a humanitarian aid supply network setting. *Industrial Marketing Management, 96*, 147–162.
69. Teece, D. J., Pisano, G., & Shuen, A. (1997). Dynamic capabilities and strategic management. *Strategic Management Journal, 18*(7), 509–533.
70. Tripathi, N., & Oivo, M. (2020). The roles of incubators, accelerators, co-working spaces, mentors, and events in the Startup development process. *Fundamentals of Software Startups*, 147–159.

71. Virkkala, P., & Saarela, M. (2024). Developing and testing a rural business incubator: A best practice framework. *European Conference on Innovation and Entrepreneurship, 19*(1), 817–825.
72. Wang, Z., He, Q., Xia, S., Sarpong, D., Xiong, A., & Maas, G. (2020). Capacities of business incubator and regional innovation performance. *Technological Forecasting and Social Change, 158*, 120125.
73. Wei, S., Zhang, Z., & Chen, X. (2021). How does business incubator motivate start-ups based on super-network: A dynamic capability perspective. *Journal of Intelligent & Fuzzy Systems*, 1–14.
74. Wright, M., Siegel, D. S., & Mustar, P. (2017). An emerging ecosystem for student start- ups. *The Journal of Technology Transfer, 42*, 909–922.
75. Wurth, B., Stam, E., & Spigel, B. (2023). Entrepreneurial ecosystem mechanisms. *Foundations and Trends in Entrepreneurship, 19*(3), 224–339.
76. Yin, R. K. (2018). *Case study research and applications* (Vol. 6). Sage.
77. Zaidi, R. A., Khan, M. M., Khan, R. A., & Mujtaba, B. G. (2023). Do entrepreneurship ecosystem and managerial skills contribute to startup development? *South Asian Journal of Business Studies, 12*(1), 25–53.
78. Zhu, Y., Tang, R. W., & Xing, K. (2023). Effective coordination and innovation-facilitating role of science parks: The place-based approach with paradoxical outcomes. *Australian Journal of Management, 48*(1), 171–195.

Knowledge Creation and Innovation: The Key Drivers of Economic Growth According to the Global Innovation Index

Norberto Semião, Luís Farinha, and Sara Nunes

Abstract This study analyzes the relationship between the dimensions of the input and output sub-indices of the Global Innovation Index (GII) in the performance of innovation ecosystems across 132 economies. Using an exploratory and quantitative approach, the study employs factor analysis and cluster analysis to interpret how the factors interact and impact the innovative performance of countries. Based on empirical data and previous studies, the research explores interactions among innovation variables, providing relevant contributions for public policy and innovation strategies. By examining the relationships between GII sub-pillars using 2023 data, the analysis offers a detailed view of innovation dynamics and opportunities for strengthening national systems. The geographic and economic diversity of the sample allows for a comprehensive perspective on global innovation capacities, contributing to an understanding of the dynamics that underpin robust and interconnected innovation ecosystems, essential for the sustainable and competitive development of economies.

Keywords Economic growth · Global innovation index · Innovation · Innovation system · Knowledge · R&D

N. Semião (✉)
Universidade da Beira Interior, Covilhã, Portugal
e-mail: norberto.semiao@ubi.pt

N. Semião · L. Farinha · S. Nunes
NECE-UBI Research Centre for Business Sciences, Covilhã, Portugal

L. Farinha · S. Nunes
Escola Superior de Gestão, Instituto Politécnico de Castelo Branco, Idanha-a-Nova, Portugal

L. Farinha
CIICESI—Centro de Inovação e Investigação em Ciências Empresariais e Sistemas de Informação, Felgueiras, Portugal

L. Farinha et al. (eds.), *Sustainable Innovations in Mechanical Engineering and Management Sciences*, Lecture Notes on Multidisciplinary Industrial Engineering,
https://doi.org/10.1007/978-3-032-30372-1_4

1 Introduction

According to Oturakci [1], the Global Innovation Index (GII) is a fundamental benchmark for assessing the performance and innovation capacity of 132 economies, using 80 indicators distributed across 21 sub-pillars, with updated data for 2023 [2]. It highlights the essential role of factors such as human capital, market sophistication, and business sophistication in driving innovative outcomes. Studies show that economies that consistently invest in innovation tend to achieve better economic results [3, 4].

This study seeks to explore the relationships between input and output elements in countries' innovation ecosystems [5], identifying critical factors for policy development and resource allocation in areas with the greatest impact [6]. Ivanova and Leydesdorff [7] emphasize that international research collaborations significantly influence innovation performance, providing valuable insights for policymakers and practitioners. As discussed by Crespo and Crespo [8], high innovation performance can be achieved through various combinations of conditions, reflected in the five GII input enablers: institutions, human capital, infrastructure, business sophistication, and market sophistication. Khedhaouria and Thurik [9] identified that integrating these conditions is essential for achieving high innovation capacity. Recent studies, such as Yu et al. [10] explore interrelations within the innovation ecosystem to interpret the GII holistically.

GII's popularity has encouraged the use of various methodologies to analyze the structure of the index. Sohn et al. [11] applied structural equation modeling to explore relationships between innovation input and output dimensions. Reis et al. [5] used quantile regression to understand this structural relationship, finding significant positive effects. Similarly, Carpita and Ciavolino [12] confirmed the influence of business sophistication on innovation within the European Union (EU-27), using data from 2012 as the reference for their study. Analyzing key innovation factors allows progress points to be identified. For example, do Couto and Silva et al. [13] introduced a novel method to combine innovation forecasting, country innovation indices, and decision analysis to identify optimal investment strategies for enhancing national innovation systems. Meanwhile, Dutta et al. [2] proposed a gap matrix based on the GII sub-pillars of human capital and business sophistication to guide strategic investments that maximize impact on innovative systems.

Thus, the purpose of this study is to investigate the role of innovation in economies at different income levels, analyzing how geopolitical factors and specific characteristics of innovation systems influence their capacity to generate economic growth.

2 Literature Review

In the context of a globalized economy, the ability to capture innovations and ideas from other regions and convert them into national value emerges as a crucial tool for economic growth [14]. Innovation undeniably plays a central role in promoting economic development. However, according to Vlasova and Saprykina [15], recent geopolitical challenges have undermined the innovative capacity and economic growth of several nations, particularly those with lower per capita income.

The literature indicates that the characteristics of innovation systems vary across countries and are associated with each nation's level of economic development. Since the late nineteenth century, research has linked innovation to economic performance [16]. Recent studies suggest that the interaction between innovation, business sophistication, and macroeconomic factors generates a process of continuous improvement: economic activity drives entrepreneurship and innovation, which in turn further stimulates the economy [17]. Additionally, researchers highlight innovation as one of the main drivers of economic development, with models aimed at promoting it, including within creative industries [18].

2.1 Global Innovation Index

The 2023 GII highlights innovative trends in 132 economies, pointing to a promising future driven by digitalization, scientific advancements, and technological development, despite current economic and geopolitical challenges [2].

Complementary research corroborates the significant influence of multiple factors on innovative performance. For instance, Oturakci [1] examines the relationship between the components of the innovation input and output sub-indices, focusing on the impact of human capital, research, business sophistication, and creative outputs, while considering the influence of countries' income levels. Gogodze [19] proposes an approach to measure national innovation systems, identifying efficient institutional capital management as crucial, particularly in low-income countries. Kawabata and Camargo Junior [20] also explore the relationship between institutional quality, innovative activity, and resulting outcomes, recognizing innovation as central to business competitiveness and national economic growth.

Grounded in the concept of innovation from the Oslo Manual [21], expanded by the European Union and OECD, the GII serves as a robust metric for assessing innovation and its economic impact. Figure 1 illustrates the main areas of the innovation ranking, offering a global perspective on the innovation environment and identifying strengths and areas for improvement.

Fig. 1 Composition of the Global Innovation Index, author's own elaboration

2.2 Economic Growth

Innovation is widely recognized as a fundamental driver of economic growth, enhancing productivity and opening new market opportunities. Schumpeter [22] views innovation as a process of "creative destruction" that drives economic growth by replacing outdated technologies with new ones.

Innovation is acknowledged as a critical engine for economic growth and improving quality of life [23]. This growing importance of innovation has intensified international competition for access to and utilization of global innovation. Recent studies confirm the significance of innovation for business competitiveness, highlighting a positive correlation between productivity and exports in firms that adopt innovation strategies [24]. Furthermore, investment in R&D is regarded as crucial for strengthening firms' competitive advantage [25].

Some studies also examine the impact of entrepreneurial contexts on economic growth, particularly in transition economies and innovation-driven economies [26]. However, it is important to acknowledge that the relationship between innovation and economic growth is not unanimous, with some studies finding no significant evidence of this connection [27].

2.3 Innovation

Bae [18] discusses high-involvement innovation, a concept fundamentally centered on creating a culture where innovation becomes part of daily life. At the national level, however, innovation modeling faces conceptual and practical challenges, requiring a comprehensive approach that considers both theoretical aspects and the specific realities of each country [2]. Cultural barriers to innovation have been identified as a key explanation for the lag of Central and Eastern European countries compared to Western Europe, as perceptions of innovation significantly impact innovative performance [28].

According to Johannessen [29], innovation is one of the main drivers of international competitiveness, efficiency, production capacity, and overall employment performance. It can be seen as a tool to assess a country's level of development. Palazzeschi, Letizia, Bucci, Ornella, Di Fabio, and Annamaria [30] argue that organizations recognize innovation as a contributor to creating competitive advantages in an increasingly challenging, dynamic, and competitive labor market. This evidence supports the perspective linking innovation and competitiveness, where the intrinsic values of national culture can foster innovation development and raise the competitiveness of both nations and organizations [31].

Khedhaouria and Thurik [9] analyzed the relationships among the five input conditions, emphasizing that their combination is crucial for a country to achieve high innovation capacity. This aspect is vital when considering scenarios where innovation is not solely the result of formal research investment but can also arise from informal processes of learning by doing [32] and the absorption of external knowledge [33].

3 Methodology

This study utilized data from the Global Innovation Index (GII), provided by the World Intellectual Property Organization, encompassing 132 countries in 2023. GII captures characteristics and trends within global and national innovation ecosystems, employing innovative approaches and metrics that enable both the evaluation of countries' innovation performance and the formulation of new policies to foster innovation [2].

Statistical analyses were conducted using the Statistical Package for Social Sciences (SPSS), version 26. Initially, a factor analysis was performed to identify associations among a potentially broad set of observed variables or indicators by reducing them to a smaller number of factors [34]. Subsequently, a hierarchical cluster analysis was applied, a multivariate statistical technique that groups samples with similar characteristics [35].

3.1 Factor Analysis

Latent variables play a key role in enriching theoretical explanations of complex behaviors that cannot be fully described by observable influences alone. Factor analysis is widely used to infer these variables from patterns of association among sets of observed variables [34].

Theoretically, the component of uniqueness can be decomposed into variability attributable to other factors, specificity, and variability attributable to error or random fluctuation. However, this decomposition is only feasible when the data matrix contains other indicators that manifest the same specificity or covariates associated with the specific component [36]. According to Hoyle and Duvall [34], one of the main concerns in factor analysis applications is determining the optimal number of factors to adequately explain the similarity among a set of indicators.

To assess the suitability of the data for factor analysis, the Kaiser-Meyer-Olkin (KMO) test and Bartlett's Test of Sphericity are considered essential [37]. The KMO index measures the adequacy of a correlation matrix by verifying whether there is sufficient common variance among variables to justify factor analysis [38]. Interpretation of the KMO index varies based on its value: Below 0.5: Unacceptable, 0.5–0.6: Inadequate, 0.6–0.7: Mediocre, 0.7–0.8: Good, 0.8–0.9: Very Good, Above 0.9: Excellent [39].

Bartlett's Test of Sphericity assesses whether the covariance matrix resembles an identity matrix, which would indicate an absence of correlations among variables [40]. However, as highlighted by Betensky [41], the "p-value" should not be interpreted in isolation; it is necessary to consider the study design and the test's applicability, including sample size and effect significance. Generally, significance values below 0.05 suggest that the data do not form an identity matrix and are approximately multivariate normal, making them suitable for subsequent analyses [42].

Factor rotation is a technique used to identify a reduced number of factors that best describe the relationships among observed variables. Various methods exist to determine the optimal number of factors to retain. The Kaiser-Guttman criterion [43, 44], for example, is widely employed and recommends retaining factors with eigenvalues greater than 1.0 ($\lambda i > 1$). Another common technique is the scree plot or Cattell [45], test which identifies the "elbow" in the graph where eigenvalues follow a descending linear trend. In practice, the results obtained through the Kaiser-Guttman criterion and scree plot are often congruent.

For factor interpretation, the varimax rotation is frequently used. This orthogonal rotation method aims for a simplified structure by maximizing the sum of squared variances within factors, which facilitates interpretation. One of the main advantages of orthogonal rotations, such as varimax, is the simplicity of results, as factor loadings directly represent the correlations between factors and observed variables [44].

3.2 *Hierarchical Cluster Analysis*

Hierarchical Cluster Analysis (HCA) employs mathematical methods to identify and group objects or individuals with similar descriptions, forming homogeneous sets [35]. This method organizes samples into clusters and subclusters, resulting in a hierarchy visualized through a dendrogram, which illustrates the organization of the samples and their relationships in a tree-like structure [46].

In this process, observations within a cluster share similar characteristics, while differing significantly from observations in other clusters. From a dataset of n individuals with information on p variables, HCA groups individuals so that those within the same cluster are as similar as possible to one another, while exhibiting significant differences from members of other clusters [47].

The goal of HCA is to organize a set of cases into homogeneous groups, ensuring high intra-group similarity and substantial inter-group differentiation. The dendrogram serves as a visual representation of these relationships, providing insights into the structure and cohesion of the clusters.

4 Results and Analysis

4.1 Factor Analysis

The objective of this analysis was to identify a restricted set of factors capable of explaining the interrelationships among a broader set of observed variables.

Bartlett's Test of Sphericity and KMO

The Kaiser-Meyer-Olkin (KMO) measure of sampling adequacy yielded a value of 0.788, indicating that the sample was suitable for factor analysis, reflecting sufficiently strong correlations between the variables. Bartlett's Test of Sphericity produced a chi-square value of 10,655.080 with a significance of 0.000 ($p < 0.05$), confirming that the variables were significantly correlated and supporting the suitability of the factor analysis for this dataset.

Factor Extraction

The analysis revealed that the first extracted factor explained 66.702% of the variance, demonstrating its relevance in explaining the variability of the data. After factor rotation, the factor loadings indicated the contribution of each variable to each factor: the first factor accounted for 32.661% of the variance in the first variable, 25.731% in the second, and 19.374% in the third.

Factor Matrix

The component matrix presented the loadings of each variable for each extracted component. The variables displayed high factor loadings in Component 1, all greater than 0.670, indicating a common underlying dimension related to the pillars of economic development and innovation. This factor is associated with elements such as investment in human capital and research, business sophistication, business excellence, knowledge and technology production, and creativity—all critical for economic growth and fostering innovative and dynamic societies.

Factor Matrix After Varimax Rotation

A Varimax rotation with Kaiser normalization was applied to simplify the interpretation of the components. Three main components were identified: Component 1, highlights the importance of Research and Development (R&D), market sophistication, and creative production for business competitiveness and technological advancement; Component 2, emphasizes the importance of a favorable institutional environment, education, and robust infrastructure, including ICTs, in supporting innovation.; Component 3, reinforces the role of institutions and business dynamics in promoting investments and the transfer of knowledge and technology.

Internal Consistency Analysis

The data provided essential statistics to assess the internal consistency of the scale, with a strong association between the input and output dimensions and the innovation capacity of a country. The high internal consistency, indicated by Cronbach's alpha values ranging from 0.952 to 0.967, demonstrated that the items were well-aligned in measuring innovation. Moreover, excluding any item did not significantly improve the consistency of the scale.

4.2 Cluster Analysis

The cluster analysis was conducted with the objective of identifying homogeneous groups within the data, classifying economies based on observed similarities and differences. The goal was to organize a set of countries into homogeneous groups such that the countries within each group were as similar as possible to each other, while being distinct from the other groups. The analysis focused on the dimensions of the GII, including "Institutions," "Human Capital and Research," "Infrastructure," "Market Sophistication," "Business Sophistication," "Knowledge and Technology Outputs," and "Creative Outputs." Based on the previous data, the study organized the 132 countries analyzed into four Global Innovation clusters, as shown in Fig. 2.

Cluster 1, 37 countries

This group is primarily concentrated in Europe, with 24 countries (64.9%), followed by Southeast Asia and Oceania, with 8 countries (21.6%). Smaller representations include North America and the Caribbean, with 2 countries (5.4%), and North Africa and Western Asia, with 3 countries (8.1%). Statistical analyses highlight significant differences between the categories, revealing varying levels of development and

		Frequency	Percent	Valid Percent	Comulative Percent
Valid	1	37	28,0	28,0	28,0
	2	56	42,5	42,5	70,5
	3	2	1,5	1,5	72,0
	4	37	28,0	28,0	100,0
	Total	132,0	100,0	100,0	

Fig. 2 Selected clusters for the study. *Source* SPSS v26

innovation impact. While institutions exhibit a high average with low variability, indicating well-developed institutional infrastructure, other areas, such as knowledge and technology, present opportunities for improvement. Strategies to boost innovation in this group should prioritize business sophistication, knowledge and technology production, and investments in human capital and research.

Cluster 2, 56 countries

The largest concentration of countries in this cluster is in Europe (26.8%), followed by North Africa and Western Asia (23.2%). The remaining regions are relatively evenly represented. The analysis reveals significant disparities between the categories: areas such as institutions show consistency, while others, such as business sophistication, show less development. The correlation with the Global Innovation Index emphasizes the importance of prioritizing business sophistication and knowledge and technology outputs for advancements in the innovation capacity of the countries in this cluster.

Cluster 3, 2 countries

This cluster is composed of two countries from Latin America, with a limited economic diversity. These countries, at different income levels, are located in specific regions of Latin America, one in Central and South America and the other in Latin America and the Caribbean. The data indicates notable challenges in areas such as market sophistication and knowledge and technology outputs, while others, such as institutions and infrastructure, demonstrate relatively stable performance. It is recommended that development strategies prioritize areas with a high correlation to innovation performance, such as market sophistication and knowledge and technology outputs, to promote balanced and effective growth.

Cluster 4, 37 countries

This group includes three countries from Central and South America (8.1%), eight from Latin America and the Caribbean (21.6%), two from North Africa and West Asia (5.4%), one from Southeast Asia and Oceania (2.7%), and twenty-three from Sub-Saharan Africa (62.2%). Public policies for this cluster should prioritize strengthening institutions, investing in human capital and infrastructure, and promoting technological and creative innovation. Strategies focused on these areas can boost economic development and the innovative capacity of the countries in the cluster, fostering an innovative and sustainable environment.

5 Discussion

The main objective of this study was to investigate the influence and the combined relationship of the input and output subindex dimensions on the innovation scores of countries. To interpret and describe the results, statistical measures from factor analysis and cluster analysis were analyzed, providing a deeper understanding of the data regarding the research problem.

5.1 Factor Analysis

The analysis of communalities helped interpret the correlations, indicating the proportion of variance explained by each variable in relation to the extracted factors. The factor analysis revealed a strong correlation between human capital and research with R&D (0.908) and a high communality (0.891), highlighting the importance of R&D investments for human capital development. Infrastructure showed a high correlation with ICTs (0.924) and ecological sustainability (0.804), suggesting the need for policies focused on technological infrastructure. Market sophistication was strongly associated with credit (0.830) and investment (0.782), emphasizing the relevance of a developed financial market. Business sophistication was correlated with skilled workers (0.935) and innovation (0.912), reinforcing the need for qualifications and collaborative networks. Knowledge and technology production exhibited high correlations with production (0.916) and the impact of knowledge (0.866), indicating an essential interconnection for innovation. Lastly, creative outputs showed high correlations with intangible assets (0.904) and creative goods/services (0.829), suggesting that policies fostering creativity could yield economic benefits.

Thus, it is concluded that the quantitative results, combined with theoretical literature, reveal the complexity and interdependence of the factors driving the innovation rankings of countries, with an emphasis on education, R&D, infrastructure, and business and institutional environments.

5.2 Cluster Analysis

The study examines the relationship between factors influencing innovation in the economies of 132 countries. Through cluster analysis, variations were identified in specific dimensions and areas where certain variables showed differentiated performances in the countries' rankings.

Statistical analysis of Cluster 1 reveals significant differences between categories, highlighting various levels of development and impact on innovation among the countries in this group. Institutions exhibit the highest average and lowest variability, indicating a strong and stable institutional infrastructure. In contrast, indicators of knowledge and technology show a lower average and higher variability, signaling areas that require greater attention and investment. The analysis clearly identifies the strengths and weaknesses in innovation. While institutions and infrastructure demonstrate stable performance, areas like knowledge and technology, and market sophistication face greater challenges. This suggests that policies and investments should focus on reducing these disparities, promoting a more balanced and effective development.

The Pearson correlation between the Global Innovation Index 2023 and its dimensions shows that all dimensions have positive and significant correlations with the innovation index. The most influential factors include business sophistication (0.851), knowledge and technology outputs (0.837), and human capital and research (0.770). Market sophistication (0.737), infrastructure (0.649), and institutions (0.521) also present important correlations, while creative outputs (0.777) reinforce the relevance of creativity in innovation.

These results emphasize the importance of a multidimensional approach to strengthen innovation capacity, with a special focus on business sophistication and human capital development. These findings provide clear strategic guidance for policymakers and leaders interested in improving innovation in their countries or organizations.

The analysis of descriptive statistics in Cluster 2 reveals significant differences between the categories, reflecting various levels of development and impact on innovation. The Pearson correlation between the Global Innovation Index 2023 Score and the various dimensions highlights important factors for global innovation.

The correlation with institutions was weak and not significant (0.083, $p > 0.05$), indicating a relatively low relevance between the global innovation index and this specific aspect. In contrast, other dimensions showed positive and significant correlations, particularly with human capital and research (0.523), infrastructure (0.668), market sophistication (0.599), business sophistication (0.694), knowledge and technology outputs (0.811), and creative outputs (0.810).

These results suggest that investments and policies focused on these areas can drive the global innovation index, providing valuable information for policymakers and business leaders in adopting effective strategies to promote innovation globally.

Cluster 3 includes only two Latin American countries with different income levels and located in distinct geographic regions: one in Central and South America and the other in Latin America and the Caribbean. Descriptive statistics indicate that areas such as institutions and infrastructure have stable performance, while market sophistication and knowledge and technology production face challenges. The comparison with Cluster 2 reveals consistent trends but highlights significant differences in market sophistication and knowledge production, indicating the need for specific strategies for more balanced development across all innovation dimensions.

The analysis of Pearson correlations between the "Global Innovation Index 2023 Score" and its dimensions shows perfect and highly significant correlations ($p < 0.01$). There is a perfect positive correlation with "Human Capital and Research," "Market Sophistication," "Knowledge and Technology Production," and "Creative Production," indicating that increases in these areas are associated with proportional increases in the innovation index. In contrast, "Institutions," "Infrastructure," and "Business Sophistication" exhibit perfect negative correlations, suggesting an inverse relationship. These perfect correlations, both positive and negative, underline the importance of these dimensions in global innovation performance, although the validity of the results may be limited by the small sample size ($N = 2$).

The study highlights critical areas for improvement, such as business sophistication and knowledge and technology production. Focusing strategies on these areas could bring significant advancements in innovation and economic development for the countries in this cluster.

The highest concentration of countries in Cluster 4 is in Sub-Saharan Africa, followed by Latin America and the Caribbean. In these contexts, public policies should focus on institutional strengthening and increasing investment in human capital and infrastructure. To enhance human capital, it is essential to direct more funding toward education and research. Additionally, business strategies need to encourage sophisticated practices and the development of efficient markets. Policies that promote creative production and technological innovation are also crucial. Strengthening institutions, human capital, infrastructure, market and business sophistication, as well as improvements in knowledge and creativity outputs, are essential for an innovative environment.

The analysis of Pearson correlations between the "Global Innovation Index 2023 Score" and its input and output dimensions reveals positive and significant correlations with almost all dimensions, except institutions. The strongest correlations were found with business sophistication (0.812), creative outputs (0.739), and infrastructure (0.665), indicating a substantial relationship with the innovation index. Other dimensions, such as human capital and research (0.493), market sophistication (0.534), and knowledge and technology outputs (0.634), also show significant positive correlations. These data highlight the importance of a holistic approach to innovation, focusing on multiple critical dimensions to achieve balanced and sustainable development.

These results provide a solid foundation for the development of targeted strategies, allowing countries and organizations to maximize their innovative potential. The effective implementation of these strategies can drive economic development and enhance the innovation capacity of countries in this group.

5.3 *Bivariate Analysis*

The bivariate analysis, including chi-square and Pearson correlation, revealed strong linear relationships between the innovation dimensions. Positive correlations with variables such as Human Capital and Research (0.925) and Infrastructure (0.909) indicate that strengthening these dimensions is associated with an increase in global innovation. The negative relationship observed between the GII Score and the Global Innovation Clusters (−0.845) suggests that the concentration of innovation does not necessarily increase innovation at the national level.

Non-parametric tests rejected the null hypothesis of a uniform distribution of dimensions across clusters, indicating significant differences in innovation characteristics. These results highlight the need for specific innovative policies for each cluster, reinforcing the validity of the model.

In summary, the study contributes to the understanding of global innovation by highlighting the importance of dimensions such as human capital, business sophistication, and infrastructure, suggesting strategic paths for the sustainable and balanced economic development of global economies.

5.4 Implications and Significance of the Findings

The comprehensive analysis of the dimensions that make up the GII reveals the key determinants of a country's innovative capacity. This study examines institutions, human capital, research, infrastructure, market and business sophistication, as well as knowledge and creativity outputs, advancing the understanding of innovation.

The positive correlations between these dimensions and the GII highlight the importance of each in fostering innovation. Strong institutions, investments in human capital and research, developed infrastructure, sophisticated business practices, and the production of knowledge and creativity are essential elements for driving innovative performance.

These results have profound implications for public policies, business strategies, and development initiatives, providing guidance on how to create a more conducive environment for innovation and economic progress. In summary, the study enriches existing knowledge by offering insights into how countries can enhance their innovative capabilities and promote sustainable long-term growth.

5.5 Limitations of the Study

This study provided valuable insights into the differences in the GII dimensions across global innovation clusters, but it has some limitations. The metrics used to evaluate each dimension may not fully capture the regional complexities. Additionally, specific characteristics of the analyzed clusters may influence the results, making it difficult to generalize the conclusions.

Future research should delve deeper into the causes and consequences of the observed discrepancies, also exploring the development of innovation clusters over time. Further studies could include new metrics, dimensions, and longitudinal data to identify trends and dynamic changes in global innovation. These approaches could enhance the understanding of the factors driving innovation and support the formulation of more effective policies for sustainable economic development.

6 Conclusion

6.1 Summary of Key Findings

The study highlighted the complexity and interdependence of factors that determine the innovation capacity of different countries. Through factor and cluster analyses, the importance of dimensions such as human capital, research and development (R&D), infrastructure, business sophistication, knowledge production, and creative outputs were emphasized.

Key findings include Factor Analysis: The extracted factors explained a significant portion of the variability in the data, with the most relevant factors linked to economic development, innovation, and business sophistication. Significant correlations between human capital, infrastructure, and technology and creativity outputs underscore the importance of public policies that encourage research, technological development, and talent cultivation. Cluster Analysis: Four global clusters were identified, showing significant differences in institutional development, market sophistication, and technological production. Advanced clusters exhibited institutional stability and strong R&D capacities, while less developed clusters faced challenges in business sophistication and infrastructure. Correlations: All dimensions showed positive and significant correlations with the Global Innovation Index (GII), confirming that improvements in these areas drive global innovation performance.

6.2 Contributions to the Field

This study advances the understanding of innovation determinants by providing: A multidimensional analysis: By examining how variables such as human capital, business sophistication, and creativity interact and impact global innovation, the study offers an integrated view of key innovation drivers. Cluster-based classification: Grouping countries into homogeneous clusters helps identify regional and contextual differences, providing a foundation for tailored policymaking. Strategic recommendations: The findings offer practical guidelines for policymakers and business leaders, emphasizing the importance of investing in human capital, infrastructure, and robust institutional environments to foster innovation.

6.3 Recommendations for Future Research

To deepen the understanding of how innovation dimensions impact economic and social development, the following are recommended: Incorporate new metrics and longitudinal data: Exploring time series and additional metrics could help identify trends and dynamic changes in innovation clusters. Analyze the impact of specific regional factors: Future studies should investigate local or regional characteristics, such as cultural and social policies, that may influence innovation outcomes. Expand the geographical and contextual scope: Including more countries and regional clusters would improve the generalizability of results and help identify global patterns. Explore variable interactions: Future research could analyze causal relationships between variables, such as the interaction between infrastructure and creative outputs, to understand how they mutually influence innovation.

These recommendations aim to deepen the understanding of innovative drivers and contribute to the formulation of more effective policies, promoting sustainable growth and global economic progress.

References

1. Oturakci, M. (2023). Comprehensive analysis of the global innovation index: Statistical and strategic approach. *Technology Analysis and Strategic Management, 35*(6), 676–688. https://doi.org/10.1080/09537325.2021.1980209
2. Dutta, S., Lanvin, B., Rivera León, L., & Wunsch-Vincent, S. (2023). Global innovation index 2023 innovation in the face of uncertainty. In Wipo. chrome-extension://efaidnbmnnnibpcajpcglclefindmkaj/https://www.wipo.int/edocs/pubdocs/en/wipo-pub-2000-2023-en-main-report-global-innovation-index-2023-16th-edition.pdf
3. Powell, W. W., & Snellman, K. (2004). The knowledge economy. *Annual Review of Sociology, 30*, 199–220. https://doi.org/10.1146/annurev.soc.29.010202.100037
4. Taalbi, J. (2017). What drives innovation? *Evidence from Economic History. Research Policy, 46*(8), 1437–1453. https://doi.org/10.1016/j.respol.2017.06.007
5. Reis, D. A., De Moura, F. R., & De Aragão Gomes, I. M. (2018). The linkage between input and output in the global innovation index. In *Proceedings of the European conference on innovation and entrepreneurship, ECIE, 2018-Septe(3)* (pp. 635–642).
6. Coutinho, E. M. O., & Au-Yong-Oliveira, M. (2024). Innovation's performance: A transnational analysis based on the global innovation index. *Administrative Sciences, 14*(2), 0–16. doi:https://doi.org/10.3390/admsci14020032.
7. Ivanova, I. A., & Leydesdorff, L. (2014). A simulation model of the triple helix of university-industry-government relations and the decomposition of the redundancy. *Scientometrics, 99*(3), 927–948. https://doi.org/10.1007/s11192-014-1241-7
8. Crespo, N. F., & Crespo, C. F. (2016). Global innovation index: Moving beyond the absolute value of ranking with a fuzzy-set analysis. *Journal of Business Research, 69*(11), 5265–5271. https://doi.org/10.1016/j.jbusres.2016.04.123
9. Khedhaouria, A., & Thurik, R. (2017). Configurational conditions of national innovation capability: A fuzzy set analysis approach. *Technological Forecasting and Social Change, 120*(October 2016), 48–58. https://doi.org/10.1016/j.techfore.2017.04.005
10. Yu, T. H. K., Huarng, K. H., & Huang, D. H. (2021). Causal complexity analysis of the global innovation index. *Journal of Business Research, 137*(August), 39–45. https://doi.org/10.1016/j.jbusres.2021.08.013
11. Sohn, S. Y., Kim, D. H., & Jeon, S. Y. (2016). Re-evaluation of global innovation index based on a structural equation model. *Technology Analysis and Strategic Management, 28*(4), 492–505. https://doi.org/10.1080/09537325.2015.1104412
12. Carpita, M., & Ciavolino, E. (2017). A generalized maximum entropy estimator to simple linear measurement error model with a composite indicator. *Advances in Data Analysis and Classification, 11*(1), 139–158. https://doi.org/10.1007/s11634-016-0237-y
13. do Couto e Silva, E., Silberglitt, R., Machado, L. C., Maia, J. M. F., & Cagnin, C. H. (2017). A portfolio analysis methodology to inform innovation policy and foresight. *Technological Forecasting and Social Change, 115*, 338–347. https://doi.org/10.1016/j.techfore.2016.06.032
14. Mahroum, S., & Al-Saleh, Y. (2013). Towards a functional framework for measuring national innovation efficacy. *Technovation, 33*(10–11), 320–332. https://doi.org/10.1016/j.technovation.2013.03.013
15. Vlasova, V., & Saprykina, A. (2023). Innovation-driven economic growth under global turbulence: How countries strengthen innovation systems to deal with threats. *Economics of Innovation and New Technology, 1–25*. https://doi.org/10.1080/10438599.2023.2276318
16. Śledzik, K. (2013). Schumpeter's view on innovation and entrepreneurship. *SSRN Electronic Journal*, (April 2013). https://doi.org/10.2139/ssrn.2257783
17. Galindo, M. Á., & Méndez, M. T. (2014). Entrepreneurship, economic growth, and innovation: Are feedback effects at work? *Journal of Business Research, 67*(5), 825–829. https://doi.org/10.1016/j.jbusres.2013.11.052
18. Bae, S. H., & Yoo, K. (2015). Economic modeling of innovation in the creative industries and its implications. *Technological Forecasting and Social Change, 96*, 101–110. https://doi.org/10.1016/j.techfore.2015.02.010

19. Gogodze, J. (2016). Mechanisms and functions within a national innovation system. *Journal of Technology Management and Innovation, 11*(4), 12–21. https://doi.org/10.4067/S0718-27242016000400003
20. Kawabata, M. K., & Camargo Junior, A. S. (2020). Innovation and institutions' quality: A comparative study between countries. *International Journal of Innovation Science, 12*(2), 169–185. https://doi.org/10.1108/IJIS-10-2019-0100
21. Collecting, G. F. O. R., & Data, I. I. (2005). Third edition ORGANISATION FOR ECONOMIC CO-OPERATION. In Communities: Vol. Third edit. http://scholar.google.com/scholar?hl=en&btnG=Search&q=intitle:Oslo+manual#0
22. Schumpeter, J. A. (1949). *The theory of economic development: An inquiry into profits, capital, credit, interest, and the business cycle* (Vol. 46). Harvard university press.
23. Atkinson, R. D., Ezell, S. J., & Stewart, L. A. (2012). The global innovation policy index (issue march).
24. Cassiman, B., Golovko, E., & Martínez-Ros, E. (2010). Innovation, exports and productivity. *International Journal of Industrial Organization, 28*(4), 372–376. https://doi.org/10.1016/j.ijindorg.2010.03.005
25. Tsai, K. H., & Wang, J. C. (2004). R&D productivity and the spillover effects of high-tech industry on the traditional manufacturing sector: The case of Taiwan. *The World Economy, 27*(10), 1555–1570. https://doi.org/10.1111/j.1467-9701.2004.00666.x
26. Gomes, S., Ferreira, J., Lopes, J. M., & Farinha, L. (2022). The impacts of the entrepreneurial conditions on economic growth: Evidence from OECD countries. *Economies, 10*(7). https://doi.org/10.3390/economies10070163
27. Afzal, A., & Gauhar, A. A. (2020). The dark side of financial innovation: Deterrent for economic growth. *Academy of Accounting and Financial Studies Journal, 24*(5), 1–11.
28. Akhmadi, S., & Tsakalerou, M. (2023). Shades of innovation: Is there an east-west cultural divide in the European Union? *International Journal of Innovation Science, 15*(2), 260–278. https://doi.org/10.1108/IJIS-01-2022-0019
29. Johannessen, J. (2009). A systemic approach to innovation: The interactive innovation model. *Kybernetes, 38*(1/2), 158–176. https://doi.org/10.1108/03684920910930330
30. Palazzeschi, L., Bucci, O., & Di Fabio, A. (2018). High entrepreneurship, leadership, and professionalism (HELP): A new resource for workers in the 21st century. *Frontiers in Psychology, 9*, 1480. https://doi.org/10.3389/fpsyg.2018.01480
31. Prim, A. L., Filho, L. S., Zamur, G. A. C., & Di Serio, L. C. (2016). The relationship between National Culture Dimensions and Degree of Innovation. *International Journal of Innovation Management, 21*(01), 1730001. https://doi.org/10.1142/S136391961730001X
32. Wright, G. (1982). An evolutionary theory of economic change. In R. R. Nelson & S. G. Winter (Eds.), *The Belknap Press of Harvard University Press, 1982. Technology in Society* (Vol. 4, pp. 315–317). https://doi.org/10.1016/0160-791X(82)90007-0
33. Abreu, M., Grinevich, V., Hughes, A., Kitson, M., & Ternouth, P. (2008). Universities, business and knowledge exchange. *Business, 64.*
34. Hoyle, R. H., & Duvall, J. (2004). Determining the number of factors in exploratory and confirmatory factor analysis. https://www.researchgate.net/publication/245904294
35. Romesburg, C. (2004). Cluster analysis for researchers. Lulu. com.
36. Bentler, P. M. (1990). Latent variable structural models for separating specific from general effects. In L. Sechrest, E. Perrin, & J. Bunker (Eds.), *Research methodology, strengthening causal interpretations of nonexperimental Data* (pp. 61–83). Department of Health and Human Services.
37. Dziuban, C. D., & Shirkey, E. C. (1974). When is a correlation matrix appropriate for factor analysis? Some decision rules. *Psychological Bulletin, 81*, 358–361.
38. Lorenzo-Seva, U., Timmerman, M. E., & Kiers, H. A. L. (2011). The hull method for selecting the number of common factors. *Multivariate Behavioral Research, 46*(2), 340–364. https://doi.org/10.1080/00273171.2011.564527
39. Wang, H., Li, X., Lei, J., & Ren, F. (2021). Quantitative analysis model of recorded data for terrorist attacks. *ACM International Conference Proceeding Series.* https://doi.org/10.1145/3469213.3471313

40. Field, A. (2013). *Discovering statistics using IBM SPSS statistics*. Sage.
41. Betensky, R. A. (2019). The p-value requires context, not a threshold. *American Statistician, 73*(sup1), 115–117. https://doi.org/10.1080/00031305.2018.1529624
42. Pallant, J. (2011). SPSS Survival Manual 4th edition. www.allenandunwin.com/spss
43. Guttman, L. (1954). Some necessary conditions for common-factor analysis*.
44. Kaiser, H. F. (1958). The Varimax criterion for analytic rotation in factor analysis*.
45. Cattell, R. B. (1966). The Scree Plot Test for the Number of Factors. Collecting, G. F. O. R., & Data, I. I. (2005). Third edition ORGANISATION FOR ECONOMIC CO-OPERATION. In Communities: Vol. Third edit. http://scholar.google.com/scholar?hl=en&btnG=Search&q=intitle:Oslo+manual#0.
46. Lee, I., & Yang, J. (2009). Common clustering algorithms.
47. King, R. S. (2015). Cluster analysis and Data mining: An introduction. *Mercury Learning and Information*.. https://books.google.pt/books?id=a4XKDgAAQBAJ

Sustainable Regional Development through Entrepreneurship Support Policies: Insights from Low-Density Areas in Northern Portugal

Anderson R. Galvão, Carla S. Marques, and Mariana Santos

Abstract The main objective of this research is to study policies to support entrepreneurship in low-density territories. To develop this research, a qualitative methodology was chosen, in which semi-structured interviews were conducted with different local actors (e.g., Municipal Councils, Intermunicipal Communities, Science and Technology Parks, incubators, business associations and Higher Education Institutes), from the districts of Vila Real and Bragança. The results show very similar perspectives regarding the weaknesses felt in both territories. However, with regard to policies to support entrepreneurship and its sufficiency, as well as the interaction between the various local actors, the views of the interviewees differ significantly. Overall, the results of this study revealed that in both the Vila Real and Bragança districts, there are some policies that aim to foster entrepreneurship and increase local attractiveness. However, these policies are insufficient to meet the real needs of the region. The originality of this study stems from the integrated analysis of policies to support entrepreneurship in low-density territories, cross-referencing the mapping of existing measures in the districts of Vila Real and Bragança with the perspective of various local actors, offering a contextualized view that is little explored in the literature.

Keywords Entrepreneurship · Low-density territories · Entrepreneurship support policies; entrepreneurial ecosystems

---

A. R. Galvão (✉) · C. S. Marques
CETRAD & University of Tras-os-Montes and Alto Douro, Vila Real, Portugal
e-mail: anderson@utad.pt

C. S. Marques
e-mail: smarques@utad.pt

A. R. Galvão
NECE—Research Centre for Business Sciences, Covilhã, Portugal

M. Santos
University of Tras-os-Montes and Alto Douro, Vila Real, Portugal

L. Farinha et al. (eds.), *Sustainable Innovations in Mechanical Engineering and Management Sciences*, Lecture Notes on Multidisciplinary Industrial Engineering,
https://doi.org/10.1007/978-3-032-30372-1_5

than being replaced by technological progress. The consistent linking of agricultural terminology with innovation-related concepts shows a deliberate strategy of technological enhancement within established productive sectors.

The prominence of *Companies*, *Attracting*, *Business*, and *Investment* reveals a municipality actively engaged in territorial competitiveness strategies designed to enhance its appeal to external economic actors. This focus on attraction and investment reflects a proactive approach to economic development that emphasizes creating favorable environments for business setup and expansion. The frequent use of business-related terminology indicates that Fundão has adopted modern territorial marketing strategies focused on building investor confidence and developing an entrepreneurial ecosystem.

The thematic prominence of *Project*, *Support*, and *Initiatives* reflects a governance philosophy characterized by proactive intervention, systematic commitment to project implementation, and a firm commitment to development. This clustering indicates that Fundão uses project-based development methods, with a structured approach to territorial improvement. The emphasis on support mechanisms underscores the understanding that successful territorial development depends on active institutional backing rather than on passive market confidence.

The presence of the words *Ecosystem*, *Institutions*, and *Producers* underscores Fundão's recognition that contemporary territorial development requires orchestrated collaboration among diverse stakeholders. This emphasis on ecosystem thinking reflects a sophisticated understanding of how innovation and development emerge through interconnected relationships rather than individual initiatives. The systematic reference to institutional networks suggests that Fundão has embraced collaborative governance models that leverage collective capabilities and shared resources.

The centrality of *Center* and *Promote* within the municipal discourse indicates Fundão's strategic positioning as both a focal point for regional activity and an active promoter of its distinctive capabilities. This dual role suggests a municipality that combines hub functions with proactive communication strategies, recognizing that territorial competitiveness requires both substantive capabilities and effective visibility within relevant networks. The addition of *Region* and *European* demonstrates Fundão's spatial awareness and its strategic orientation toward multi-scalar development approaches. This linguistic emphasis on trans-local connectivity suggests recognition that contemporary municipal competitiveness depends upon integration with broader territorial networks and opportunities. The European reference particularly indicates alignment with supranational development frameworks and funding mechanisms.

The prominence of *Quality* and *Sustainable* reflects Fundão's commitment to development models that prioritize long-term viability and excellence rather than short-term growth maximization. This emphasis indicates a municipality that has adopted modern development trends focusing on environmental responsibility and qualitative improvement alongside growth.

The convergence of these thematic elements portrays Fundão as a highly developed municipality that has successfully combined traditional agricultural strengths with modern technological capabilities while staying aware of broader competitive

1 Introduction

Over time, entrepreneurship has been recognized as a key driver of development and economic growth at global, national, and regional levels. It is closely linked to the introduction of innovations into the economy, as any entrepreneurial initiative can result in the creation and transformation of businesses, directly impacting the development of local communities [1]. This relationship highlights that entrepreneurship is not only an economic phenomenon, but also a social one, influencing the vitality and resilience of the regions in which it operates.

In recent decades, several countries have developed policies to encourage the creation of new companies with the aim of promoting local economic growth [2, 3]. These policies can act directly or indirectly, through structural, fiscal or economic measures, which influence the level of entrepreneurship in a given region [4]. In Portugal, for example, the concentration of economic opportunities on the coast highlights significant territorial inequality. While these areas offer favorable conditions for creating businesses, inland territories face specific limitations and challenges that make it difficult for entrepreneurs to establish themselves [5].

The growing concern for the balanced development of Low-Density Territories (LDTs) has placed entrepreneurship at the center of local and national policies, in an attempt to economically revitalize these regions. Understanding the factors that promote business creation therefore becomes fundamental, since the expansion of entrepreneurship positively influences local economic vitality [6]. However, much of the recent research continues to focus on urban or metropolitan contexts, where entrepreneurial ecosystems are denser and financial, human, and technological resources are more available [7].

In contrast, low-density territories face significant structural challenges, such as geographic isolation, a shortage of qualified human capital, and a lower diversity of support networks, factors that can limit the effectiveness of policies to promote entrepreneurship [8]. The literature still lacks studies that assess the suitability and impact of innovation policies adapted to these contexts, including differentiated instruments such as rural incubators, local microfinancing programs, or policies to encourage interterritorial cooperation [9]. This gap highlights the need for empirical research focused on the effectiveness of specific measures for low-density territories, in order to support the formulation of policies better tailored to their real needs.

The main objective of this research is to study policies to support entrepreneurship in two low-density territories in Portugal.

To develop the study, semi-structured interviews were conducted with 19 local actors from the district of Bragança and Vila Real (low-density regions). The information collected from the interviews was subjected to content analysis and data coding using NVIVO software.

This study constitutes a relevant contribution to deepening knowledge about entrepreneurship in low-density territories, by carrying out an integrated analysis of entrepreneurship support policies and the perceptions of local actors in the districts of Vila Real and Bragança. By highlighting how structural, demographic, and institutional factors influence entrepreneurial dynamism, this work offers a solid empirical contribution to the debate on the need to adapt public policies to the specificities of peripheral territories. At the same time, it proposes a collaborative governance model that values collaboration between municipalities, higher education institutions, business associations, and incubators as an essential condition for strengthening local ecosystems. From a theoretical point of view, it reinforces the relevance of the concept of entrepreneurial ecosystem applied to low-density contexts, expanding the existing literature that has mainly prioritized urban and metropolitan spaces.

2 Literature Review

2.1 Entrepreneurship in Low-Density Territories

LDTs are characterized as places that have lower population density or per capita income rates, compared to national averages, which are normally associated with municipalities or regions in the interior of a country [8]. Furthermore, LDTs have a low relational density, that is, the interaction between people and institutions is practically non-existent, either due to the lack of both or the limited supply by the competent entities [10]. These territories, mainly in the countries located in southern Europe, still have low academic qualifications, as well as a business fabric that is not very entrepreneurial and innovative, which causes an obstacle to their economic development [11]. LDTs are typically characterized by an aging population and depopulation, largely caused by rural exodus or migratory flows of younger populations. Therefore, demographic changes have severe consequences for the development of LDTs, as they cause an imbalance in economic activity [12].

For Cappelleras et al. [13], the problem of population density can influence the development of a territory, as this issue will affect the generation of companies and their growth and, consequently, their survival rate. Higher density regions present a greater diversity of resources and employment opportunities, providing a greater impact on business growth, contrasting with the reality evident in LDT [14]. Although LDTs retain weaknesses such as those previously mentioned, and these hinder their establishment as development hubs, they currently present a range of components that are not automatically related to the problem of underdevelopment and geographic isolation [13]. This impulse arises due to the position demarcated by government entities, who believe that LDTs can become a center of economic development [15].

In this sense, entrepreneurship can make an important contribution to achieving growth and economic development in a region, because it enables job creation, good market dynamics, knowledge sharing and innovation [16]. According to Bosma and Levie [17] and Müller [18], entrepreneurship has become an essential means of boosting the economic development of LDTs, as it allows for a huge margin of innovation and job creation, which are some of the major needs of these territories. Entrepreneurship and companies created in LDTs contribute significantly to the distinction and preservation of local natural resources, heritage or others of equal importance, in order to support economic development, but taking into account the regional valorization of these territories (e.g., [19–21]).

In this sense, government policies play a fundamental role in developing these territories, in order to encourage entrepreneurship [22]. Defining strategies and programs that suit each territory contributes significantly to the development of LDTs, in order to more easily identify their needs [23, 24].

2.2 *Entrepreneurship Support Policies*

Governments and policymakers, who are increasingly aware of the benefits of entrepreneurship in society and the economy, have initiated a strategy of increasing measures and programs to promote it [25, 26]. Policies to support entrepreneurship aim to directly affect entrepreneurial activity in a variety of locations, increasing their competitiveness and creating better conditions for entrepreneurs and the population in general [27, 28].

According to Parker [29], the various types of policies assumed greater relevance in territories marked by high unemployment rates and low economic development. Policies can play a significant role in job creation measures and in the economic and social development of the population of these territories [30]. The territory where entrepreneurship emerges is crucial, given that it can influence the birth and death rate of various companies, due to the level and number of business opportunities that arise there, as well as the impact that these can have on the availability and quality of services and resources existing in that territory [31]. For these reasons, the way to mitigate these economic, social and business inequalities is to stimulate economic growth through entrepreneurship, and more specifically, through policies to support it [32]. These will be more effective when there is a prior analysis of the opportunities and challenges of each location, as this allows policymakers to act in accordance with the reality presented, whether in economic, institutional or social terms [33, 34].

In this sense, governments play a fundamental role in stimulating the level of entrepreneurship in a given region, as they often seek ways to establish themselves as competitive hubs, with the aim of attracting new projects to their territories, in addition to supporting local companies [35]. The government has multiple tools at its disposal to promote entrepreneurship [36], among which we can highlight the enactment of a series of tax and/or economic incentives and benefits, by way of example, this could be a reduction in certain municipal taxes, or the rental of a common commercial space, in order to create an environment conducive to the development of entrepreneurial activity and, consequently, attract more investment to the location in question [37]. This can also provide direct financial support to companies that meet the requirements established by the same, in order to speed up the process of establishing a business in the location, since access to this type of support is quite time-consuming for an entrepreneur or a company that wants to establish itself in that location [38]. In addition to this obstacle, the issue of administrative or bureaucratic processes is considered one of the main barriers to boosting entrepreneurial activity, and it is necessary to drastically reduce this problem by the numerous competent government institutions in order to achieve better business dynamics [32, 39]. According to Bruton et al. [25], governments must ensure the proper functioning of their entrepreneurial activity, always encouraging the creation of new business opportunities through the implementation of public measures and programs, which must consider the context of the territory, that is, the opportunities and challenges inherent to it.

In fact, policies to support entrepreneurship combined with LDT must follow a line of constant promotion of attracting people to these territories, given that, as previously presented, entrepreneurship is seen as a tool for the development of any region [40, 41]. Therefore, the development of strategies and policies must remain a concern for all governments, both at local and national levels, in addition to clearly requiring care on the part of the entire entrepreneurial ecosystem, to identify all the opportunities and challenges in each territory where it is located [42, 43].

3 Methodology

3.1 Characterization of the Territory

The districts of Vila Real and Bragança are located in the inland northern region of Portugal, integrating NUTS II "Norte" and NUTS III "Alto Tâmega" and "Douro" (Vila Real) and "Terras de Trás-os-Montes" (Bragança). Both are low-density territories, with extensive rural areas, low population density, pronounced demographic aging and economic structures marked by a predominance of primary sector activities and small trade and service companies.

The Vila Real district is comprised of 14 municipalities, covers 4328 km^2, and had, according to the 2021 Census, 185,695 inhabitants. It stands out for its strong ties to agriculture and winemaking (Douro Demarcated Region) and for its dispersed network of small and medium-sized businesses. Among the municipalities with the largest number of businesses are Vila Real, Chaves, and Valpaços. The territory, however, presents different dynamics, with municipalities closer to major highways registering a higher business birth rate.

The district of Bragança, with 12 municipalities, covers 6608 km^2 and has 122,804 inhabitants (2021 Census), being the least populous in the Trás-os-Montes and Alto Douro region. Its economy is linked to agribusiness, forestry and livestock farming, with a growing focus on nature and cultural tourism. The municipality of Bragança is the main urban hub, concentrating population, services, and businesses, followed

Table 1 Characterization of the territory

Indicator	Vila Real	Bragança
Number of municipalities	14	12
Total area (km^2)	4328	6608
Resident population	185,695	122,804
Population density (inhab./km^2)	~ 42.9	~ 18.6
Most populous municipality	Vila Real (49,571 inhabitants)	Bragança (34,582 inhabitants)
Total number of companies (2020)	25,612	23,566
Municipality with the most companies (2020)	Vila Real (6870 companies)	Bragança (6343 companies)

by Mirandela and Macedo de Cavaleiros. Its proximity to the Spanish border lends it strategic importance for cross-border trade flows.

Both districts share structural challenges, such as the loss of young populations, difficulties in retaining skilled workers, a weak business community, and the need to strengthen rail transport links. However, they benefit from a strategic location relative to Spain, relatively low living costs, and the potential for increasing the value of endogenous resources, which opens up opportunities for policies to encourage entrepreneurship and attract investment (Table 1).

Source INE [44, 45].

3.2 *Type of Study*

With the main objective of this research being to study policies to support entrepreneurship in low-density territories, a qualitative methodology was chosen through interviews with local actors in the regions under study. Qualitative methodology has been gaining interest among entrepreneurship researchers, since the qualitative technique allows gathering information that leads to results that certainly could not be measured or translated into numbers [46]. According to Yin [47], qualitative methods should be used when there is a need to analyze in depth certain processes, certain characteristics of organizations and/or the vision or experience of individuals.

To develop this study, a survey of entrepreneurship support policies was first carried out in all municipalities in the districts of Vila Real and Bragança, as well as a set of secondary data, such as the existence of entrepreneurship support offices, industrial zones, among others, which allowed us to observe what type of policies or programs were in force in these locations. This survey aimed to determine which municipalities had a set of policies to support entrepreneurship, as well as which entities implemented them and helped residents set up their businesses.

Subsequently, semi-structured interviews were conducted with Municipal Councils, Business Associations, Incubators, Higher Education Institutions (HEIs), Intermunicipal Communities (CIMs) and Science and Technology Parks, located in the territories under study. The interviews were conducted with people who were nominated by the institutions themselves and who were linked to entrepreneurship.

3.3 *Data Collection and Analysis*

In order to collect primary data capable of capturing the complexity of the phenomenon under study, semi-structured interviews were conducted based on a previously prepared script. This guide included open-ended questions, organized into four sections: interviewee characterization, low-density territories, policies to support entrepreneurship, and main challenges. Each section included three to four central questions, allowing for deeper understanding of the context and capturing different perspectives from participants.

Nineteen officials from various local institutions were interviewed, including city councils, business associations, intermunicipal communities (CIM), science and technology parks, incubators and higher education institutions (HEIs). The interviews took place between September and December 2024, being conducted in person or by videoconference, depending on the availability of the participants, with an average duration of approximately 60 minutes. Table 2 presents the characterization of the interviewees, showing that 11 interviews were carried out in the district of Vila Real and 8 in the district of Bragança.

After collecting data, it was organized, analyzed and processed. All interviews conducted were transcribed and compiled into a single document, allowing for a first comprehensive reading of the material. Then, the data extracted from the transcripts, as well as the information from complementary documents, were subjected to content analysis using NVIVO 15.0 software. This process involved organizing sources and coding information, resulting in the creation of tree nodes that allowed content to be labeled and branched in a structured way. It is important to emphasize that all the information analyzed reflects the context existing during the period in which the interviews were conducted.

Table 2 Characterization of interviewees

	Institution	Position	Age	Gender	Qualifications
Vila Real District					
Interviewee VR1	Business association	President	41	Masculine	Bachelor's degree in computer systems engineering
Interviewee VR2	Business association	Management technique	37	Feminine	Master's in management
Interviewee VR3	Intermunicipal community	1st secretary	60	Masculine	Bachelor's degree in communication
Interviewee VR4	Intermunicipal community	1st secretary	53	Masculine	PhD in computer science
Interviewee VR5	Incubator	Responsible for corporate reception	44	Feminine	PhD in agriculture and sustainability
Interviewee VR6	Development association	Secretary-general	60	Masculine	Bachelor's degree in management
Interviewee VR7	Municipality	Chief of staff of the presidency	41	Masculine	PhD in geography
Interviewee VR8	Municipality	Councilor	45	Masculine	Bachelor's degree in management
Interviewee VR9	Municipality	Division head	59	Feminine	Bachelor's degree in animal engineering
Interviewee VR10	Municipality	President	53	Masculine	Bachelor's degree in management
Interviewee VR11	University	Vice-rector	58	Masculine	PhD in electrical engineering
Bragança District					
Interviewee B 1	Business association	Superior technique	51	Feminine	Bachelor's degree in philosophy
Interviewee B 2	Intermunicipal community	Secretary-general	60	Masculine	Master in rural development
Interviewee B 3	Development association	Director	61	Masculine	Bachelor's degree in philosophy
Interviewee B 4	Development association	Coordinator	39	Masculine	PhD in agricultural production chains
Interviewee B 5	Municipality	Councilor	52	Masculine	Bachelor of Laws
Interviewee B 6	Municipality	Office coordinator	40	Masculine	Master's in management
Interviewee B 7	Municipality	Chief of staff of the presidency	41	Masculine	Master's degree in law
Interviewee B 8	University	Senior technician, entrepreneurship office	48	Masculine	Bachelor's degree in accounting and administration

4 Results

4.1 Low Density Territories

4.1.1 Main Problems

Low-density territories present structural problems aggravated by recent migratory flows, which have intensified depopulation, aging and low qualifications of the population. They suffer asymmetries compared to other regions, with fewer job opportunities, a predominance of the primary sector and an underdeveloped business fabric, limiting innovation and entrepreneurship [8, 48]. With regard to the districts of Vila Real and Bragança, Fig. 1 presents the main problems that interviewees mentioned throughout the data collection.

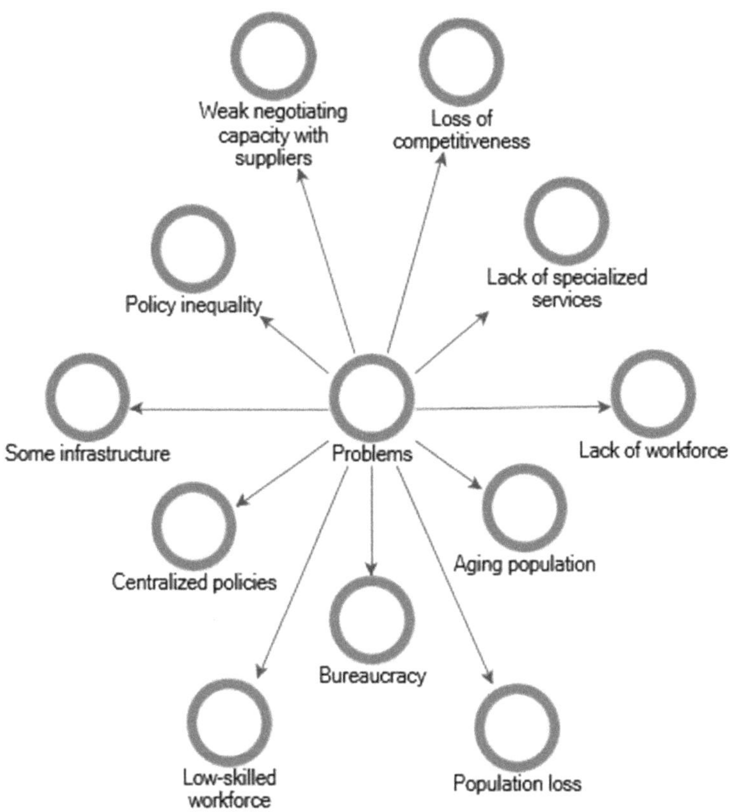

Fig. 1 Main problems of LDT (Vila Real and Bragança)

In general, interviewees highlighted population loss and an aging population as the main problems in both districts. In addition to this issue, there has been a significant loss of labor, especially skilled labor, as the majority have decided to leave the region in search of better living conditions. At a structural level, interviewees highlight problems with some infrastructure, combined with the limited availability of specialized services and the bureaucracy on which the region depends, given its highly centralized political structure.

In this sense, the following excerpts provide a detailed look at the main challenges that the interviewees highlighted in the two districts.

> **Interviewee VR1:** "It seems very clear to me that the primary obstacle in this region, and perhaps throughout the interior, stems from the inequality of our country's policies, which prevents the entire territory from being equitable. We're talking about a territory, and our country is 300 kilometers wide, so talking about the interior or the coastline in 300 kilometers these days is a bit absurd. Because the interior of Portugal is the coast of Europe. [...] Therefore, the lack of services in the interior compared to the coast is an obstacle to the settlement of people. Without people, there can be no businesses. And there can be no businesspeople and no entrepreneurs. [...]"

> **Interviewee VR4:** "[...], for example, is the infrastructure part. We can say that we have highways, but we don't have trains. [...]"

> **Interviewee B5:** "In low-density areas, it's not easy, on the one hand, to attract investment and retain people. [...] One of the biggest problems is the lack of skilled labor. The country is unbalanced. We have a coastline overcrowded with people, and the interior is underpopulated. Therefore, the challenge here is settling people so that investors can later emerge and settle them as well. This is common across all municipalities."

> **Interviewee B8:** "Demographic desertification, so to speak, leads to the absence of high-quality services, namely education, culture and medical care. [...]."

In general, the interviewees mentioned the difficulty of retaining and attracting people to the two districts under study, in addition to the issue of the aging population and its low qualifications. Interviewee B5 highlights the clear imbalance between inland and coastal regions, which has led to a lack of investment in these areas and the depopulation they have experienced in recent decades. Therefore, the demographic problem felt in Vila Real and Bragança also causes losses in various sectors essential to the local populations, as interviewee B8 states. It is also important to highlight that interviewee VR1 focuses on the region's difficulties in the inequality of national public policies, which end up stagnating its development. This interviewee emphasizes that the implementation of public policies that are not suited to the reality of the region itself does not allow for the reversal of the current situation, and logically does not transfer sufficient investments to the business sector, for example. Furthermore, the same interviewee refers to the heterogeneity that exists between the coast and the interior and how centralized public policies devalue the region itself, as mentioned by interviewee VR4.

4.1.2 Main Benefits

Low-density territories, despite the structural challenges they face, offer significant benefits that can be strategic for local development. Figure 2 presents the main benefits of TLD identified by the interviewees.

Figure 2 shows that the interviewees mention several benefits of both districts, such as their strategic location, as some municipalities in the region under study border Spain, and they are also centrally located to Portugal and Spain. Furthermore, a large number of respondents highlighted the better quality of life, safety, and low cost of living as the main reasons for these territories, as they are the main reasons for people to settle there, along with other factors, such as employment.

In contrast, the possession of numerous endogenous resources that cluster in the Trás-os-Montes region is also highlighted, in addition to the existence of a significant number of markets yet to be explored, compared to other regions of the country, which leads to low competition in some sectors, allowing for the maintenance of a very attractive market in the region, as can be seen in the excerpts below:

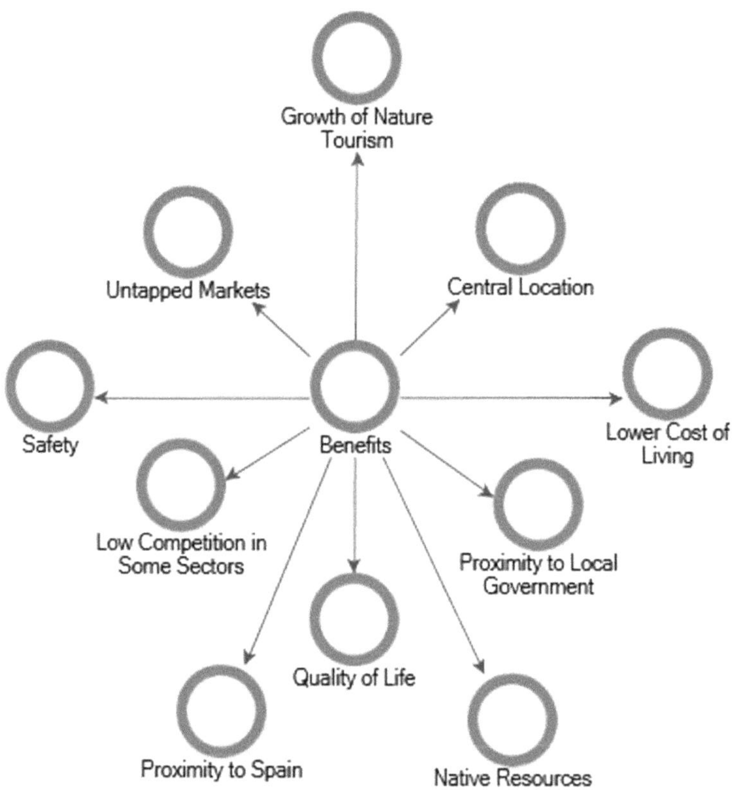

Fig. 2 Main benefits of LDT (Vila Real and Bragança)

Interviewee VR1: "[…] The fact that our geographical location is quite good allows us to have a quality of life that is well above average, with a lower cost compared to large centers and where we can travel from our home to our workplace in just a few minutes. […]"

Interviewee VR6: "We have an advantage, as we are a border area, very close to Spain. Furthermore, we have low competition in some sectors and some untapped markets […]."

Interviewee B2: "The great benefits of our region are: first, safety, then low prices, for example in housing, and then, as we currently have good accessibility, because we are approximately 1 hour from the airport and the same distance from Lisbon and Madrid. We're peripheral to the Portuguese capital, but we're very central to Spain. We're very close to Galicia, and now with the new access roads, we're very close to Castile and León, very close to Zamora, very close to Salamanca. So, imagine a company that wants to export by road to the European continent; it's more centrally located here than in Lisbon."

Interviewee B8: "The region's major utilities, the first, which comes from its natural characteristics, that is, from a set of products and services endogenous to the region, namely, has to do with agroindustry. And the other benefit, for now, is the average spending level for families and young people who decide to settle in the region. We're still relatively competitive compared to the coast, particularly regarding the main housing costs we face today."

4.2 Entrepreneurship in Low-Density Territories

Entrepreneurship in low-density territories plays a strategic role in economic and social development, contributing to job creation, the valorization of endogenous resources, and population retention. In these contexts, marked by structural weaknesses, geographic isolation, and a shortage of skilled labor, entrepreneurial initiatives can revitalize traditional sectors, promote innovation, and explore uncompetitive market niches.

According to those interviewed, territorial cohesion policies are fundamental for progress in entrepreneurship in LDT. In this sequence, interviewee VR2 states that there must be a positive differentiation regarding young people from the interior, since the region significantly lacks a young population. This idea aims to promote youth entrepreneurship and investment in these regions, so that they can be trained, with the aim of retaining a younger and more qualified population in these places, as interviewee VR5 asserts. From this perspective, interviewee VR11 asserts that these regions can be a great opportunity for entrepreneurs to invest in, given that there is enormous accessibility throughout the entire Trás-os-Montes and Alto Douro region, allowing individuals to develop their activities in this region and be just a few kilometers from major centers in Portugal and Spain.

From another perspective, interviewees B4 and B7 highlight that support from the central government is insufficient, as it does not efficiently address the region's depopulation and does not provide incentives for new companies to establish themselves there. Interviewee B4 also adds that entrepreneurs must deeply analyze the region and the possible return they will have in it, in order to be profitable for the entrepreneurs, due to all the challenging issues that LDTs face daily, as interviewee B7 mentions.

Interviewee VR2: "Considering that the region has an inverted demographic pyramid and that its main need is youth, it is not possible to continue, impassively and serenely, to watch young people being forced to leave without having the opportunity to settle locally. It is therefore essential to promote youth entrepreneurship and support investment, through a clearly positive differentiation for young people in the interior. A community that lacks the capacity to motivate, recognize, and retain its young people is a community with no future. [...]."

Interviewee VR5: "[...] For example, the search for these types of territories by technology-based companies has led some people to settle here. When they come, they bring their families, and we are automatically helping to combat depopulation. [...]."

Interviewee B4: "I know that in terms of PRR support, there has been some investment to help the population settle in the territory. Initially, these supports were probably inefficient or weak, and perhaps not attractive to entrepreneurs. And then, when making their investment here in the region, entrepreneurs also have to understand what they'll be working on, what they'll be exploring, and what the return will be, because it's not enough to just arrive and have the ambition to open a company and hire workers. The type of service they provide must be efficient and deliver results."

Interviewee B7: "In both aspects simultaneously. Firstly, due to all the constraints that low-density territories face and the lack of true territorial cohesion policies. The State must send positive signals and create incentives that encourage companies to establish themselves in these regions and territories."

4.3 Entrepreneurial Ecosystems

4.3.1 Policies to Support Entrepreneurship

Entrepreneurship support policies aim to create favorable conditions for the creation and growth of companies, stimulating innovation, competitiveness and economic development. They may include tax incentives, financial support, training programs, administrative simplification and the provision of support infrastructure, such as incubators and industrial zones [25, 49]. When adapted to territorial specificities, these policies have greater potential to respond to local needs, promoting the establishment of entrepreneurs, economic diversification and job creation, especially in more vulnerable contexts, such as low-density territories [50]. Figure 3 shows the interviewees' opinions on the policies to support entrepreneurship in the countries under study.

According to the interviews carried out, it was possible to see that there are two distinct opinions. One group believes that there are policies to support entrepreneurship in the region. Interviewees report that incubators are crucial to establishing entrepreneurial activity in the region, as they provide entrepreneurs with more affordable physical space and are responsible for assisting with business structuring and developing business plans. Additionally, incubators are a fundamental part of identifying appropriate financing for each entrepreneur, as well as identifying how current public policies can benefit them. Other notable initiatives include financial incentives, reduced-price land, entrepreneur support offices, entrepreneurship incentive programs, municipal tax reductions, business park expansion, and agricultural exploration programs.

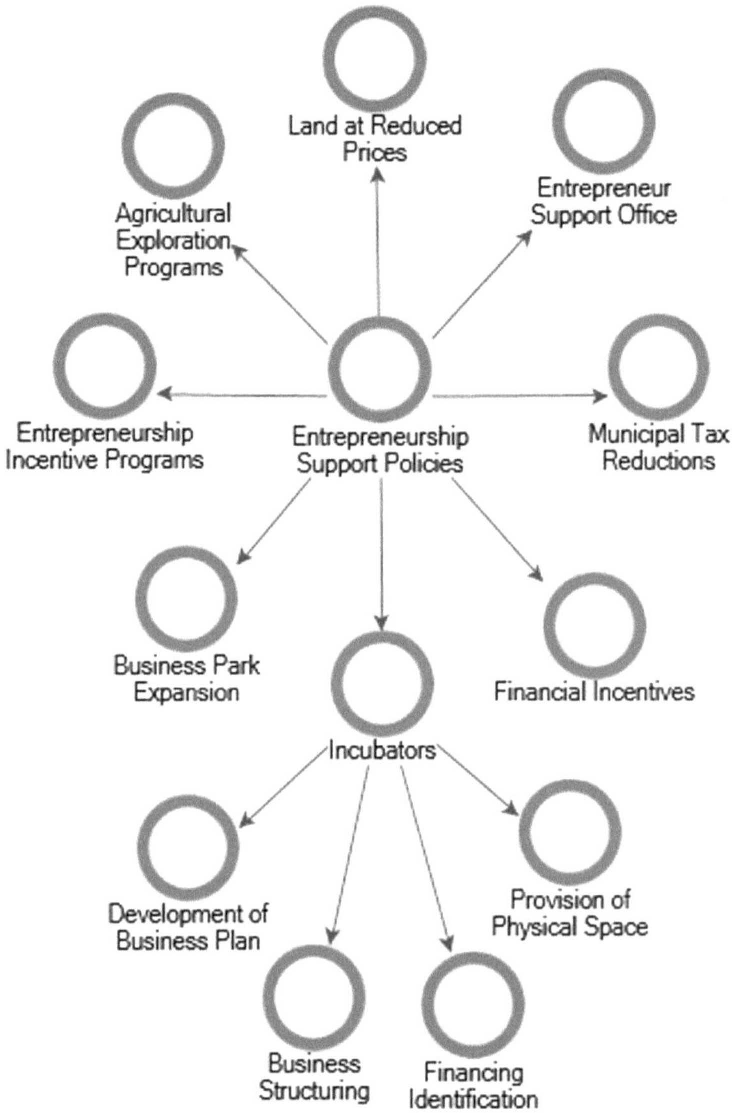

Fig. 3 Policies to support entrepreneurship in LDT (Vila Real and Bragança)

Interviewee B5: "We offer two types of support: entrepreneurship support for commerce, and entrepreneurship support for those who want to settle in our Industrial Zone. At the local commerce level, we have a regulation called "Regulation of Incentives for Traditional Commerce", which, in short, has three supports. Support through a subsidy for those looking to open a business. A second type of support is available for those looking to modernize or implement marketing initiatives for their businesses. Finally, there is a third support for those entrepreneurs who eventually resort to Community Funds. An application for a Community

Fund does not include the full investment, and the City Council also contributes a certain amount, also as a non-refundable contribution, to the non-shared portion. [...].”

Interviewee B6: “This office has a specialized technician who provides all kinds of support to young entrepreneurs. We welcome people who genuinely want to launch their own business. We provide all kinds of support. We also have a partnership with a Polytechnic Institute that provides us with some support.”

Interviewee VR10: “We seek to make the municipality more attractive for investment and the establishment of new companies that can generate jobs and, therefore, contribute to the settlement of our population. These supports include the provision of land for companies to set up in the Business Park at reduced costs; a 5-year exemption from IMI (Municipal Property Tax) for companies that establish themselves in the municipality and create jobs; and an exemption from IMT (Municipal Property Tax) for properties acquired by companies for the purpose of carrying out their activities. […].”

Interviewee VR5: “There are several support competitions related to entrepreneurship, from the municipality and the incubator. We support the creation and structuring of the business plan, and we also help with the search for financing from funds, banks, or other sources. We also have our incubator, whose success rate is quite high, the mortality rate of companies there is quite low, and these are mainly [...]”.

However, some interviewees have a different opinion, stating that policies to support entrepreneurship are practically non-existent, and the few that exist are mostly implemented at the national level, which does not address the region’s real problems. Interviewee VR6 argues that in these territories there is a profound lack of training focused on entrepreneurship, which means that the culture for it is poorly rooted, leading individuals to not be properly informed about everything inherent to entrepreneurship, including support. Regarding this last question, interviewee B1 states that there are currently some training courses that enable potential entrepreneurs to be trained, however, specialized technical support is non-existent, often falling short of meeting the real needs of entrepreneurs. The same interviewee also states that only a few municipalities have financial support, however the information that is published for their target audience is extremely scarce and ambiguous, which leads individuals to believe that there is no type of support in the entrepreneurial sphere. In this sequence, interviewee B8 claims that direct financial incentive policies are non-existent, that is, the monetary support that could be provided to entrepreneurs is not carried out directly by local entities, since they do not have sufficient conditions to do so. Therefore, entrepreneurs are highly dependent on the guidelines and support imposed by central and European authorities, especially with regard to direct support, that is, monetary incentives to start their businesses.

Interviewee B1: “There’s zero support in terms of development, training, and specific consulting. However, some municipalities offer some financial support for starting a business, but you have to call them, because there’s nothing on the website. It’s not easy to access information about support in most municipalities. [...]”

Interviewee VR6: “[…] I think there is a lack of training for entrepreneurship here, and what we call culture for development. [,,,]”

Interviewee B8: “There is no direct financial incentive policy […].”

On the other hand, there is an interviewee who states that the policy of incentives for entrepreneurship is demarcated by each municipality. Each municipality defines and implements its own support, which causes disarticulation among numerous local or regional entities, resulting in a lack of joint action for the proper implementation and implementation of policies to support entrepreneurship.

> **Interviewee VR3**: "Support for entrepreneurship is typically something each municipality tries to promote on its own. The policy isn't global. What municipality A implements isn't implemented by municipality B. Each municipality is different; some receive more support than others. This is especially true in business reception areas, and that's where municipalities have the opportunity to help, because the business reception area or Industrial Zone belongs to the municipality. [...]."

Table 3 presents a brief systematization of the mapping of policies to support entrepreneurship in the territories under study.

Table 3 was developed from an exhaustive mapping of policies to support entrepreneurship in the municipalities of Vila Real and Bragança districts, based on information collected, mostly, from the official websites of the Municipal Councils, as well as from business associations, incubators and higher education institutions. On a smaller scale, secondary sources, such as local or regional news, were also used. This survey allowed us to identify the presence or absence of different types of initiatives and support infrastructure, including entrepreneur support offices, incubators, business associations, industrial zones, training programs, financial support, tax benefits, and other elements relevant to stimulating entrepreneurial activity.

The analysis revealed significant heterogeneity among municipalities: while some offer a diverse range of support, combining physical structures, tax incentives, and training programs, others demonstrate significant deficiencies, offering only one or two measures. Business associations and industrial zones are common features in most municipalities, but direct financial support and tax benefits are less common. It was also observed that municipalities that most clearly and frequently publicize their incentives also tend to be those with the most active policies. Conversely, the absence or scarcity of information, often unclear or incomplete, generally corresponds to municipalities with fewer supports, located primarily in the interior of the Trás-os-Montes region.

These results confirm what some interviewees, such as B1 and VR3, pointed out: even within low-density territories themselves, there is significant regional disparity, with each municipality defining and implementing its own policies to promote entrepreneurship. This scenario highlights the need to strengthen not only the quantity and diversity of implemented policies, but also their coordination and dissemination, in order to enhance the attractiveness and effectiveness in stimulating entrepreneurial activity.

4.3.2 Role of Local Actors

Local stakeholders play a crucial role in supporting entrepreneurship, especially in low-density areas where resources and opportunities are more limited. Entities

Table 3 Mapping of policies to support entrepreneurship in the territories under study

Municipalities	Entrepreneur Support Offices	Business Incubators	Business Associations	Industrial Zones	Training Programs	Financial Support	Tax benefits	Other important aspects
Vila Real	No	Yes	Yes	Yes	Yes	No	No	Yes
Alijó	Yes	Yes	Yes	Yes	No	Yes	Yes	Yes
Boticas	Yes	Yes	Yes	Yes	No	Yes	Yes	Yes
Chaves	Yes	Yes	Yes	Yes	No	No	No	Yes
Mesão Frio	No	No	Yes	Yes	No	No	No	Yes
Mondim de Basto	No	No	Yes	Yes	No	Yes	Yes	Yes
Montalegre	Yes	Yes	Yes	Yes	Yes	No	No	No
Murça	Yes	Yes	Yes	Yes	No	Yes	Yes	Yes
Peso da Régua	No	No	Yes	Yes	No	No	Yes	No
Sabrosa	No	No	Yes	Yes	No	No	No	No
Ribeira de Pena	No	Yes	Yes	Yes	No	No	No	No
Valpaços	No	Yes	Yes	Yes	Yes	No	No	Yes
Vila Pouca de Aguiar	Yes	Yes	Yes	Yes	Yes	No	Yes	Yes
Santa Marta de Penaguião	No	No	Yes	Yes	No	No	Yes	No
Bragança	Yes	Yes	Yes	Yes	No	No	Yes	Yes
Alfândega da Fé	Yes	Yes	Yes	Yes	Yes	No	No	Yes
Miranda do Douro	Yes	No	Yes	Yes	Yes	No	Yes	Yes
Vimioso	Yes	Yes	Yes	Yes	Yes	Yes	Yes	Yes

(continued)

Table 3 (continued)

Municipalities	Entrepreneur Support Offices	Business Incubators	Business Associations	Industrial Zones	Training Programs	Financial Support	Tax benefits	Other important aspects
Macedo de Cavaleiros	Yes	No	Yes	Yes	No	Yes	Yes	Yes
Mirandela	Yes	Yes	Yes	Yes	No	Yes	Yes	Yes
Carrazeda de Ansiães	No	Yes	Yes	Yes	No	Yes	Yes	Yes
Vinhais	No	No	Yes	Yes	Yes	No	No	Yes
Vila Flor	Yes	Yes	Yes	Yes	No	No	No	No
Freixo de Espada à Cinta	No	No	Yes	Yes	No	Yes	Yes	No
Mogadouro	Yes	No	Yes	Yes	No	Yes	Yes	Yes
Torre de Moncorvo	Yes	No	Yes	Yes	No	Yes	Yes	Yes

such as city councils, business associations, higher education institutions, incubators, and science and technology parks can create favorable conditions for business creation and consolidation by offering technical, infrastructural, training, and financial support [51]. Cooperation between these stakeholders is essential to build a solid entrepreneurial ecosystem, capable of attracting investment, retaining talent and promoting innovation, thus contributing to the economic and social development of the region [52, 53].

The results of the interviews show that all interviewees consider the existence of interaction between the various local actors. The interviewees report that in recent years, joint work has been carried out between a wide range of local entities, with the aim of promoting entrepreneurship and initiatives that make companies and their production profitable, and empower local commerce itself, in addition to training, and the implementation of joint applications to boost various local areas, as mentioned by interviewees B4, VR3 and VR7. Interviewee B3 reiterates the same idea, stating that without interaction between various local actors, it would not be possible to implement policies to promote entrepreneurship and boost the region itself.

> **Interviewee B4**: "There has been ongoing collaboration involving various institutions to provide the best support for business owners and companies and to stimulate entrepreneurship. The City Council has worked in partnership with the Intermunicipal Community, regional Higher Education Institutions, the Business Association, the Agricultural Cooperative, the Development Association, among others, to promote entrepreneurship through various projects and initiatives that not only stimulate entrepreneurship but also seek to contribute to greater profitability for companies and a reduction in their production costs."

> **Interviewee B3:** "There has always been interaction, because I think the greatest added value that has occurred is precisely the fact that this interaction exists. [...]."

> **Interviewee VR3**: "That's been happening. I don't know if it's produced much, but it has. There have been seminars. The Intermunicipal Community itself has organized entrepreneurship seminars. [...]."

> **Interviewee VR7:** "This interaction has been ongoing. There are several examples of joint applications aimed at promoting local commerce. An application was prepared in collaboration with the municipality and the Business Association for the "Digital Neighborhoods." There has been collaboration on European projects such as the INTERREGE EUROPE RATIO, among other examples. [...]."

However, some interviewees demonstrate that the interaction between local actors is not consistent and that they believe that it is essential to have a close connection with them. Interviewee B1 reports that the spirit of cooperation is non-existent between the various entities, which causes a strategic disparity between them, as interviewee VR1 points out. He declared that it was urgent for local actors to follow the same line of reasoning, otherwise there would be no possibility of any kind of local or regional progress, which would ultimately increase the challenges evident in the territory.

> **Interviewee B1**: "I'm not aware of anything that's been done. Now, why? I can't explain it either. We think too much about ourselves. Our goal is to have our projects, finance the infrastructure, pay the expenses, always thinking about how this might work for the companies because we work for the companies. We think for ourselves; there's no such spirit of collaboration, of cooperation. [...]."

Interviewee VR1: "There have always been natural and institutional relationships between these institutions, there have been meetings, debates, and some conversations. Unfortunately, there usually isn't much continuity after the meeting, and this was important. I think it would be necessary to bring everyone together and define the strategic line. But defining the strategic line involves knowing how to propose and listen, because we're talking about universities, incubators, companies, and associations here, and perhaps we all have different focuses, and things will only work when the strategy reflects everyone's focus, not just the focus of a few. [...]."

4.4 Proposals to Improve the Attractiveness of Low-Density Territories for Entrepreneurship

The attractiveness of low-density areas for entrepreneurship is based on a combination of factors that can represent competitive opportunities, despite structural challenges. Figure 4 shows the interviewees' proposals for improving the attractiveness of the regions under study.

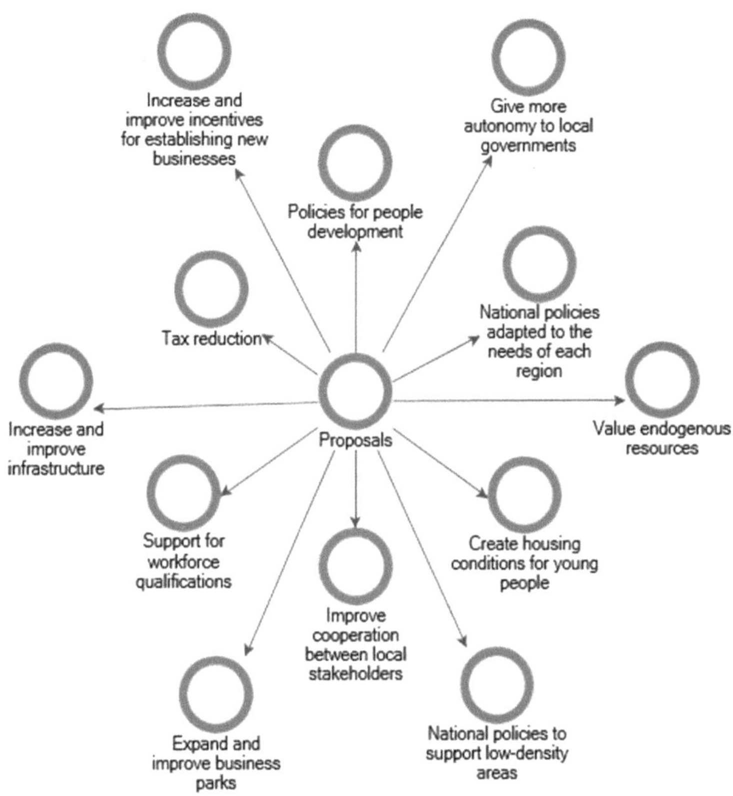

Fig. 4 Proposals to improve the attractiveness of LDT (Vila Real and Bragança)

According to the interviews, the proposals are divided into four main groups: fiscal measures, housing and population measures, business measures, and national measures. In the first group, the policy of tax reduction for companies is noteworthy, so that with the existence of positive discrimination for LDT, people and companies increasingly seek out these territories. In the second group, respondents value the existence of policies that keep the population in LDT, such as the creation of habitable conditions for young people and the existence of support for workforce qualifications. In the third group, interviewees emphasized the need to improve incentives for establishing new businesses through the expansion and improvement of business parks, the enhancement of infrastructure to welcome new entrepreneurs to the region, and the progression of interaction between local actors, in order to consolidate the best strategies for the region's development. The last group includes the creation of a series of national policies adapted to the reality of the territory itself, ultimately offering greater autonomy to local authorities, and in a global way (both on the part of local and central authorities) the existence of an authentic valorization of the endogenous products of each region.

Interviewee B5: "Attracting new entrepreneurs and attracting people to settle in the region isn't achieved through municipal policies alone. Municipalities don't have the authority or financial capacity alone; this policy is a cross-cutting one, and it's the central administration, the government, that should truly implement positive policies, positive discrimination for the interior, not the municipalities. [...]"

Interviewee B2: "[...] We have to have the capacity to think, although we are a small country, it is a very diverse country, since Alentejo is different from Trás-os-Montes, so we need political courage to create policies adapted to the region. The support measures implemented here are all the same for the country. [...]."

Interviewee VR9: "Reduction to the legal minimum of municipal taxes applied (IMI, exemption from fees applied for licensing housing or industrial construction) and increase in the limit of IRS deductions for 10 years. [...]."

Interviewee VR10: "It was crucial that the Central Government do the same, implementing genuine positive discrimination measures for these territories that would contribute to the establishment of companies. It's clear that establishing companies in these regions will require reducing their production costs, as this is the only way to overcome the distance to raw materials, major markets, and export centers."

Interviewee VR2: "A huge effort has been made to attract, for example, footwear companies and other businesses in other sectors. But it's been difficult because tax measures don't keep pace with government support at the central level. And often at meetings held in Lisbon, because we have a small population, we don't count for anything. [...]."

In this sense, in general, the interviewees state that the territory urgently needs national public policies that discriminate the territory in a positive way, in order to attract people to settle in these places, as well as companies, since the policies implemented at the municipal level are insufficient, as reported by interviewees B5 and VR2. Interviewees B2 and VR10 state that it is imperative that the central government implement appropriate policies for LDTs, since all implemented policies and programs currently follow the same conditions and incentives as High-Density Territories (HDTs), which do not have the same shortcomings or challenges as the former. Therefore, all these measures are insufficient and unrealistic because they are not adapted to the territory.

Interviewees VR9 and VR10, who emphasize the need for tax incentive policies, such as reducing property taxes (IMI) and licensing fees, among others, encourage large companies to settle in these territories when comparing the tax levels of the HDTs. Interviewee VR10 also argues that a reduction in production costs should be foreseeable so that companies don't feel the impact of being distanced from large centers.

On the other hand, interviewee B8 emphasizes that measures must be implemented to create habitable conditions for the population wishing to settle in these territories, such as price controls and the restoration of vacant houses and the historic center itself, leading to a significant increase in demand and also in the housing supply, which is quite scarce throughout the region.

5 Conclusion

With the aim of this study, we analyzed the policies to support entrepreneurship in two low-density territories in Portugal (the districts of Vila Real and Bragança). We found that entrepreneurship is central to the economic and social development of these regions, contributing to demographic revitalization, productive diversification and the valorization of endogenous resources. Despite the opportunities associated with the strategic location, the competitive cost of living, and the wealth of local resources, constraints persist, such as an aging population, a shortage of skilled labor, infrastructural weaknesses, and the lack of adapted public policies.

Existing policies are limited in scope, relying on national and European initiatives, and poorly adapted to local specificities. Measures such as reduced land prices in industrial zones, the creation of incubators and support offices, and some tax benefits, while significant, have not been sufficient to reverse depopulation or attract sustained investment, compounded by the difficulty in accessing clear and comprehensive information.

The role of local actors is heterogeneous: in some cases, there is cooperation between municipalities, higher education institutions, business associations, and incubators, generating joint initiatives and training; in others, isolated action prevails, weakening the entrepreneurial ecosystem. Evidence confirms that collaborative governance is crucial for policy effectiveness and territorial competitiveness.

Given these limitations, it is essential to adopt differentiated policies and positive discrimination, with attractive tax measures, incentives for setting up businesses, programs to retain qualified young people, and strategies to enhance endogenous resources. It is equally essential to improve living conditions, modernize infrastructure, and promote technology-based and creative sectors, creating a more favorable environment for innovation and investment.

This study contributes to the existing literature on entrepreneurship in LDT by deepening the understanding of the specificities that shape the effectiveness of support policies in these contexts. By comparatively analyzing the districts of Vila Real and Bragança, the research offers empirical evidence on how structural, demographic, and institutional factors influence entrepreneurial attractiveness and dynamism. This contribution is particularly relevant in an area of research that remains largely unexplored, given that most studies on policies to promote entrepreneurship focus on urban and metropolitan contexts. By bringing the perspective of multiple local stakeholders, the study enriches the academic debate on entrepreneurial ecosystems in peripheral regions, highlighting the importance of adapting policies to territorial realities.

On a practical level, the results offer valuable insights for policymakers working in low-density areas. The evidence collected suggests that policies to support entrepreneurship should be more differentiated and territorially adapted, incorporating positive discrimination mechanisms that compensate for the structural disadvantages of these territories. At the local level, municipalities can benefit from creating joint strategies and stronger collaborative networks between city councils, higher education institutions, business associations, incubators, and science and technology parks, maximizing resources and avoiding dispersion of efforts. For entrepreneurship in the regions studied, the study highlights the importance of concrete measures such as administrative simplification, the provision of land and incubation spaces at reduced costs, the implementation of competitive tax benefits, and the creation of programs to retain qualified young people. Additionally, strengthening housing conditions and enhancing endogenous resources emerge as key elements in attracting and retaining entrepreneurs, contributing to more balanced and sustainable economic development.

The main limitations of this study are its qualitative nature and the sample size restricted to the districts of Vila Real and Bragança, which limits the generalizability of the results. Furthermore, data collection occurred at a specific point in time, reflecting only the policies and perceptions prevailing during that period. Finally, the difficulty in accessing clear and up-to-date official information on support for entrepreneurship limited the comparative analysis between all municipalities.

Future research could expand the geographic scope to other low-density territories in Portugal and abroad, allowing for more comprehensive comparative analyses. It will also be important to adopt mixed methodologies, combining qualitative and quantitative approaches, to validate and deepen the conclusions. Additionally, longitudinal studies could monitor the evolution of policies and entrepreneurial ecosystems, assessing their effective impact over time.

Funding This work is supported by national funds, through the FCT–Portuguese Foundation for Science and Technology under the projects UID/04011/2025 (https://doi.org/10.54499/UID/04011/2025), UID/GES/04630/2020 (https://doi.org/10.54499/UID/04630/2025) and project BOLDER, ref. 2024.17994.PEX (https://doi.org/10.54499/2024.17994.PEX).

References

1. Ribeiro, M. I., Fernandes, A., Matos, A., & Cabo, A. (2016). Empreendedorismo, Inovação e Desenvolvimento Local: As micro e pequenas empresas do interior norte de Portugal. *Revista de Empreendedorismo e Gestão de Micro e Pequenas Empresas, 1*(3), 34–53. http://hdl.handle.net/10198/13745
2. Ciccarino, I. D. M., Borges, A., & Viveiros, D. S. D. (2017). Um estudo sobre o empreendedorismo incentivado por políticas públicas de assistência social no município do Rio de Janeiro. *Cadernos de Gestão e Empreendedorismo, 6*(1), 15–29. https://doi.org/10.32888/cge.v6i1.12724
3. Stam, E., & van de Ven, A. (2021). Entrepreneurial ecosystem elements. *Small Business Economics, 56*(2), 809–832.
4. Verheul, I., Wennekers, S., Audretsch, D., & Thurik, R. (2002). An eclectic theory of entrepreneurship: Policies, institutions and culture. In *Entrepreneurship: Determinants and policy in a European-US comparison* (pp. 11–81). Springer US.
5. Vaz, D., & Nofre, J. (2017). Conhecimento, Criatividade e Novas Dinâmicas Urbanas: Repensar os Territórios de Baixa Densidade em Portugal. *Revista Portuguesa de Estudos Regionais, 49*, 77–88. https://www.review-rper.com/index.php/rper/article/view/498/414
6. Lee, S. Y., Florida, R., & Acs, Z. (2004). Creativity and entrepreneurship: A regional analysis of new firm formation. *Regional Studies, 38*(8), 879–891.
7. Tekic, A., & Tekic, Z. (2025). Resource endowments as antecedents of high-impact entrepreneurship: Embracing the complexity of entrepreneurial ecosystems. *Journal of Business Research, 197*, 115444.
8. Almeida, J., & Daniel, A. D. (2025). Addressing the distinctive features of entrepreneurial ecosystems in low-density territories. *Journal of Entrepreneurship and Public Policy.*
9. Pitelis, C. N., Teece, D. J., & Yang, H. (2024). Dynamic capabilities and MNE global strategy: A systematic literature review-based novel conceptual framework. *Journal of Management Studies, 61*(7), 3295–3326.
10. Galvão, A. R., Marques, C. S., Monteiro, C., & Mascarenhas, C. (2024). The creative industries' role in low-density territories' sustainability: The Douro region. *SAGE Open, 14*(2). https://doi.org/10.1177/21582440241258412
11. Keryan, T., Veréb, V., Madureira, L., Bento, R., Ramos, L., Marques, C., & Dax, T. (2025). Is low-density a viable territorial category for enhancing EU cohesion policy implementation? *European Planning Studies, 33*(4), 491–511.
12. Stam, E. (2015). Entrepreneurial ecosystems and regional policy: A sympathetic critique. *European Planning Studies, 23*(9), 1759–1769.
13. Cappelleras, J. L., Contín-pilart, I., Larraza-kintana, M., & Martin-sanchez, V. (2015). Regional population density and entrepreneurial growth aspirations: The moderating role of individual human capital. *Druid15*, 3–25. https://silo.tips/download/regional-population-density-and-entrepreneurial-growth-aspirations-the-moderatin
14. Fritsch, M., Brixy, U., & Falck, O. (2006). The effect of industry, region, and time on new business survival: A multi-dimensional analysis. *Review of Industrial Organization, 28*(3), 285–306.
15. Pezzini, M. (2001). Rural policy lessons from OECD countries. *International Regional Science Review, 24*(1), 134–145. https://doi.org/10.1177/016001701761013024
16. Meyer, N., & Jough, J. D. (2018). The importance of entrepreneurship as a contributing factor to economic growth and development: The case of selected European countries. *Journal of Economics and Behavioral Studies, 10*(4), 287–299.
17. Bosma, N. S., & Levie, J. (2010). *Global Entrepreneurship Monitor: 2009 Global Report. Babson College.* Universidad del Desarrollo, Reykjavík University, London Business School. https://dspace.library.uu.nl/handle/1874/209570
18. Müller, S. (2016). A progress review of entrepreneurship and regional development: What are the remaining gaps? *European Planning Studies, 24*(6), 1133–1158.

19. Kolehmainen, J., Irvine, J., Stewart, L., Karacsonyi, Z., Szabó, T., Alarinta, J., & Norberg, A. (2015). Quadruple helix, innovation and the knowledge-based development: Lessons from remote, rural and less-favoured regions. *Journal of the Knowledge Economy, 7*(1), 23–42.
20. Siemens, L. (2015). "We moved here for the lifestyle": A picture of entrepreneurship in rural British Columbia. *Journal of Small Business & Entrepreneurship, 27*(2), 121–142.
21. Stathopoulou, S., Psaltopoulos, D., & Skuras, D. (2004). Rural entrepreneurship in Europe. *International Journal of Entrepreneurial Behavior & Research, 10*(6), 404–425.
22. Ahmad, A. R., WFW, Y., Noor, H. M., & Ramin, A. K. (2012). Preliminary study of rural entrepreneurship development program in Malaysia. *Journal of Global Entrepreneurship, 2*(1), 1–8.
23. Miles, M. P., & Morrison, M. (2018). An effectual leadership perspective for developing rural entrepreneurial ecosystems. *Small Business Economics, 54*(4), 933–949.
24. Pato, M. L., & Teixeira, A. A. C. (2014). Twenty years of rural entrepreneurship: A bibliometric survey. *Sociologia Ruralis, 56*(1), 3–28.
25. Bruton, G. D., Ahlstrom, D., & Li, H. (2010). Institutional theory and entrepreneurship: Where are we now and where do we need to move in the future? *Entrepreneurship Theory and Practice, 34*(3), 421–440.
26. Carballo, I. E., Belloni, C. M., Amorós, M. L., & Fracchia, E. L. (2017). Emprendedorismo y políticas públicas. Una introducción a la literatura. *Revista Perspectivas de Políticas Públicas, 7*(13), 37–88.
27. Li, Z., Jintana, S., & Jotaworn, S. (2025). Empowering the next generation of entrepreneurs: Intersections of education, policy, and business model innovation. *Cogent Education, 12*(1), 2542392.
28. Lundström, A., & Stevenson, L. A. (2005). *Entrepreneurship policy: Theory and practice* (Vol. 9). Springer. https://link.springer.com/content/pdf/10.1007/b104813.pdf
29. Parker, C. S. (2009). When politics becomes protest: Black veterans and political activism in the postwar south. *The Journal of Politics, 71*(1), 113–131.
30. Barca, F., Brezzi, M., Terrible, F., & Utili, F. (2005). Measuring for decision making: Soft and hard use of indicators in regional development policies. *Statistics, Knowledge and Policy*, 50.
31. Potter, J. (2005). Entrepreneurship policy at local level: Rationale, design and delivery. *Local Economy, 20*(1), 104–110.
32. Castaño, M. S., Méndez, M. T., & Galindo, M. Á. (2016). The effect of public policies on entrepreneurial activity and economic growth. *Journal of Business Research, 69*(11), 5280–5285.
33. Ascani, A., Crescenzi, R., & Iammarino, S. (2012). Regional economic development. A Review. *SEARCH, WP01/03*, 2–26. http://www.ub.edu/searchproject/wp-content/uploads/2012/02/WP-1.3.pdf
34. Murray, A. T. (2024). A location analytics perspective of regional science at a crossroad. *Papers in Regional Science, 103*(4), 100034.
35. Quttainah, M. A. (2015). Entrepreneurship, innovation and public policy: A survey of literatures. *International Journal of Economics & Management Sciences, 4*(279) https://www.semanticscholar.org/paper/Entrepreneurship%2C-Innovation-and-Public-Policy%3A-A-Quttainah/b1df337aade12ed5d77ee0bfb8a798c24d4462da
36. Gnyawali, D. R., & Fogel, D. S. (1994). Environments for entrepreneurship development: Key dimensions and research implications. *Entrepreneurship Theory and Practice, 18*(4), 43–62.
37. Cohen, B. (2005). Sustainable valley entrepreneurial ecosystems. *Business Strategy and the Environment, 15*(1), 1–14.
38. Henrekson, M., & Stenkula, M. (2010). Entrepreneurship and public policy. In *Handbook of entrepreneurship research: An interdisciplinary survey and introduction* (pp. 595–637). Springer New York.
39. Tsai, W. H., & Kuo, H. C. (2011). Entrepreneurship policy evaluation and decision analysis for SMEs. *Expert Systems with Applications, 38*(7), 8343–8351.
40. Acs, Z. J., & Szerb, L. (2007). Entrepreneurship, economic growth and public policy. *Small Business Economics, 28*(2), 109–122.

41. Stephens, H. M., Partridge, M., & Faggian, A. (2013). Innovation, entrepreneurship and economic growth in lagging regions. *Journal of Regional Science, 53*(5), 778–812.
42. Sroka, W., Filipiak, T., & Barczyk-Ciula, J. (2024). The role of the local government in fostering entrepreneurship–evidence from municipalities in the Kraków metropolitan area. *Economic and Regional Studies, 17*(1), 21–57.
43. Tamásy, C. (2006). Determinants of regional entrepreneurship dynamics in contemporary Germany: A conceptual and empirical analysis. *Regional Studies, 40*(4), 365–384.
44. INE (2022a). *População residente (N°) por local de residência (à data dos Censos de 2021), Sexo, Grupo etário, e Naturalidade; Decenal.* Instituto Nacional de Estatística. https://www.ine.pt/xportal/xmain?xpid=INE&xpgid=ine_indicadores&contecto=pi&indOcorrCod=0011628&selTab=tab0
45. INE (2022b). *Empresas (N°) por localização geográfica (NUTS—2013) e Dimensão; Anual.* Instituto Nacional de Estatística. https://www.ine.pt/xportal/xmain?xpid=INE&xpgid=ine_unid_territorial&menuBOUI=13707095&contexto=ut&selTab=tab3
46. Perren, L., & Ram, M. (2004). Case-study method in small business and entrepreneurial research: mapping boundaries and perspectives. *International small business journal, 22*(1), 83–101. https://doi.org/10.1177/0266242604039482
47. Yin, R. K. (2015). *Qualitative research from start to finish. Guilford publications*
48. Reynolds, J. (2022). Building an entrepreneurial ecosystem to catalyze innovation in rural places. In *Building rural community resilience through innovation and entrepreneurship* (pp. 31–48). Routledge.
49. Thai, M. T. T., & Turkina, E. (2014). Macro-level determinants of formal entrepreneurship versus informal entrepreneurship. *Journal of Business Venturing, 29*(4), 490–510.
50. Audretsch, D. B., Grilo, I., & Thurik, A. R. (2007). *Handbook of research on entrepreneurship policy*. Edward Elgar Publishing.
51. Acs, Z., Åstebro, T., Audretsch, D., & Robinson, D. T. (2016). Public policy to promote entrepreneurship: A call to arms. *Small Business Economics, 47*, 35–51.
52. Bischoff, K., Volkmann, C. K., & Audretsch, D. B. (2018). Stakeholder collaboration in entrepreneurship education: An analysis of the entrepreneurial ecosystems of European higher educational institutions. *The Journal of Technology Transfer, 43*(1), 20–46.
53. Parmar, B. L., Freeman, R. E., Harrison, J. S., Wicks, A. C., Purnell, L., & Colle, S. D. (2010). Stakeholder theory: The state of the art. *The Academy of Management Annals, 4*(1), 403–445.

The Relationship Between Open Innovation and Public Funding in Portuguese SMES: Impact on Innovative Capacity and Collaboration within Innovation Ecosystems

Norberto Semião, Sara Nunes, and Luís Farinha

Abstract This study investigates the relationship between open innovation factors and publicly funded Research and Development Innovation (RDI) projects in small and medium-sized enterprises (SMEs) in Portugal, with an emphasis on the impact of public funding on these companies' innovative capacity. The methodology was based on a quantitative statistical analysis applied to a dataset obtained from innovative Portuguese companies. The empirical results reveal that financial support facilitates the adoption of open innovation practices, promoting collaboration with universities and research centers, which enhances SMEs' capabilities by allowing them to overcome resource limitations. Additionally, a positive relationship was identified between public funding and the strengthening of collaborative networks, improving access to external knowledge and competitiveness in the global market. The conclusions underscore the relevance of targeted public policies that can reduce barriers for organizations in establishing partnerships and creating an inclusive innovation ecosystem. This study provides essential data for formulating strategies to support innovation and sustainable growth for SMEs in Portugal.

Keywords SME · Innovation · Innovation management · Innovation ecosystem · Open innovation

---

N. Semião (✉)
Universidade da Beira Interior, Covilhã, Portugal
e-mail: norberto.semiao@ubi.pt

N. Semião · S. Nunes · L. Farinha
NECE-UBI Research Centre for Business Sciences, Covilhã, Portugal

S. Nunes · L. Farinha
Instituto Politécnico de Castelo Branco, Escola Superior de Gestão, Idanha-a-Nova, Portugal

L. Farinha
CIICESI—Centro de Inovação e Investigação em Ciências Empresariais e Sistemas de Informação, Felgueiras, Portugal

L. Farinha et al. (eds.), *Sustainable Innovations in Mechanical Engineering and Management Sciences*, Lecture Notes on Multidisciplinary Industrial Engineering,
https://doi.org/10.1007/978-3-032-30372-1_6

1 Introduction

In 1996, the European Commission adopted the first common definition of small and medium-sized enterprises (SMEs), applied across the European Union. On May 6, 2003, a new recommendation was approved to reflect economic changes. SMEs are the engine of the European economy, serving as an essential source of employment, entrepreneurship, and innovation, and are crucial for fostering competitiveness and job creation. This new definition came into force on January 1, 2005 [16].

According to Bruque and Moyano [10], SMEs have been the driving force behind economic growth and technological progress in many countries through their capacity to innovate in a rapidly changing and increasingly competitive global market. Thus, the importance of SMEs should not be underestimated, given their critical role in a modern economy [54]. For Taques et al. [57], innovation can be an important source of competitive advantage for companies, enabling the creation of new products or services and the improvement of existing ones. Schumpeter [53] views innovation as a process of "creative destruction" that drives economic growth by replacing old technologies with new ones.

According to Thanawala [58], Schumpeter assigned a central role to innovation in economic development, characterizing it as the introduction of new products, production processes, organizational structures, or markets, leading to "creative destruction" that replaces old structures with more efficient and competitive solutions. Davidsson [18], in turn, highlights the role of innovation in entrepreneurship, asserting that the social and economic context shapes entrepreneurs' innovative potential, making it crucial for value creation and sustainable development. Chesbrough [13], through the concept of Open Innovation, describes how innovative companies utilize innovative networks and external partnerships to develop more comprehensive solutions and integrate complementary resources. This networked management model facilitates access to knowledge and technology, fostering the development of networking skills [11].

SMEs heavily depend on external entities and cooperate with partners to improve their performance [41]. Cooperation with external organizations to access technology is an important competitive advantage. SMEs use network partnerships to compensate for weaknesses, reduce transaction costs and risks, and exchange knowledge and skills. Therefore, network relationships are critical to the performance of SMEs [38]. According to Radziwon and Borgers [49], SMEs face challenges in managing open innovation due to their dependence on external partners and misalignment between their business models and business ecosystems. Inter-company collaboration is essential to overcome these obstacles. The article discusses how open innovation is linked to knowledge management and categorizes inter-company connections, helping businesses choose the best mechanisms to transfer knowledge and maximize outcomes [47].

Small businesses, due to their limited resources, may face difficulties in meeting the demands required to implement innovation projects [1, 8]. Furthermore, while innovation can bring significant benefits, it also comes with a set of challenges and risks that companies must manage [5, 21]. Nohria and Gulati [45] argue that large companies, in contrast, have the capacity to absorb failures in innovative projects, whereas for small companies, a failure could represent an existential risk.

This study, based on evidence and using a questionnaire administered to Portuguese companies, aims to investigate the importance of innovation for SMEs and the challenges they face in its implementation, especially in the current economic and technological context. This exploratory research seeks to deepen the understanding of the reality of innovation in Portuguese SMEs, analyzing whether the ability to innovate differs between companies of different sectors and sizes. The results aim to contribute to the discussion on innovation in SMEs, bridging the gap between theory and practice and paving the way for future research.

2 Literature Review

Creativity and innovation are essential for organizational success [6]. Bessant [9] discusses high-involvement innovation, a concept fundamentally centered on fostering a culture where innovation becomes part of everyday life. Innovation capability requires a collaborative and open culture, supported by incentives that reward challenges to current practices [55]. According to Laursen and Salter [37], innovation models indicate that many innovative companies have shifted how they seek new ideas, with a notable increase in the use of external networks across various company sizes [27]. As business realities evolve, employees' behavior and actions must adapt accordingly [20].

In the era of "open innovation" [13], companies increasingly rely on external sources of innovation, leveraging ideas, knowledge, resources, and external networks critical to SME innovation success. Open innovation allows SMEs to share risks and access knowledge and partnerships, particularly in robust industrial ecosystems [42]. Innovation is recognized as a key factor for long-term success and survival in organizations. According to Palazzeschi et al. [48], organizations acknowledge that innovation contributes to competitive advantage in an increasingly challenging and dynamic labor market.

Limited resources and reduced R&D capacity make networks indispensable for SMEs to mitigate uncertainties in developing and utilizing new technologies [19]. The concept of the innovation ecosystem has gained prominence in the literature on strategy, innovation, and entrepreneurship, though it is applied in various and often contradictory ways [25]. The term "ecosystem" has become common in strategy discussions over the past two decades, reflecting an interest in interdependencies between organizations [3]. According to Koenig [34] and Moore [44], an innovation ecosystem consists of an intentional community of economic actors co-evolving, aligning values, and expanding capabilities through collective innovation. Adner [2] states that innovation ecosystems foster collaborative agreements, enabling companies to integrate individual offerings into customer-focused activities, thereby receiving positive feedback on the value and performance of innovations that are unattainable in isolation. An innovation ecosystem reflects the dynamics between various stakeholders—such as companies, customers, and government entities—whose interactions influence innovation performance [37].

Few studies have examined the implementation of open innovation in SMEs compared to large enterprises. However, Spithoven et al. [56] note that while resource scarcity in SMEs is a barrier, it also motivates them to seek external knowledge. According to Cohen and Levinthal [15], firms' absorptive capacity can be a significant factor when comparing SMEs and large enterprises in the context of innovation. In Portugal, a study by Julião et al. [30] reveals that SMEs exhibit high receptivity to open innovation implementation, driven primarily by process improvement, with decision factors varying based on company size. Existing studies on open innovation in SMEs reveal significant challenges in effectively implementing mechanisms to promote open innovation [50]. Leiponen and Helfat [40] highlight a broad and valuable range of information sources—such as customers, suppliers, competitors, universities, public research organizations, consultants, and professional and industrial associations frequently used by companies. Katila and Ahuja [32], along with Laursen and Salter [37], emphasize the heavy reliance of companies on external information sources, integrating them with internal innovation sources.

Keyword co-occurrence analysis by Sabando-Vera et al. [52] reveals a growing trend in publications about open innovation in SMEs, highlighting central themes such as company performance, R&D networks, business management, business models, capabilities, and knowledge transfer. Recent studies compiled by Vanhaverbeke et al. [63] explore topics such as SME collaboration with crowdsourcing platforms [7], the adoption of open innovation to support frugal development [29], SME involvement in crowdfunding [23], and the role of key individuals entrepreneurial leaders—in supporting open innovation adoption [4].

Innovation strategy is widely discussed in research literature [14, 33], but strategic innovation management still faces theoretical gaps and conflicting predictions, requiring systematic reviews to guide future research and management decisions [33]. SMEs increasingly favor open innovation models over closed ones, focusing on partnerships along supply chains. Evidence shows that these partnerships enhance short-term innovation and help sustain SMEs' long-term innovation success [14, 62]. According to Haahti et al. [26], cooperative strategies are crucial for SMEs to enhance their knowledge of export markets and improve export performance. Additionally, evidence indicates that innovation and exports mutually reinforce each other, creating a virtuous cycle where innovation drives exports, and knowledge gained in foreign markets fosters further innovation and growth [24]. García-Morales et al. [22], argue that leadership is fundamental to innovation outcomes, with variations depending on company size. In large organizations, transformational leadership is positively associated with organizational innovation [22, 31, 61]. Conversely, smaller, less complex organizations benefit more from transactional leadership in achieving innovation in management [61]. Modern leadership emphasizes channeling employee energy collaboratively, prioritizing inspiration and engagement over issuing orders [35]. Participative leadership increases trust, commitment, and employee valuation of managers, leading to greater motivation for task completion [64].

SME managers face strategic decisions about appropriating returns on innovative investments that differ from those of larger firms, prioritizing speed and confidentiality over patents [28, 36]. Exceptions include SMEs collaborating with universities, typically science based or R&D-focused SMEs, for whom patents remain a relevant mechanism for capturing innovation profits [39]. Holgersson [28] highlights that patents serve as crucial signaling mechanisms for attracting clients and securing venture capital investments in such contexts.

Since Schumpeter's analysis, finance has been seen as a vital part of the innovation processes; authors like Mazzucato [43] and O'Sullivan [46] argue that supporting the real economy requires reducing companies' dependence on financial capital and restructuring the financial sector to ensure the long-term resources necessary for innovation. Small companies face financing constraints, while larger companies investing in R&D have easier access to external funding [60]. The high perception of technological uncertainty by external investors regarding SMEs makes it more difficult for them to access external financing. Above all, Trönnberg and Hemlin [59] and Zhang et al. [65] mention that SMEs face more obstacles in obtaining bank loans compared to large companies. Exporting can alleviate these constraints by signaling stable cash flows and improving creditors' willingness to finance strategic, innovative investments critical to SME competitiveness [12, 51, 60]. Investments in fixed assets, such as machinery and equipment, are crucial for driving innovation and SME performance in export markets. However, access to financing these investments remains challenging for young, innovative SMEs seeking growth and export opportunities [42].

3 Methodology

The present study adopted a quantitative approach, using a structured questionnaire with 58 questions as the main data collection instrument. The aim was to investigate Research, Development, and Innovation (RDI) practices in Portuguese companies. The questionnaire was developed based on the "Innovation Scoring" tool from COTEC Portugal [17] and supported by relevant literature. It was subsequently reviewed by experts to ensure the validity and reliability of the items.

In addition to socio-demographic dimensions, the questionnaire covered several essential aspects of organizational innovation, including strategy, organization, processes, enablers, impact, open innovation, and funded R&D projects. Each dimension was subdivided to enable a detailed analysis of factors associated with RDI management. The questions were structured using 5-point Likert scales, multiple-choice questions, and open-ended fields, allowing for a mixed-methods analysis (quantitative and qualitative) of the data and providing an in-depth understanding of innovative practices and outcomes within organizations.

For the present study, the focus was specifically on the dimensions of Open Innovation and Funded R&D Projects, given their relevance to analyzing collaboration practices and open innovation in Portuguese companies. The Open Innovation dimension was explored to understand how companies engage in external innovation initiatives, as well as the benefits, challenges, and capacity-building opportunities arising from these practices. The Funded R&D Projects dimension focused on partnerships with universities and research institutions, as well as support from funding entities to promote R&D key elements for innovation in Portugal. The analysis of these dimensions provided specific insights into openness to innovation and strategic collaborations in the Portuguese context.

Data collection took place between August 22 and September 25, 2024, with the questionnaire being sent via email directly to the selected companies. The data obtained was analyzed using IBM SPSS Statistics 26.0 software, employing descriptive statistics to summarize general information and inferential techniques to examine correlations between variables. The analysis aimed to identify trends and calculate new variables influencing the implementation of R&D and Innovation (RDI), to maximize the benefits and competitive potential of open innovation in Portuguese companies.

Variables were defined based on the dimensions of Open Innovation and Funded R&D&I Projects to capture critical factors for understanding collaboration and innovation practices in Portuguese companies.

Open Innovation Collaboration Factor: Comprising 21 items, this reflects the organizational strategy of establishing external partnerships to drive innovation, encompassing activities such as market prospecting, investing in startups, and participating in consortia. The organization prioritizes collaborations with various stakeholders, with a special focus on international partnerships, aiming to expand knowledge, increase global visibility, and attract external investments—essential factors for strengthening innovation capacity and competitiveness.

Benefits and Challenges in Open Innovation Factor: With 18 items, this addresses the gains and challenges faced in collaborative innovation. Benefits include increased innovation capacity, agility in market response, and strengthening internal competencies. Challenges involve cultural resistance to external innovation, issues of trust and intellectual property protection, and the complexity of managing partnerships. Careful management of these aspects is essential to maximize returns and ensure security in open innovative initiatives.

Opportunities, Capacity Building, and Development in Open Innovation Factor: Including 22 items, this focuses on identifying and exploiting strategic collaborations that promote innovation. It includes participation in networks and clusters, networking events, and collaborative platforms. The organization invests in acceleration and incubation programs, open challenges, hackathons, and partnerships with universities and R&D centers. Capacity-building actions focus on areas such as innovation management, design thinking, and agile methodologies, fostering continuous talent development and innovation.

Entities in the Context of Funded R&D&I Projects Factor: Represented by 17 items, this reflects the impact of publicly funded projects on business development and innovation. It includes the acquisition and commercialization of patents, licensing of technological solutions, and creation of new collaboration and subcontracting agreements. Strengthened alliances among partners, increased collaboration with universities, the emergence of spinoffs, and absorption of external knowledge are also observed. Funded projects provide financial support and encourage the attraction of qualified talent.

Collaboration with Universities and R&D Institutes in Funded R&D&I Projects Factor: Formed by 16 items, this emphasizes the importance of information and knowledge sources for innovation. It includes integrating diverse sources, both internal and external, such as universities, R&D institutes, and government entities. Collaboration with suppliers, customers, competitors, consultants, and R&D firms, as well as participation in conferences and sectoral associations, enhances knowledge exchange. Additional sources, such as technical literature and sectoral regulations, contribute to continuous updates, strengthening competitiveness and compliance with standards.

These factors enable an in-depth analysis of collaboration practices and open innovation in Portuguese companies, providing insights into the impacts and challenges of implementing RDI (Research, Development, and Innovation). Additionally, they help identify trends and variables that influence competitiveness and organizational performance in the context of innovation.

The target population of the study comprised Portuguese companies from various sectors. The questionnaire was distributed to an initial sample of 7000 companies, covering a wide range of business activities. The study obtained 113 valid responses, which constitute the sample analyzed. Despite the 1.61% response rate, the diversity in sectors, sizes, and locations ensures adequate representativeness for the study's objectives, reflecting the diversity of the Portuguese business landscape.

In terms of sectors, most companies belong to Manufacturing Industries (34.5%), followed by Trade (15.0%) and Consulting and Scientific Activities (13.3%). Regarding the company's age, 30.9% have been operating for over 40 years, indicating market maturity and experience, while a small portion (1.8%) represents recent businesses with 1–2 years of activity. Geographically, the distribution is also broad, ensuring diverse business contexts.

Concerning company size, 17.7% are micro-enterprises, the most representative group is small enterprises at 37.2%, followed by 23.9% medium-sized enterprises, and 21.2% large companies, demonstrating a heterogeneous mix of organizational structures. In terms of sales volume, 42.4% of the companies report values between €1 million and €10 million, while 20.4% have sales exceeding €50 million, indicating a balanced distribution in financial capacity.

This sample provides a detailed and representative view of RDI management practices in Portuguese companies, offering relevant data to assess the implementation of RDI practices and their impacts on innovation, sustainability, and competitiveness in the Portuguese context.

4 Results and Analysis

To analyze the results obtained across different dimensions, the meaning of the variables used to explore open innovation dynamics was calculated, identifying relevant trends and correlations. The descriptive and inferential statistical techniques applied provided a comprehensive and detailed view of the factors impacting the implementation of RDI, as well as the benefits and challenges faced by organizations. The results aim to clarify the relationships between innovation factors and the competitive potential of companies, offering practical insights for effective innovation management.

Table 1 presents the Pearson correlation coefficients between the created factors. All correlations shown are statistically significant at the 0.01 level, suggesting strong robustness among the components analyzed and the existence of significant relationships between these factors.

The analysis of the correlations obtained highlights the strong positive and significant association between the factors, all of which are significant at the 0.01 level. It is observed that Collaboration is strongly associated with the other factors, especially with Entities (correlation of 0.719), suggesting that a collaborative environment strengthens partnerships and development. Benefits and Challenges are more strongly correlated with Training and Development Opportunities (0.705), highlighting that facing challenges creates growth opportunities.

Contacts with Universities and R&D Institutes show strong correlations with Training Opportunities (0.648), indicating that academic support is crucial for innovation. Overall, the data suggests that a collaborative environment and institutional involvement are essential to maximize benefits, overcome challenges, and promote organizational development.

Table 1 Pearson correlation coefficients between the factors created

	Collaboration	Benefits and challenges	Oppourtunities, empowerment, and development	Entities	Contacts with universities and R&D institutes
Collaboration	1				
Benefits and challenges	0.567**	1			
Oppourtunities, empowerment, and development	0.656**	0.705**	1		
Entities	0.719**	0.504**	0.509**	1	
Contacts with universities and R&D institutes	0.500**	0.572**	0.648**	0.435**	1

**Correlations is significant at the 0.01 level (2-tailed)

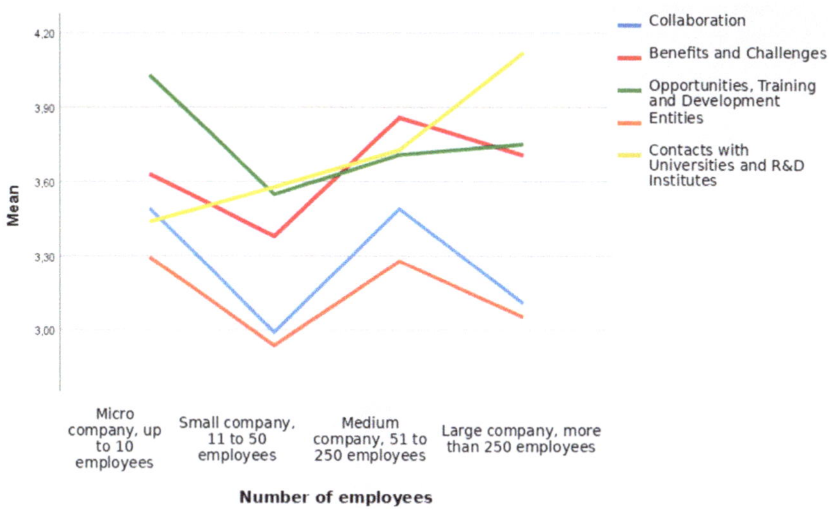

Fig. 1 Average values of the dimensions under analysis based on the number of employees in the companies surveyed

Figure 1 contains the average values of the analyzed dimensions based on the number of employees in the companies surveyed. According to the analysis, differences in various dimensions based on the number of employees in the companies were investigated. The results of the Kruskal-Wallis tests showed that for the dimensions of Collaboration, Benefits and Challenges, Opportunities, Training and Development, and Entities, no statistically significant differences were found between different company sizes, suggesting consistency in the responses regardless of the number of employees. However, for the dimension of Contacts with Universities and R&D Institutes, a statistically significant difference was observed ($p = 0.002$), with a tendency for higher scores in larger companies.

This result indicates that as the number of employees increases, the frequency and intensity of contact with educational and research institutions also tends to grow, possibly due to a greater capacity for investment in innovation and development. These data suggest a correlation between the size of the company and its tendency to interact with universities and R&D institutes, possibly driven by innovative needs and greater resource availability.

Figure 2 shows the average values of the dimensions under analysis based on companies that receive public funds to finance Research and Development (R&D) activities or other projects. The significance level considered for the tests adopted was the same as in the previous analysis. The results of the Kruskal-Wallis tests revealed statistically significant differences for the dimensions of Collaboration ($p = 0.014$), Opportunities, Capacity, and Development ($p = 0.014$), Entities ($p = 0.030$), and Contacts with Universities and R&D Institutes ($p = 0.011$), suggesting that access to public funds is associated with a significant variation in these dimensions. In

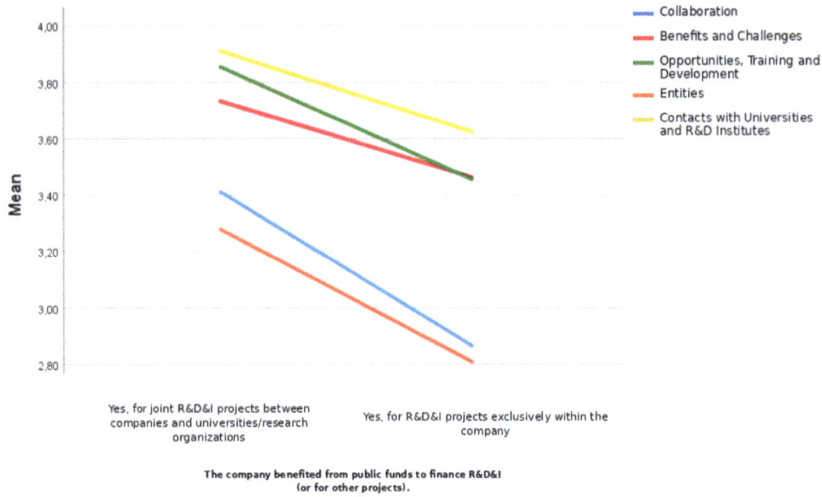

Fig. 2 Average values of the dimensions under analysis based on the benefits of public funds to finance R&D&I of the surveyed companies

contrast, for the dimension of Benefits and Challenges, no significant differences were identified, indicating that access to public funding does not significantly affect this dimension.

Following the same trend as the previous analysis of Contacts with Universities and R&D Institutes, the greatest statistically significant difference was observed, with a tendency for higher scores in companies that received funding for collaborative projects with universities and research organizations. This result suggests that access to public funds for partnership-driven R&D projects may promote more frequent and intense engagement with educational and research institutions, as opposed to purely internal projects, which, although receiving funding, show less intensity of interaction in these areas.

These data points to a correlation between the type of public funding and the likelihood of companies establishing contacts with universities and R&D institutes, highlighting that collaborative projects supported by public funds are essential for innovation and development in the business context.

5 Discussion

The literature review highlights that creativity and innovation are essential for the success of SMEs in competitive markets, and that open innovation allows these companies to overcome resource limitations, thereby increasing their capacity to innovate. This empirical study confirms that factors such as collaboration, the benefits and challenges of open innovation, capacity-building opportunities, and engagement with external entities are consistently perceived across companies of different sizes, validating the importance of external collaboration for SMEs, regardless of their size. However, a significant difference was found in engagement with universities and research institutes, with larger companies showing a greater propensity to establish contacts with these institutions. This result suggests that, although all companies' value open innovation, SMEs face additional barriers in accessing research institutions, highlighting the need for support policies that promote their integration into innovative ecosystems.

The results of this research confirm and enhance the findings in the literature regarding the impact of public funding on innovation in Portuguese companies. It was found that public funding has a significant effect on the dimensions of Collaboration, Opportunities, Capacity Building and Development, Entities, and Contact with Universities and R&D Institutes, suggesting that this type of support can strengthen companies' ability to innovate and collaborate within innovative ecosystems. In particular, the data show that funding facilitates interaction with academic entities, enabling companies to access external resources and knowledge, a key practice for open innovation and the creation of strategic partnerships. However, no significant differences were observed in the Benefits and Challenges dimension, indicating a common perception among companies regarding the obstacles to innovation, regardless of access to funding.

These results emphasize the importance of public funding for the development of innovation in companies, suggesting that public policies should promote not only access to funds but also the strengthening of collaboration networks to increase the competitiveness and sustainability of these companies in the market.

5.1 Implications and Significance of the Findings

The study provides valuable insights into the influence of public funding on innovation in Portuguese companies. The empirical analysis indicates that open innovation, driven by financial support, is essential for strengthening the innovative capabilities of SMEs, helping them overcome resource constraints and integrate into collaborative networks with universities and research institutions.

The results emphasize the importance of public funding as a driver of innovation, facilitating external collaborations, promoting capacity building and development, and strengthening connections with academic entities. These factors are crucial for the competitiveness of organizations in a dynamic global market, aligning with existing literature that highlights the relevance of access to financial resources for participation in innovative ecosystems.

In addition to confirming existing knowledge on the subject, the study highlights significant differences between companies of various sizes in their engagement with research institutions. This finding reinforces the need for public policies aimed at specifically supporting organizations, focusing on reducing barriers to partnerships with universities and research centers. It is recommended that such policies include financial incentives, mentoring programs, and platforms that facilitate interaction between companies and research entities, fostering a collaborative and more equitable environment.

The study also points out limitations that suggest the importance of future research exploring additional factors influencing open innovation in companies, going beyond quantitative variables and incorporating qualitative aspects. Longitudinal studies, for example, could provide a more dynamic view of the interactions between funding, innovation, and collaboration over time, helping policymakers adapt strategies to changes in the business environment.

In summary, the study clarifies the relationship between open innovation and public funding in SMEs and provides input for public policies aimed at strengthening the innovative capacity of these companies in Portugal. By fostering an inclusive innovation environment, it is possible not only to improve the competitiveness of organizations but also to contribute to the country's sustainable economic development.

5.2 Limitations of the Study

This research is limited to companies in the Portuguese context, which restricts the generalization of the results to other countries or types of companies, especially larger ones. Innovation dynamics and interactions with research institutions may differ significantly across regions and sectors.

Another important point is the absence of a longitudinal approach. The study relies on cross-sectional data, which prevents the analysis of the evolution of the effects of public funding on innovation over time. A longitudinal approach could provide a broader view of collaboration dynamics and the ongoing impact of funding, allowing the identification of trends and changes over time.

Finally, the predominance of quantitative data limits the understanding of qualitative aspects of interactions and the impacts of collaborations. Including a qualitative analysis, through interviews or case studies, would enrich the understanding of the relationships between companies and research institutions, providing insights into the challenges and opportunities of partnerships.

6 Conclusion

6.1 *Summary of Key Findings*

This study highlights the significant impact of public funding on innovation in Portuguese SMEs. Key findings include Collaborative Impact: Public funding strengthens collaboration with universities and research institutions, enhancing SMEs' ability to access external resources and expertise. Correlations Across Dimensions: Strong positive correlations were found between collaboration, capacity building, and development opportunities, emphasizing the role of partnerships in fostering innovation. Company Size and Collaboration: Larger companies exhibit greater propensity for engagement with academic entities, whereas SMEs face additional barriers, underscoring the need for targeted support. Uniform Challenges: Perceptions of innovation challenges, including trust and cultural resistance, are consistent across companies, regardless of funding access.

6.2 *Contributions to the Field*

The study provides critical insights into the dynamics of open innovation and public funding in SMEs by: Validating the role of financial support in promoting integration into innovation ecosystems. Highlighting the importance of collaboration networks as a mechanism to overcome resource limitations. Offering empirical data that supports the development of public policies tailored to reduce barriers for SMEs in accessing innovative ecosystems. Extending the understanding of how public funding can bridge the gap between SMEs and research institutions, fostering knowledge transfer and strategic partnerships.

6.3 *Recommendations for Future Research*

To build on the findings of this study, future research should: Adopt Longitudinal Approaches: Investigate the evolution of public funding's impact on innovation over time, providing a dynamic perspective on funding and collaboration trends. Incorporate Qualitative Analyses: Utilize interviews or case studies to explore the nuanced interactions and challenges of partnerships between SMEs and research institutions. Expand Geographical Scope: Analyze how public funding impacts SMEs in other countries or regions to identify cross-contextual insights and practices. Explore Sectoral Variations: Investigate how specific industries respond to public funding and innovation incentives to tailor policies more effectively.

This research lays the groundwork for a deeper understanding of the interplay between innovation, collaboration, and public policies, driving both theoretical advancements and practical applications in innovation ecosystems.

References

1. Acs, Z. I., & Audretsch, D. B. (1988). Innovation in small and large firms: An experimental analysis. *AER, 78*, 678–690.
2. Adner, R. (2006). Match your innovation strategy to your innovation ecosystem. *Harvard Business Review, 84*(4), 98-107+148. https://www.scopus.com/inward/record.uri?eid=2-s2.0-33645683761&partnerID=40&md5=591278e960fef6e4e747d99a5684badf
3. Adner, R. (2017). Ecosystem as structure: An actionable construct for strategy. *Journal of Management, 43*(1), 39–58. https://doi.org/10.1177/0149206316678451
4. Ahn, J. M., Minshall, T., & Mortara, L. (2018). How do entrepreneurial leaders promote open innovation adoption in small firms? In *Researching open innovation in SMEs* (pp. 137–177). World Scientific.
5. Amoroso, S., Moncada-Paternò-Castello, P., & Vezzani, A. (2017). R&D profitability: The role of risk and Knightian uncertainty. *Small Business Economics, 48*(2), 331–343. https://doi.org/10.1007/s11187-016-9776-z
6. Anderson, N., Potočnik, K., & Zhou, J. (2014). Innovation and creativity in organizations: A state-of-the-science review, prospective commentary, and guiding framework. *Journal of Management, 40*(5), 1297–1333. https://doi.org/10.1177/0149206314527128
7. Anderson, R. W., Acur, N., & Corney, J. (2017). How do SMEs use open innovation when developing new business models? In *Researching open innovation in SMEs* (pp. 179–209). World Scientific. https://doi.org/10.1142/9789813230972_0006
8. Bart, N. (1994). Innovation and diffusion in small firms: Theory and evidence. *Small Business Economics, 6*, 327–347.
9. Bessant, J. R. (2003). *High-involvement innovation: Building and sustaining competitive advantage through continuous change*. Wiley.
10. Bruque, S., & Moyano, J. (2007). Organisational determinants of information technology adoption and implementation in SMEs: The case of family and cooperative firms. *Technovation, 27*(5), 241–253. https://doi.org/10.1016/j.technovation.2006.12.003
11. Bullinger, H. J., Auernhammer, K., & Gomeringer, A. (2004). Managing innovation networks in the knowledge-driven economy. *International Journal of Production Research, 42*(17), 3337–3353. https://doi.org/10.1080/00207540410001695970
12. Carnahan, S., Agarwal, R., & Campbell, B. (2010). The effect of firm compensation structures on the mobility and entrepreneurship of extreme performers. *Business, 920*(October), 1–43. https://doi.org/10.1002/smj
13. Chesbrough, H. W. (2003). *Open innovation: The new imperative for creating and profiting from technology*. Harvard Business School.
14. Clausen, T., Pohjola, M., Sapprasert, K., & Verspagen, B. (2012). Innovation strategies as a source of persistent innovation. *Industrial and Corporate Change, 21*(3), 553–585. https://doi.org/10.1093/icc/dtr051
15. Cohen, W. M., & Levinthal, D. A. (1990). Absorptive capacity: A new perspective on learning and innovation. *Administrative Science Quarterly, 35*(1), 128–152. https://doi.org/10.2307/2393553
16. Commission, E. (2003). *The new SME definition: User guide and model declaration*. https://www.eusmecentre.org.cn/wp-content/uploads/2022/12/SME-Definition.pdf.
17. COTEC, P. (2017). *Manual Prático de Apoio ao Preenchimento do Innovation Scoring (COTEC Portugal) 2017*. https://www.innovationscoring.pt/

18. Davidsson, P. (2016). Researching entrepreneurship. *International journal of Entrepreneurial Behaviour & Research, 33*(6). https://doi.org/10.1007/978-3-319-26692-3
19. Diez, J. R. (2002). Metropolitan innovation systems: A comparison between Barcelona, Stockholm, and Vienna. *International Regional Science Review, 25*(January), 63–85.
20. Dobni, C. B. (2008). Measuring innovation culture in organizations. *European Journal of Innovation Management, 11*(4), 539–559. https://doi.org/10.1108/14601060810911156
21. Eisenhardt, K. M., & Martin, J. A. (2000). Dynamic capabilities: What are they? *Strategic Management Journal, 21*(10–11), 1105–1121. https://doi.org/10.1002/1097-0266(200010/11)21:10/11<1105::AID-SMJ133>3.0.CO;2-E
22. García-Morales, V. J., Jiménez-Barrionuevo, M. M., & Gutiérrez-Gutiérrez, L. (2012). Transformational leadership influence on organizational performance through organizational learning and innovation. *Journal of Business Research, 65*(7), 1040–1050. https://doi.org/10.1016/j.jbusres.2011.03.005
23. Giudici, G., & Rossi-Lamastra, C. (2017). Crowdfunding of SMEs and Startups: When open investing follows open innovation. In *Researching open innovation in SMEs* (pp. 377–396). World Scientific. https://doi.org/10.1142/9789813230972_0012
24. Golovko, E., & Valentini, G. (2011). Exploring the complementarity between innovation and export for SMEs' growth. *Journal of International Business Studies, 42*(3), 362–380. https://doi.org/10.1057/jibs.2011.2
25. de Gomes, L. A. V., Facin, A. L. F., Salerno, M. S., & Ikenami, R. K. (2018). Unpacking the innovation ecosystem construct: Evolution, gaps and trends. *Technological Forecasting and Social Change, 136*, 30–48. https://doi.org/10.1016/j.techfore.2016.11.009.
26. Haahti, A., Madupu, V., Yavas, U., & Babakus, E. (2005). Cooperative strategy, knowledge intensity and export performance of small and medium sized enterprises. *Journal of World Business, 40*(2), 124–138. https://doi.org/10.1016/j.jwb.2005.02.003
27. Hagedoorn, J. (2002). Inter-firm R&D partnerships: An overview of major trends and patterns since 1960. *Research Policy, 31*(4), 477–492. https://doi.org/10.1016/s0048-7333(01)00120-2
28. Holgersson, M. (2013). Patent management in entrepreneurial SMEs: A literature review and an empirical study of innovation appropriation, patent propensity, and motives. *R and D Management, 43*(1), 21–36. https://doi.org/10.1111/j.1467-9310.2012.00700.x
29. Hossain, M. (2017). Adoption of open innovation by small firms to develop frugal innovations for inclusive development. In *Researching open innovation in SMEs* (pp. 115–135). World Scientific. https://doi.org/10.1142/9789813230972_0004
30. Julião, J., Ferreira, I., & Gaspar, M. (2022). Why do SMEs implement open innovation? The case of Portugal. *International Journal of Innovation, Management and Technology, 13*(3), 58–63. https://doi.org/10.18178/ijimt.2022.13.3.922
31. Jung, D. D., Wu, A., & Chow, C. W. (2008). Towards understanding the direct and indirect effects of CEOs' transformational leadership on firm innovation. *The Leadership Quarterly, 19*(5), 582–594. https://doi.org/10.1016/j.leaqua.2008.07.007
32. Katila, R., & Ahuja, G. (2002). Something old, something new: A longitudinal study of search behavior and new product introduction. *Academy of Management Journal, 45*(6), 1183–1194.
33. Keupp, M. M., Palmié, M., & Gassmann, O. (2012). The strategic management of innovation: A systematic review and paths for future research. *International Journal of Management Reviews, 14*(4), 367–390. https://doi.org/10.1111/j.1468-2370.2011.00321.x
34. Koenig, G. (2013). *Business ecosystems revisited. Understanding business ecosystems: How firm succeed in the New World of convergence, Brussels, De Boeck* (pp. 69–83).
35. Lampikoski, K., & Emden, J. B. (1999). *Managing innovatively: Exploit creative resources.* WSOY.
36. Lanjouw, J. O., & Schankerman, M. (2004). Protecting intellectual property rights: Are small firms handicapped? *The Journal of Law and Economics, 47*(1), 45–74.
37. Laursen, K., & Salter, A. (2006). Open for innovation: The role of openness in explaining innovation performance among U.K. manufacturing firms. *Strategic Management Journal, 27*(2), 131–150. https://doi.org/10.1002/smj.507

38. Lechner, C., Dowling, M., & Welpe, I. (2006). Firm networks and firm development: The role of the relational mix. *Journal of Business Venturing, 21*(4), 514–540. https://doi.org/10.1016/j.jbusvent.2005.02.004
39. Leiponen, A., & Byma, J. (2009). If you cannot block, you better run: Small firms, cooperative innovation, and appropriation strategies. *Research Policy, 38*(9), 1478–1488. https://doi.org/10.1016/j.respol.2009.06.003
40. Leiponen, A., & Helfat, C. E. (2010). Innovation objectives, knowledge sources, and the benefits of breadth. *Strategic Management Journal, 31*(2), 224–236. https://doi.org/10.1002/smj.807
41. Lin, F., & Lin, Y. (2016). The effect of network relationship on the performance of SMEs. *Journal of Business Research, 69*(5), 1780–1784. https://doi.org/10.1016/j.jbusres.2015.10.055
42. Love, J. H., & Roper, S. (2015). SME innovation, exporting and growth: A review of existing evidence. *International Small Business Journal: Researching Entrepreneurship, 33*(1), 28–48. https://doi.org/10.1177/0266242614550190
43. Mazzucato, M. (2013). Financing innovation: Creative destruction vs. destructive creation. *Industrial and Corporate Change, 22*(4), 851–867. https://doi.org/10.1093/icc/dtt025
44. Moore, J. F. (1996). *The death of competition: Leadership and strategy in the age of business ecosystems*. Harper Business.
45. Nohria, N., & Gulati, R. (1996). Is slack good or bad for innovation? *Academy of Management Journal, 39*(5), 1245–1264.
46. O'Sullivan, M. (2006). 240 finance and innovation. In J. Fagerberg & D. C. Mowery (Eds.), *The Oxford handbook of innovation* (p. 0). Oxford University Press. https://doi.org/10.1093/oxfordhb/9780199286805.003.0009.
47. Öberg, C., & Alexander, A. T. (2019). The openness of open innovation in ecosystems—Integrating innovation and management literature on knowledge linkages. *Journal of Innovation and Knowledge, 4*(4), 211–218. https://doi.org/10.1016/j.jik.2017.10.005
48. Palazzeschi, L., Bucci, O., & di Fabio, A. (2018). Re-thinking innovation in organizations in the industry 4.0 scenario: New challenges in a primary prevention perspective. *Frontiers in Psychology, 9*(Jan), 1–6. https://doi.org/10.3389/fpsyg.2018.00030
49. Radziwon, A., & Bogers, M. (2019). Open innovation in SMEs: Exploring inter-organizational relationships in an ecosystem. *Technological Forecasting and Social Change, 146*(May), 573–587. https://doi.org/10.1016/j.techfore.2018.04.021
50. Radziwon, A., Chesbrough, H., Vanhaverbeke, W., & West, J. (2024). The future of open innovation. In *The Oxford handbook of open innovation* (pp. 914–934). https://doi.org/10.1093/oxfordhb/9780192899798.013.57.
51. Riding, A., Orser, B. J., Spence, M., & Belanger, B. (2012). Financing new venture exporters. *Small Business Economics, 38*(2), 147–163. https://doi.org/10.1007/s11187-009-9259-6
52. Sabando-Vera, D., Yonfa-Medranda, M., Montalván-Burbano, N., Albors-Garrigos, J., & Parrales-Guerrero, K. (2022). Worldwide research on open innovation in SMEs. *Journal of Open Innovation: Technology, Market, and Complexity, 8*(1). https://doi.org/10.3390/joitmc8010020.
53. Schumpeter, J. A. (1942). *Capitalism, Socialism and Democracy*. George Allen.
54. Şen, T. (2008). *Total quality management in e-commerce* (p. 132).
55. Skarzynski, P., & Gibson, R. (2008). *Innovation to the core: A blueprint for transforming the way your company innovates*. Harvard Business Press.
56. Spithoven, A., Vanhaverbeke, W., & Roijakkers, N. (2013). Open innovation practices in SMEs and large enterprises. *Small Business Economics, 41*(3), 537–562. https://doi.org/10.1007/s11187-012-9453-9
57. Taques, F. H., López, M. G., Basso, L. F., & Areal, N. (2021). Indicators used to measure service innovation and manufacturing innovation. *Journal of Innovation and Knowledge, 6*(1), 11–26. https://doi.org/10.1016/j.jik.2019.12.001
58. Thanawala, K. (1994). Schumpeter's "theory of economic development" and development economics. *Review of Social Economy, 52*(4), 353–363. http://www.jstor.org/stable/29769751

59. Trönnberg, C. C., & Hemlin, S. (2014). Lending decision making in banks: A critical incident study of loan officers. *European Management Journal, 32*(2), 362–372. https://doi.org/10.1016/j.emj.2013.03.003
60. Ughetto, E. (2008). Does internal finance matter for R&D? New evidence from a panel of Italian firms. *Cambridge Journal of Economics, 32*(6), 907–925. https://doi.org/10.1093/cje/ben015
61. Vaccaro, I. G., Jansen, J. J. P., van den Bosch, F. A. J., & Volberda, H. W. (2012). Management innovation and leadership: The moderating role of organizational size. *Journal of Management Studies, 49*(1), 28–51. https://doi.org/10.1111/j.1467-6486.2010.00976.x
62. Vahter, P., Love, J. H., & Roper, S. (2014). Openness and innovation performance: Are small firms different? *Industry and Innovation, 21*(7–8), 553–573. https://doi.org/10.1080/13662716.2015.1012825
63. Vanhaverbeke, W., Frattini, F., Roijakkers, N., & Usman, M. (2018). *Researching open innovation in SMEs*. World Scientific.
64. Yukl, G. (2006). *Leadership in organizations, 9/e*. Pearson Education India.
65. Zhang, X., & mei, Song, Z. lin, & Zhong, Z. (2016). Does "small bank advantage" really exist? Evidence from China. *International Review of Economics and Finance, 42*, 368–384. https://doi.org/10.1016/j.iref.2015.10.009

Institutionalization of the Entrepreneurship and Innovation Program at Federal University of Technology—Paraná, Brazil

C. S. Siqueira and M. Jacometti

Abstract The objective of this article was to identify the institutions established in the Entrepreneurship and Innovation Program (Proem) of the Federal University of Technology—Paraná (UTFPR), Brazil. To this end, the qualitative research strategy was used, in which documentary research was adopted as a method. We sought to describe how Proem was created, starting in 1997, and its institutionalization. To investigate the case, the sociological perspective of institutional theory was adopted to support the description of the institutional context in which Proem is immersed, considering the Triple Helix theory approach. When investigating the documents, thematic content analysis was used, which identified two final categories: interactions of institutional spheres and institutionalization of Proem. As a result, it was observed that Proem's operations are aligned with the logic of the Triple Helix model and have the potential for institutionalization over time. The importance of Proem in structuring technology-based startup companies to ensure their permanence in the market is highlighted, making it necessary to carry out further research to assess the degree of institutionalization and the effectiveness of the program.

Keywords Institutions · Triple Helix · Institutionalization · Entrepreneurship

1 Introduction

Entrepreneurial activity is an important vector of the economy, adding new interactions and promoting development. Such activity aims to break the current economic cycle by creating new markets, products, processes, raw materials, and organizational forms [1, 2]. In the context of the Knowledge Society [3], competitiveness and innovation of companies are fundamental for the dissemination of technologies, communication, and information [4].

C. S. Siqueira · M. Jacometti (✉)
Universidade Tecnológica Federal do Paraná, Curitiba, PR, Brazil
e-mail: jacometti@utfpr.edu.br

L. Farinha et al. (eds.), *Sustainable Innovations in Mechanical Engineering and Management Sciences*, Lecture Notes on Multidisciplinary Industrial Engineering,
https://doi.org/10.1007/978-3-032-30372-1_7

Intrinsic to this socioeconomic panorama, universities are at the center of actions. Integrated economy, in addition to universities, includes institutional participants such as industries and governments, which characterize the Triple Helix [3–5]. Thus, universities have become important contributors to entrepreneurial ecosystems through research and workforce education [6].

With exponential technological advancement and the evolution of consumer habits, innovation has become a key factor in the success and competitiveness of organizations. Technology-based companies, in particular, play an important role in creating and adopting innovative solutions, which promote changes in traditional industry and pave the way for new markets and services. Therefore, the journey from the initial idea to the market can be permeated by challenges, especially for start-up companies, where there is a lack of resources, contacts, and expertise. Innovation incubators promote an ecosystem that supports and accelerates the growth of emerging startups [7–9].

The Brazilian business reality is predominantly made up of micro and small companies (MSCs), the majority of which do not innovate and, when faced with crises, do not plan, and become uncompetitive [10, 11]. The importance of structuring these companies to ensure their permanence in the market is highlighted, given that low productivity arises, among other aspects, from the lack of innovation [10]. MSCs have few employees and make up around 77.8% of new entrants into the Brazilian market, being formed only by partners or owners. Of these, 15.0% represent the salaried workforce, while 82.2% belong to surviving companies [12].

Recognizing this panorama, UTFPR launched, in 1997, the Entrepreneurship and Innovation Program (Proem), which seeks to spread entrepreneurial culture and form startup companies. Thus, mirroring the Triple Helix model, there are partnerships with the public and private sectors. The relevance of UTFPR's innovative environments is linked to the regional and economic development of the cities where it is based, as each Campus plans its courses according to the needs of the region in which it is located, demonstrating the University's commitment to the local community. The interiorization of UTFPR seeks to provide opportunities and access to courses and areas of activity in all regions covered by the thirteen campuses in the Paraná [13].

As the university adopts a more entrepreneurial approach, its conduct also influences the surrounding institutional environment [14]. The institutionalization, which has a social nature, implies obligations that guide behavior such as: regulations, norms, values, practices, and expectations [15]. Institutions are regulative, normative and cognitive structures [16] that regularize the behaviors of social actors, experiences and routine practice that, eventually, can offer opportunities for agency and change, therefore helping to understand processes and social representations [17]. To explain institutional changes, the theory focuses on the actors' ability to reflect [18].

There is concern in identifying how institutions are established [19]. In this article, the institutionalization of Proem was explored and how this process occurred at the level of social actors, when implementing innovation environments at UTFPR. Therefore, the purpose was to identify the institutions recommended and established in the UTFPR Proem, since its creation. Etzkowitz introduced, in the 1990s, the concept of the "triple helix", which was designed to designate the complex relationship between the university-industry-government spheres. In this structure, institutions operate as independent entities, capable of interacting and playing mutual roles, thus creating a social configuration highly favorable to innovation. The Triple Helix forms trilateral networks and hybrid organizations with actions encouraged but not controlled by the government [20].

2 Literature Review

2.1 Triple Helix Fundamentals

The 1970s represent a milestone, in which industrialized countries prioritized knowledge as the basis of their economies, as well as the diffusion of communication and information technologies [4]. A new pact was established between the Academy and the Society. It is understood that public financing for higher education institutions is linked to their direct contribution to economic development. The Academy, by its nature, presents an environment conducive to innovation due to the gathering of knowledge and intellectual capital, as it has potential for research and students are a source of future entrepreneurs [14, 20].

Thus, Etzkowitz introduced, in the 1990s, the concept of the "triple helix", which was conceived to designate the complex relationship between the university-industry-government spheres. In this structure, institutions operate as independent entities, capable of interacting and playing mutual roles, thus creating a social configuration highly favorable to innovation [21]. University, Government and Industry form trilateral work networks and hybrid organizations with actions encouraged but not controlled by the government [20, 22], as shown in the Fig. 1.

The first level of the Triple Helix encompasses the internal transformation occurring in each of the helices, as well as the establishment of new relationships between them. Each of the helices has its own purpose, but they begin to "act as the role of the other" [23]. In other words, universities develop facilities like incubators; companies build internship and training programs and sponsor scientific conferences; and governments define public policies that encourage partnerships [14, 23, 24].

The second dimension consists of influence of one helix on the other. For example, in the development of government policies and programs that encourage universities to establish relationships with the productive sector [6, 14, 23].

In the third dimension, the author explains that the increasing interaction between institutional spheres has the effect of generating new structures with each of the helices and how the actors' actions can be understood in terms of the result of a complex set of institutional pressures. And, finally, the fourth dimension of this model represents the response to the interactions of institutional spheres: entrepreneurial dynamics as the basis of the economy [14, 16, 23].

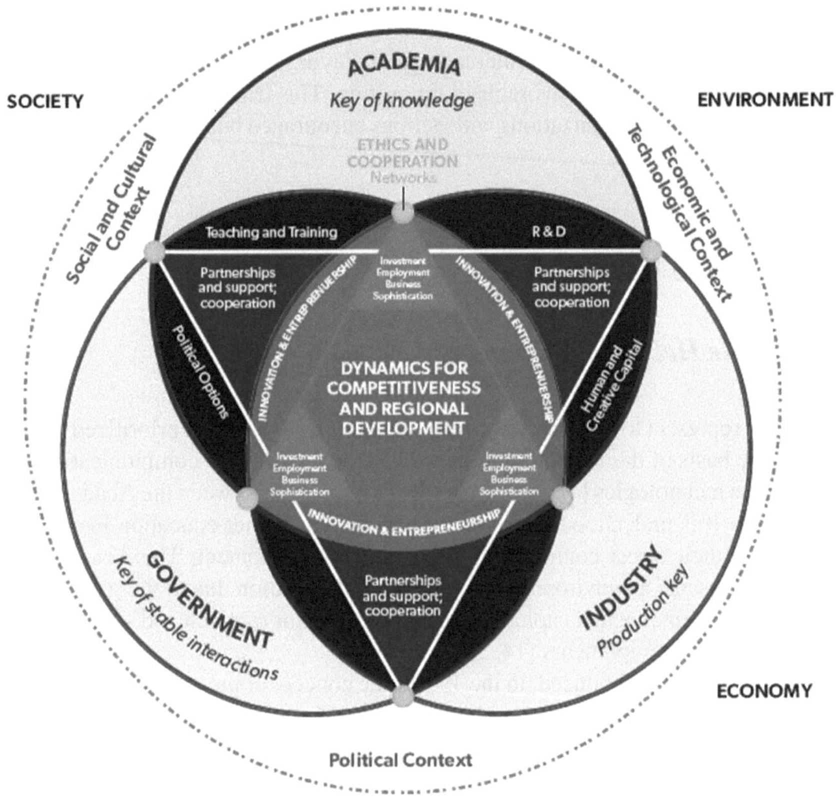

Fig. 1 Triple Helix model and its institutional context. *Source* Retrieved from https://triple-helix.uff.br/about-triple-helix/, last accessed 2023/10/30

One of the effects of this knowledge-based economy was the emergence of scientific and technological parks [4, 25]. In the 1950s, the first parks appeared: Stanford Research Park (1951), Cornell Business and Technology Park (1952) and Research Triangle Park (1959) [25, 26]. Universities foster innovation through strong interaction with affiliated companies which, consequently, reinforces the dynamics of the local economic environment [4, 25].

Technology Park is an organizational structure that promotes the fluidity of knowledge between the university, research and development in companies, institutions, industry, and the market [4, 25]. It also seeks to simplify the process of creating and developing companies through innovation incubators, in addition to providing services and physical space to enterprises, where the objective is to promote and revitalize local and regional economies, as well as the development of goods and services with technology aggregate [4, 8, 27].

Technology-based companies use scientific-technological knowledge as a fundamental element in their operations, as well as entrepreneurial training, while they exist as hybrid spaces in the context of the Triple Helix [4, 24].

Innovation incubators seek to reduce barriers to entering the market, as well as increasing the chances of success for these technology-based companies [9]. The authors explain that start-up companies are offered numerous resources, such as financing, mentoring, work space, access to a network of contacts and specialized facilities. Thus, leveraging the resources added to the expertise and disposition in the incubator, companies can develop their ideas, refine their business models, and gain access to markets, consumers, and investors [9].

The beginning of the incubator movement occurred in Brazil in 1984, when the National Council for Scientific and Technological Development (CNPq) created a program to help technology parks and incubators, which, in 1987, created the National Association of Entities Promoters of Innovative Enterprises (Anprotec) [4].

In Brazil, there are also efforts aimed at institutional consolidation and improvement aiming to promote S, T&I in the country. Law n° 13,243/2016, which modifies, among others, Law n° 10,973/2004 (known as the Innovation Law), confirms this commitment [25]. In relation to technical and financial support, there are actions by CNPq and Brazilian Micro and Small Business Support Service (Sebrae), which has supported the creation of incubators since 1991. From the 2010s onwards, Anprotec developed a specific program to support the creation and consolidation of incubated companies called the Reference Center for Supporting New Enterprises (Cerne) [4, 28]. Cerne established guidelines and metrics to create a reference model for the management and evaluation of incubators and technology parks in Brazil. Anprotec "has been operating to improve the innovation ecosystem in Brazil [...] in partnership with Sebrae, which aims to promote the improvement of incubator results in quantitative and qualitative terms, structured into four levels of maturity" [28].

2.2 *Institutional Theory and Macro X Micro Duality*

Within the sociological institutionalist thought focused on the study of organizations, in the late 1970s, a proposal was presented, within the macro scope, addressing the influence of institutions [29]. Another emphasis was given to the micro scope in the formation of institutions, which studied the power of cognitive beliefs that anchor behavior, that is, social knowledge, once institutionalized, exists as a fact, as part of objective reality [30]. There were advances in theory in the macro sphere regarding the environment, whose perspective became dominant in social institutionalism [15, 16]. Therefore, the purpose of organizations is not solely to structure themselves to achieve efficiency in their activities, but rather to do so by virtue of their symbolic effects, which result in the fact that organizations respond to tangible and intangible influences from the environment.

The Weberian notion of bureaucratic rationality to analyze the way in which organizations become more homogeneous through institutionalized practices was widely used [15]. It is noteworthy that this process, isomorphism, usually occurs because of uncertainties, and is considered a type of self-defense and attempt at survival by organizations operating in the organizational field [29, 31–34].

In this view, formal organizations have constructed myths, which consist of (1) being impersonal, rational, pursuing social purposes as if they were technological, and (2) the need for a highly institutionalized environment. Both myths are legitimized, taken for granted and socially expected. The rules define new organizations, redefine existing ones, and specify the means to copy each other through isomorphy [29].

The analysis of the configuration of organizational fields, with the aim of understanding the similarities in the structures and practices of organizations, proved to be very relevant [15]. Nevertheless, the authors' main legacy lies in their approach to institutionalization, through the three widely recognized mechanisms of isomorphism. In this way, the authors discuss three types of institutional pressures that can cause isomorphic processes and, consequently, conformity and legitimation, in organizations that are part of the organizational field, namely: (1) coercive isomorphism, (2) normative isomorphism and (3) mimetic isomorphism.

Coercive isomorphism occurs when organizations face pressures, whether formal or informal, imposed by others with which they have a dependent relationship. These transformations are the result of imposition and authority, resulting in a growing homogeneity between organizations in the same field [15, 29].

The State is one of the main actors in the organizational field that exerts significant pressure on organizations, aiming to encourage coercive isomorphism. These pressures are often applied through the formulation of public programs and policies to obtain legitimacy [15, 35].

On the other hand, organizations find themselves compelled to adhere to imposed standards due to the need to obtain resources and incentives arising from the acceptance of these programs and policies [33].

This mechanism may not be the result of conscious choices, as the adoption of rules may occur with the aim of preserving the legitimacy of an organization. And this force can be subtle, because when exercised by different interested parties, such as other organizations and the media, it forms an environment where the strategy of using information brings credibility [36].

Normative isomorphism, in turn, derives from professionalization [33], in the sense of defining working conditions and methods to legitimize the autonomy of the profession. No matter how different the professionals are within the same company, they are like other professionals existing in other organizations [15, 37, 38].

Isomorphic mimetic pressure arises in hectic industrial periods, under conditions of uncertainty. They look for stability by looking to their leaders for role models and answers [36]. Mimicry often occurs without the company taken as a model being aware of the fact or even wanting to be imitated by competitors. In mimetic isomorphism, innovative companies can serve as an example to be followed, since the adoption of innovations can strengthen the legitimacy of an organization, while demonstrating the implementation of better working conditions [15].

As presented, institutional isomorphism is an advantageous phenomenon for organizations, as similarity facilitates inter-organizational transactions and favors their functioning through the incorporation of socially accepted rules, even achieving external recognition of their field. Thus, some benefits of isomorphic identity can be listed: stability, legitimation, status, reputation, and access to resources [36].

Legitimacy, in turn, is fundamental for the longevity of organizations, which try to adapt their structures and practices to current environmental values. Therefore, the search for legitimacy and conformity to standards causes organizations to adopt lines of action taken for granted and rationalized in society, which are related to the maintenance or change of existing cultural values and practices. This means that institutions can renounce competitiveness and efficiency to achieve recognition and legitimacy in the institutional context in which they are established [15].

2.3 *Institutionalization*

Institutionalization is a phenomenon that can be understood both as a process and as a variable [30]. It is the process by which individual actors communicate what is socially accepted as reality and, at some point throughout this cognitive process, the meaning of an action can be perceived as more or less integrated with the reality taken for granted. To maintain compliance, organizations that mirror institutional rules tend to increase their formal structure (bureaucracy) to protect themselves from the uncertainties generated by technology [29]. Such processes occur less as a form of competition and more as results of mimetic processes [33].

The restrictions imposed by institutions define moral, legal, and cultural limitations. However, it is essential to also recognize that they support activities and actors, providing direction and resources. And even though they provide stability and order, they can undergo changes, both incremental and revolutionary. Therefore, one should not reflect on institutions as a "property" or state of an existing social order, but also as a "process" [16, 39].

Institutional theory does not deny, nor is it inconsistent with change. On the contrary, several institutions converge isomorphically, which implies a movement from one position to another [40]. Institutional diffusion occurs in three stages: habitualization, objectification and sedimentation [39]. And it allows the organization to define when and at what speed the institutionalization process will occur. In turn, the authors [40] were based on the seminal article by [39], incorporating the stages of diffusion, within a model with six stages.

The first stage (I) occurs when events or leaps destabilize established practices (precipitating shocks). Leaps can take the form of social unrest, technological disruptions, competitive discontinuities, or regulatory changes [40]. The authors further explain that these changes precipitate (stage II) the entry of new participants (deinstitutionalization), the rise of existing actors, or local entrepreneurship. In this way, its effect is to disrupt the stable consensus at the socially constructed field level, introducing new ideas and thus the possibility of change.

Stage III, called pre-institutionalization, occurs when a new concept is adopted and organizations innovate independently, seeking technically viable solutions to locally perceived problems [40]. Some organizations wait and wait to better understand this new concept and decide whether and when they will adopt it. Thus, decision making normally occurs with the question: "will this help or hurt?" [39].

Reinstitutionalization (i.e., from stage III to stage VI) is not a matter of direct transition. Rather, it concerns practices that are institutionalized or not. Actors follow existing institutional "scripts". Therefore, for new practices to be widely adopted, they need to be "theorized". Stage IV of theorizing involves the development and specification of abstract categories and elaboration of these categories in detail [40].

Theorization involves two main tasks: the specification of a general organizational failure for which a local innovation is a solution or treatment and the justification of the innovation. Diffusion (step V) occurs only if new ideas are convincingly presented as more appropriate than existing practices [39]. Objectification happens when the concept is spreading and companies begin to imitate the behavior [39]. Therefore, the diffusion stage occurs as the innovation is disseminated and becomes an object, gaining social consensus about its pragmatism [40].

Organizations imitate behaviors because they anticipate benefits. In this sense, pragmatic legitimacy is linked to new ideas and new economic forms [40]. Suggest that legitimacy is given from the diffusion stage. However, in a more regular scenario, it is not common for legitimacy to be based exclusively on financial return. On the contrary, new ideas need to be justified and aligned with current regulations before being disseminated.

Thus, stages IV and V are based on the semi-institutionalization phase of Tolbert and Zucker [39] who explain that this sedimentation occurs when the concept is already institutionalized and completely disseminated by the organization's members. Therefore, if the company has not yet joined, it may suffer penalties, such as negative public opinion, for example.

Finally, total reinstitutionalization (which characterizes institutional change) occurs when adoption is so dense that its ideas come at the level of cognitive legitimation, as they are taken for granted and naturally adopted. Once institutionalized, ideas can survive for generations and are accepted without criticism as the only possible form of behavior [40]. If reinstitutionalization does not occur, we have what the authors call fads. Therefore, it is essential to analyze the dimensions of the institutional environment to obtain a more in-depth understanding of how institutions influence the actions of participants within this scenario [15].

3 Methodology

This research collected data exclusively from documents, aiming to extract information contained in them to understand the phenomenon. It is a procedure that employs methods and techniques for seizure and documents analysis of different natures. Approach is considered documentary when it is the only qualitative strategy, being used as independent method [41]. Obtaining data involved research on the UTFPR site and its institutional documents. To complement and enrich the information, documents were examined considering that these records reflect the organization's system and structure, covering a variety of sources, including: (1) UTFPR homepage, (2) Institutional Development Plan for 2017 and 2023, (3) Anprotec/Sebrae Good Practices Manual, (4) Proem Processes Manual [42], and (5) newspaper or internet articles.

The data analysis technique used was content analysis [43], which integrates both theoretical and practical exploration, assuming a particularly significant role in the context of social investigations [44]. Furthermore, the meaning interpreted by the author may coincide with or differ from the meaning perceived by readers, highlighting the plurality of meanings. Thus, content analysis, when examining texts, not only processes the literal form, but also enables a broad and multifaceted understanding, recognizing subjectivity and interpretative diversity [44].

However, in qualitative content analysis, thematic analysis allows data to be investigated considering a theme, organizing ideas to answer the research problem [45]. The researcher seeks to align the method of investigation and analysis more cohesively with the objectives of his research. Therefore, we opted for thematic content analysis, as in this work, there is a greater demand for the interpretation of other elements of the field, such as a specific context, the identification of common meanings, in which a semantic analysis would not offer the necessary sufficiency to the case [45].

3.1 Analysis Categories

The documents incorporated into the sample need to be representative and relevant to the objectives of the analysis, comprehensively including the field to be investigated [44]. In this way, from the analyzed documents mentioned in the previous section, the intention of analyzing the PDIs within the temporal difference of ten years stands out.

When selecting documents, four criteria were used [41]: (1) authenticity; (2) credibility, since the documents are public data and there are no questions regarding their origin; (3) representativeness, as they are routine protocols or initiatives within the public system in which they are integrated; and (4) significance, since they are official documents and have clarity and objectivity in their purposes. The pre-analysis was conducted with the purpose of organizing the preliminary ideas presented by the theoretical framework, systematizing initial ideas, and creating indicators for the interpretation of the information gathered [43, 46]. During the analysis, we sought to categorize the information into themes or groups that contributed to the understanding of the content underlying the speeches [46], as shown in Table 1.

The development of categories occurred as proposed by Braun and Clarke [47], following the selection of material and familiarization with the data. The coding was carried out based on the research question and the literature review, which,

Table 1 Analysis category progression summary

Initial categories	Categories intermediaries	Finals categories
Entrepreneurial culture	Development and entrepreneurship	Interactions in the institutional spheres
Development (technological, economic, and social)		
Entrepreneurship		
Public power	Triple helix	
Public-private initiative		
Technologic innovation		
Incubator	Innovation habitats	Institutionalization of Proem
Technologic park		
Process	Implementation of Proem	
UTFPR Proem		
Institutional actors		

when corroborated by the observed themes, constituted tables and thematic maps. Subsequently, the themes were refined [48].

Thematic analysis was performed using Atlas.ti24 software [49], which facilitated control of the analysis of various documents. As limitation, primary data was not yet sought to triangulate with the results obtained in the documentary analysis. It is not possible, for example, to assess the institutional effectiveness of Proem and whether there are signs of decoupling [50, 51]. This limitation must be overcome by seeking to carry out in-depth interviews with the main actors involved in the construction of the program, as well as direct observations at the Institution.

4 Results

4.1 Interactions of Institutional Spheres

An entrepreneurial University plays a crucial role in a knowledge-based economy, being a significant driver of socioeconomic development [24]. However, the Brazilian economy is not yet knowledge-based, it relies heavily on commodities and there is a growing understanding that Brazilian universities cannot be seen as separate from the rest of the economic world. Science is currently closer to market than years ago [14].

UTFPR contributes to the environment in which it operates. Part of its mission is dissemination of knowledge to the industry and its transformation into commercial assets through its ownership and profitable use [14]. In the UTFPR PDIs of 2013 and 2023, regional development and its role in society is present in the mission [13, 52].

There is also the aim of being a catalytic agent to promote coordination with public authorities and the private sector, by transferring technology through its technology parks and business incubators.

UTFPR's mission, in 2013, included being innovative, however, in 2023, it did not present "being innovative" in its explicit text, but rather as a concept on which the mission was built. Entrepreneurship and innovation began to have implicit importance in 2023, as they became values of the University [13]. In the 2013 PDI, there are goals stipulating the encouragement of innovation, development of entrepreneurial culture and increase in innovative actions. In this way, the PDI documents show the encouragement of entrepreneurship at UTFPR, where it seeks to achieve, in the cognitive scope of individuals, transformation actions, in a cultural-cognitive conception of institutions [52]. Thus, actors who connect with cultural beliefs tend to feel competent and connected. A cultural-cognitive conception of institutions highlights the central role played by the socially mediated construction of common framework of meanings.

If the behavior of some universities was considered unprecedented when they began to take an entrepreneurial direction, the reality for them today is very different, that is, the expectation for them to act entrepreneurially is established [14]. This is what is observed between the two UTFPR PDIs analyzed. The change in their roles represents changes in values and norms, which are influenced by various institutional pressures, including the entrepreneurial turn of some universities.

Proem was planned and created in 1997, in Curitiba, when UTFPR was still Cefet-PR, and its implementation intended to provide access to programs, events and entrepreneurship and innovation actions for students, employees and graduates of the Institution, as well as for the external community [53].

In essence, Proem seeks to disseminate entrepreneurial culture and support the development of innovative technology-based ventures [42]. To this end, it has set up the following structures: (1) Technological Hotel, (2) Innovation Incubator, (3) Business Accelerator, (4) Technological Scientific Park, and (5) Junior Companies [53], to operate on all UTFPR Campuses, however, not all of them have this structure.

The maintained initiative provides resources necessary for its operation. Proem operates on the thirteen UTFPR Campuses, and seeks to encourage the creation of sustainable enterprises that contribute to the economic development of the region [53]. Connections between university, industry, and government, constituting a Triple Helix, play a fundamental role in driving knowledge-based socioeconomic growth and development [24].

It is pertinent to view entrepreneurial transition as the main means that drives the promotion of innovation [14]. However, the drive to establish a Triple Helix at the regional level is originating from "regional innovation enablers" and "regional innovation promoters" [24]. And, in accordance with the UTFPR PDIs of 2013 and 2023 [13, 52], policies for the management and technological, economic, and social development of its surroundings were established, as well as the understanding of the needs and demands of the community to constitute the processes academics.

A challenge, when developing regional innovation, is that often only the outcome is targeted, while the process is neglected [24]. Thus, demonstrating how the process seeks to be well structured, Proem promotes several actions to disseminate entrepreneurial culture, such as lectures, meetings, testimonials from participants in innovative environments [54]. And, when identifying inefficiencies on a Campus, it recognizes the importance of Sebrae's actions in the University space to assist in the process of implementing Proem's innovation habitats.

It was identified that interactions between partners promote innovation and institutionalization of practices [24]. The entrepreneurial turn at UTFPR depends on institutional work and can be understood as result of a confluence of internal and external forces that are shaped through a historical and recursive interaction between regulatory, normative, and cultural-cognitive pressures, jointly derived from each actor of Triple Helix.

4.2 Institutionalization of Proem

Practices that were previously unconventional in their institutional contexts, over time, become legitimate and institutionalized, shaping their own institutional environment. The entrepreneurial experience of some prestigious universities, such as Stanford University, has contributed to changes in the way Society perceives the role of universities. The development of its transformation started to influence other universities to act in a similar way, adapting their structures and strategies to also be more entrepreneurial [14].

In view of the above, Proem at UTFPR is an important agent for this transformation to occur, as it plays a crucial role in promoting an entrepreneurial mentality and operates in conjunction with other institutional structures, such as Junior Companies and the extension curriculum [13]. In relation to the incubation process at Proem, the aim is to develop in entrepreneurs the maturity necessary for sustainable innovative development, resulting in profitable businesses that have an impact on the environment in which they operate. This process is divided into three stages: pre-incubation, incubation, and post-incubation [53–55].

Pre-incubation, also known as technological hotel, consists of verifying, testing, and validating the business proposal or solution and this stage lasts up to twenty-four months. Therefore, if there is demand or interest in the solution, and the team is prepared to open a business, the project is qualified by evaluating its business plan. This stage, like the others, has specific regulations [53, 55]. The incubation phase itself is dedicated to supporting, qualifying, and monitoring the enterprise with innovative potential and a technological base. And, finally, the post-incubation phase is the process of accelerating the venture, in which the graduate of the incubation program becomes an associate. In both periods, qualifications, consultancy, advice, mentoring, events, access to markets and investors are provided. And in relation to the duration of these two stages, they can last up to sixty months [55].

The processes described above are ways of standardizing and legitimizing Proem across all UTFPR Campuses, constituting a high potential for institutionalizing the program. In this way, it is understood that institutions are not just a property or an existing social order, but rather a process, including institutionalization and deinstitutionalization [16, 39]. It is important to recognize that institutions are guides and resources for acting, as well as for prohibiting and limiting actions [16].

In the first version of the Proem Manual, restructured in 2006, an attempt was made to establish standardized guidelines for incubation processes, describing the routines, to guide the agents responsible for the incubators. There are countless procedures, scripts, sheets, guidance guides, notices, covering details such as badge making, even a table specifying the process of disseminating entrepreneurial culture [42].

In addition to being understood as a process and guide within the Institution, the documentation analyzed, in accordance with Scott's three institutional pillars, fits into the normative pillar, as it represents rules that introduce a prescriptive, evaluative, and mandatory dimension at the University. Normative systems include both values and norms. Values are conceptions of what is preferred or desirable, along with the construction of standards to which existing structures or behaviors can be compared and evaluated, as we can see in the 2023 PDI [13]. Standards specify how things should be done [16]. The Proem documents [42] specify its implementation through example documents to be used in different situations to implement and expand.

Normative systems are seen as limitations imposed on social behavior, but at the same time they empower and facilitate social action. They confer rights as well as responsibilities [16]. Proem Manual, in addition to making it possible to implement an incubator on UTFPR campuses that did not yet have one, was a facilitator for entrepreneurial and innovative culture to be maintained and processes managed.

However, in complex situations, individuals or organizations may find themselves faced with competing leadership. Actors faced with conflicting regulatory requirements and standards often have difficulty acting, as compliance with one standard compromises the regulatory support of other bodies [16]. As the 2013 and 2023 UTFPR PDIs are premised on regional development [13, 52], it is questioned what the understanding of the needs and demands of communities is, and whether these fit into normative standards. Some values and norms are applicable to all members of community, while others apply only to some selected types of actors or positions [16].

Thus, from the point of view of the cognitive pillar [16], it is important to mention that without an investigation with Proem actors, it is not possible to identify to what extent there is an effective culture and practices clearly copied between the different UTFPR Campuses. In this context, it is only possible to assume that it occurs, considering the actions of management.

5 Conclusions

The present study aimed to identify the institutions of Proem developed at UTFPR. To this end, documentary research with a qualitative approach was conducted, which provided a first impression of how UTFPR perceives its innovation environments and how it proposes their development, to then be able to understand how it acted in implementing the Program. The reality found clarifies that there is an intention for UTFPR maintains an entrepreneurial and innovative culture. Since 1997, it has been structuring innovation habitats and, more recently, it has been diversifying with innovation hubs, innovation and technological centers, and maker spaces. In this way, it is possible to identify the institutions recommended by Proem.

In relation to S, T&I public policies, the Federal Government's actions occur through of the Ministries of Education and Science, Technology, and Innovation, CNPq, among others. In legislation on the subject, the State Government acts through the State Secretary of Science, Technology, and Innovation, and in each municipality where UTFPR operates, support varies according to the City Hall's engagement. Industry has the participation of entities linked to the Federation of Industries of Paraná, Anprotec/Sebrae and companies in general. This conjunction corroborates the Triple Helix theory and as a result, if universities incorporate an entrepreneurial and innovative perspective, their actions impact the institutional context [16].

The documentary analysis highlighted an environment with potential for institutionalization, because there is a process of a socio-technical nature that establishes a set of standards, such as: regulations, norms, values, expectations, and practices [14]. Scott's question stands out, where standards are adopted by some organizations, but not by others in similar situations [16, 29].

This study identified elements that can be explored in future studies to better understand the phenomenon in universities. In the present case, we intend to investigate Scott's question regarding the UTFPR Proem, in other words, why some structures, standards and practices were adopted by some campuses and others not, since they are in similar situations and under the auspices of a single management.

Finally, it is highlighted that the main limitations of the study are related to the need for a counterpoint with the feasible reality and how, from a cognitive point of view, intentions are applied in the organization's routine.

References

1. Schumpeter, J. A. (1961). *Capitalismo, socialismo e democracia*. Fundo de Cultura.
2. Nascimento, T. C., Cherobim, A. P. M. S., & Mendonça, A. T. B. B. (2019). The influence of venture capital on innovation in Brazilian startups. *Cadernos EBAPE.BR, 17*(4), 1016–1030. https://doi.org/10.1590/1679-395174721x
3. Ranga, M., & Etzkowitz, H. (2013). Triple helix systems: An analytical framework for innovation policy and practice in the knowledge society. *Industry & Higher Education, 27*(3), 237–262. https://doi.org/10.5367/ihe.2013.0165
4. Amaral, M. (2015). Management and assessment of innovation environments. *Triple Helix, 2*(19), 1–20. https://doi.org/10.1186/s40604-015-0030-5
5. Jepperson, R. L., & Meyer, J. W. (2021). *Institutional theory: The cultural construction of organizations, states, and identities*. Cambridge University Press.
6. Lahikainen, K., Kolhinen, J., Ruskovaara, E., & Pihkala, T. (2019). Challenges to the development of an entrepreneurial university ecosystem: The case of a Finnish University Campus. *Industry and Higher Education, 33*(2), 96–107. https://doi.org/10.1177/0950422218815806

7. Kollmann, T., Stöckmann, C., Niemand, T., Hensellek, S., & Cruppe, K. (2021). A configurational approach to entrepreneurial orientation and cooperation explaining product/service innovation in digital vs. non-digital startups. *Journal of Business Research, 125*, 508–519. https://doi.org/10.1016/j.jbusres.2019.09.041
8. Sánchez Ocampo, E., Iacono, A., & Leandro, F. R. (2019). Gestão da inovação em empresas de base tecnológica: um estudo de caso em empresas incubadas. *Innovar, 29*(74), 71–84. https://doi.org/10.15446/innovar.v29n74.82062
9. Van Rijnsoever, F. J., Van Weele, M. A. V., & Eveleens, C. P. (2017). Network brokers or hit makers? Analyzing the influence of incubation on start-up investments. *International Entrepreneurship and Management Journal, 13*(2), 605–629. https://doi.org/10.1007/s11365-016-0416-5
10. Organization for Economic Cooperation and Development. (2018). *Oslo manual of the organisation for economic cooperation and development: Guidelines for collecting, reporting, and using data on innovation* (4th ed.). OECD.
11. Zen, A. C., Machado, B. D., López, A. I. J., Borges, M. C., & Menezes, D. C. (2017). Rota da inovação: uma proposta de metodologia de gestão da inovação. *Revista de Administração Contemporânea, 21*(6), 875–892. https://doi.org/10.1590/1982-7849rac2017170151
12. Instituto Brasileiro de Geografia e Estatística (2023). Demografia das empresas e estatísticas de empreendedorismo: 2021. IBGE, Coordenação de Cadastro e Classificações, Rio de Janeiro
13. Universidade Tecnológica Federal do Paraná. (2023). *Plano de desenvolvimento institucional da Universidade Tecnológica Federal do Paraná: 2023–2027*. EDUTFPR. https://www.utfpr.edu.br/documentos/reitoria/documentos-institucionais/pdi
14. Amarante, J. M., & Crubellate, J. M. (2020). Pressões institucionais, trabalho institucional e o desenvolvimento do direcionamento empreendedor de universidades. *Revista de Administração Contemporânea, 24*(2), 119–133. https://doi.org/10.1590/1982-7849rac2020170383
15. Dimaggio, J. P., & Powell, W. W. (2005). A gaiola de ferro revisitada: Isomorfismo institucional e racionalidade coletiva nos campos organizacionais. *Revista de Administração de Empresas, 45*(2), 74–89. https://periodicos.fgv.br/rae/article/view/37123
16. Scott, W. R. (2014). *Institutions and organizations* (4th ed.). Sage Publications.
17. Perkmann, M., Phillips, N., & Greenwood, R. (2022). Institutional arbitrage: How actors exploit institutional difference. *Organization Theory, 3*(2), 1–20. https://doi.org/10.1177/26317877221090313
18. Goldenstein, J., & Walgenbach, P. (2021). How on earth did this happen? The relationship of practical consciousness and institutional evolution. *Organization Theory, 2*, 1–22. https://journals.sagepub.com/doi/pdf/10.1177/26317877211020324
19. Greenwood, R., Oliver, C., Sahlin, K., & Suddaby, R. (Eds.). (2008). *Introduction of the sage handbook of organizational institutionalism* (pp. 1–46). Sage. https://doi.org/10.4135/9781849200387.n1
20. Etzkowitz, H. (2003). Innovation in innovation: The triple helix of university-industry-government relations. *Social Science Information Sur Les Sciences Sociales, 42*(3), 293–337. https://journals.sagepub.com/doi/10.1177/05390184030423002
21. Zhou, C. (2008). Emergence of the entrepreneurial university in evolution of the triple helix: The case of Northeastern University in China. *Journal of Technology Management in China, 3*(1), 109–126. https://doi.org/10.1108/17468770810851539
22. Etzkowitz, H., & Leydesdorff, L. (2000). The dynamics of innovation: From national systems and mode 2 to a triple helix of university-industry-government relations. *Research Policy, 29*(2), 109–123. https://doi.org/10.1016/S0048-7333(99)00055-4
23. Etzkowitz, H. (2011). The triple helix: Science, technology, and the entrepreneurial spirit. *Journal of Knowledge-Based Innovation in China, 3*(2), 76–90. https://doi.org/10.1108/17561411111138937
24. Etzkowitz, H., & Zhou, C. (2017). *The triple helix: University-industry-government innovation and entrepreneurship* (2nd ed.). Routledge. https://doi.org/10.4324/9781315620183
25. Faria, A. F., Battisti, A. C., Sediyama, J. A. S., Alves, J. H., & Silvério, J. A. (2021). *Parques tecnológicos do Brasil*. NTG/UFV.

26. Serviço Brasileiro de Apoio às Micro e Pequenas Empresas. (2020). Ecossistemas de empreendedorismo inovadores e inspiradores. Sebrae
27. Mota, P. G., Silva, A. L. B., & Limongi, R. F. C. (2022). The use of agile methodologies and their contribution to innovation of the business model: A study of multiple cases in the context of incubators and startups. *Iberoamerican Journal of Entrepreneurship and Small Business, 11*(2), 1–12. https://doi.org/10.14211/ibjesb.e2170
28. Associação Nacional de Entidades Promotoras de Empreendimentos Inovadores. (2018). *Cerne é apresentado como modelo de referência no World Incubation Summit*. Anprotec. Retrieved from https://anprotec.org.br/site/tag/cerne/
29. Meyer, J., & Rowan, B. (1977). Institutionalized organizations: Formal structure as myth and ceremony. *American Journal of Sociology, 83*(2), 340–363. https://doi.org/10.1086/226550
30. Zucker, L. G. (1977). The role of institutionalization in cultural persistence. *American Sociological Review, 42*(5), 726–743. https://doi.org/10.2307/2094862
31. Selznick, P. (1996). Institutionalism old and new. *Administrative Science Quarterly, 41*(2), 270–277. https://doi.org/10.2307/2393719
32. Machado-da-Silva, C. L., & Fonseca, V. S. (1993). Homogeneização e diversidade organizacional: uma visão integrativa. In *17° Encontro Anual da Anpad* (Vol. 9, pp. 147–159).
33. Duarte, D. V., & Tavares, B. (2012). Institucionalismo e determinações governamentais: evidências de isomorfismo em empresas de base tecnológica. In *3° Encontro de Administração Pública e Governo*. ENAPG.
34. Souza, R. P., Ramalho, L. D. F., & Guerreiro, R. (2019). Características dos sistemas de custos gerenciais de subsidiárias alemãs operando no Brasil: uma análise sob a perspectiva do isomorfismo e do decoupling. *Revista Universo Contábil, 15*(3), 61–82.
35. Machado-da-Silva, C. L., Fonseca, V. S., & Crubellate, J. M. (2010). Estrutura, agência e interpretação: elementos para uma abordagem recursiva do processo de institucionalização. *Revista de Administração Contemporânea, 4*, 77–107.
36. Struckell, E., Ojha, D., Patel, P. C., & Dhir, A. (2022). Strategic choice in times of stagnant growth and uncertainty: An institutional theory and organizational change perspective. *Technological Forecasting and Social Change, 182*, 1–18.
37. Larson, M. S. (1977). *The rise of professionalism: A sociological analysis*. University of California Press.
38. Collins, R. (1979). *The credential society: An historical and sociology of education and stratification*. Columbia University Press.
39. Tolbert, P. S., & Zucker, L. G. (1996). The institutionalization of institutional theory. In S. R. Clegg, C. Hardy, & W. R. Nord (Eds.), *The handbook of organization studies: Theory & method* (Vol. 1, pp. 175–190). Sage.
40. Greenwood, R., Suddaby, R., & Hinings, C. (2002). Theorizing change: The role of professional associations in the transformation of institutionalized fields. *Academy of Management Journal, 45*(1), 58–80. https://doi.org/10.2307/3069285
41. Kripka, R. M. L., Scheller, M., & Bonotto, D. L. (2015). Pesquisa documental na pesquisa qualitativa: conceitos e caracterização. *Revista de Investigaciones, 14*(2), 55–73. https://www.researchgate.net/publication/280924900
42. Universidade Tecnológica Federal do Paraná. (2006). *Manual de procedimentos do Proem*. UTFPR. https://nuvem.utfpr.edu.br/index.php/s/DekSi3spYunkNy1
43. Bardin, L. (2016). Análise de conteúdo. 3rd reimp. Edições 70, São Paulo
44. Moraes, R. (1999). Análise de conteúdo. *Revista Educação, 22*(37), 7–32. https://edisciplinas.usp.br/pluginfile.php/4125089/mod_resource/content/1/Roque-Moraes_Analise%20de%20conteudo-1999.pdf
45. Barbosa, M. A. S., Silva, M. R., & Nunes, M. S. C. (2017). Pesquisa qualitativa no campo estudos organizacionais: explorando a análise temática. In: 41° Encontro da Associação Nacional de Pós-Graduação e Pesquisa em Administração. Anpad. *São Paulo*.
46. Silva, A. H., & Fossá, M. I. (2015). T.: Análise de conteúdo: exemplo de aplicação da técnica para análise de dados qualitativos. *Qualit@s Revista Eletrônica, 17*(1), 1–14. https://www.fepiam.am.gov.br/wp-content/uploads/2020/06/2113-7552-1-PB.pdf

47. Braun, V., & Clarke, V. (2006). Using thematic analysis in psychology. *Qualitative Research in Psychology, 3*(2), 77–101. https://doi.org/10.1191/1478088706qp063oa
48. Souza, L. K. (2019). Pesquisa com análise qualitativa de dados: conhecendo a análise temática. *Arquivos Brasileiros de Psicologia, 71*(2), 51–67.
49. ATLAS.TI Homepage. *Qualitative data analysis*. Release 7.6.0. Berlin, Atlas.ti GmbH. http://www.atlasti.com. Last accessed April 11, 2024
50. Jastram, S. M., & Foersterling, J. (2024). Clarifying ends in institutional decoupling theory: Insights from social impact research. *Journal of Business Economics, 94*, 501–523. https://doi.org/10.1007/s11573-023-01170-8
51. Kern, A., Laguecir, A., & Leca, B. (2018). Behind smoke and mirrors: A political approach to decoupling. *Organization Studies, 39*(4), 543–564.
52. Universidade Tecnológica Federal do Paraná. (2014). *Plano de desenvolvimento institucional da UTFPR: 2013–2017*. UTFPR. https://www.utfpr.edu.br/documentos/reitoria/documentos-institucionais/pdi
53. PROEM Homepage. https://portal.utfpr.edu.br/inovacao/empreendedorismo-2/empreendedorismo-1. Last accessed 15 March, 2024.
54. Universidade Tecnológica Federal do Paraná. (2015). Conselho de Relações Empresariais e Comunitárias. Resolução n° 02/15, de 30 set. Coemp. https://nuvem.utfpr.edu.br/index.php/s/H8QmENiALdZuwF8
55. SPRINT—Incubadora da UTFPR Homepage. https://portal.utfpr.edu.br/inovacao/empreendedorismo-2/incubadora-de-inovacoes-da-utfpr. Last accessed 28 July 2023.

Interactions between Innovation Capacity and Innovation Ecosystems: A Bibliometric Review

Nilvane Boehm Manthey, Carlos Ricardo Rossetto, Jorge Cunha, Fernando Cesar Lenzi, Gustavo Behling, and Everton Luis Pellizzaro De Lorenzi Cancellier

Abstract This study presents a systematic literature review aimed at understanding how existing research on Innovation Ecosystems addresses the theme of Innovation Capacity. Our methodology involved a comprehensive search of the Scopus database for publications spanning 2010–2023, utilizing "innovation ecosystem" and "innovation capacity" as core keywords. Subsequent keyword co-occurrence analysis, executed in R, helped map the thematic structure of the field. The findings illustrate a dynamic evolution in the conceptual and practical integration of these two domains, identifying predominant research themes and outlining a robust agenda for future studies. Ultimately, this review underscores the growing strategic imperative for collaboration within innovation ecosystems and highlights the crucial need for organizations to develop a deeper understanding of effective mechanisms to enhance their innovation capacity for sustainable competitive advantage.

Keywords Innovation capacity · Innovation ecosystems · Systematic literature review · Bibliometric analysis · Collaboration

1 Introduction

Innovation capability, a multifaceted concept encompassing essential dimensions for organizational development and sustainability, has long been a cornerstone of strategic management theory. Pioneering work by Schumpeter [49] underscored innovation as a driver of economic advancement, encompassing diverse forms from new products to organizational methods. Miller and Friesen [39] further highlighted its multidimensional nature, stressing its pervasive influence across organizational

C. R. Rossetto (✉) · J. Cunha · F. C. Lenzi · G. Behling
Universidade do Vale do Itajaí, Itajaí, Brazil

N. B. Manthey · E. L. P. De Lorenzi Cancellier
Universidade do Estado de Santa Catarina, Florianópolis, Brazil

L. Farinha et al. (eds.), *Sustainable Innovations in Mechanical Engineering and Management Sciences*, Lecture Notes on Multidisciplinary Industrial Engineering,
https://doi.org/10.1007/978-3-032-30372-1_8

levels, from product development to leadership strategy. Such a vital and multifaceted concept naturally draws attention from diverse fields, including engineering, economics, psychology, and administration [26].

Concurrently, the concept of innovation ecosystems has emerged as a crucial lens for understanding how various actors and institutions interact to foster innovation within a defined environment [1, 27]. These ecosystems are recognized as pivotal contexts for facilitating technological advancements and market growth. **However, despite the individual prominence of both innovation capability and innovation ecosystems in scholarly discourse, a systematic and integrated understanding of their conceptual intersection and mutual influence remains nascent**. While some studies touch upon elements of this relationship, there is a **lack of comprehensive synthesis detailing how innovation capability is explicitly framed and integrated within the broader theoretical framework of innovation ecosystems across various research streams**. This fragmented view hinders a holistic comprehension of how organizations can effectively cultivate their internal innovation capacities by leveraging the external dynamics of their surrounding ecosystems.

Addressing this critical gap, this systematic literature review aims to provide a structured overview of how innovation capability is discussed, framed, and integrated within studies on innovation ecosystems. Theoretically, this study contributes by offering a consolidated conceptual map of this intersection, identifying dominant perspectives and emerging themes, thus advancing our understanding of this complex interplay. Practically, it will inform managers, policymakers, and practitioners on critical areas for intervention to strategically enhance organizational innovation capacity within an ecosystem context, ultimately fostering competitive advantage and sustainable growth.

Specifically, this research seeks to: (1) analyze the theoretical frameworks linking Innovation Ecosystems and Innovation Capability; (2) identify the main research topics and trends at this intersection; and (3) propose a comprehensive agenda for future research. To achieve these objectives, a systematic literature review was conducted using the Scopus database. The search strategy involved the keywords "innovation ecosystem" and "innovation capability" (title, abstract, and keywords fields), covering publications from 2010 to 2023. A keyword co-occurrence analysis, performed using R software, was then utilized to map the thematic areas and reveal underlying conceptual structures.

The remainder of this paper is structured as follows: Section 2 details the methodological approach. Section 3 presents the findings from the systematic review and co-occurrence analysis. Section 4 discusses the theoretical and practical implications, and Sect. 5 concludes with an agenda for future research.

2 Theoretical Framework Innovation Capacity

Organizational innovation is the introduction into the organization of new methods for business management in the workplace and/or in the relationship between the company and external agents, such as a new product, service, technology, production process, new administrative structure or system, or a new plan or program [13]. The concept of innovation capability predates the notion of dynamic capacity [11] and refers to a company's ability to continuously adapt and improve by developing new products and services, new production methods, identifying new markets, and strategically aligning its innovative orientation with organizational behaviors and processes [61].

More recently, the ongoing digital transformation has significantly broadened the understanding of innovation capability, introducing concepts such as 'digital innovation capability' which encompasses the organizational ability to leverage digital technologies for developing new processes, products, and business models [37].

The multidisciplinary nature of innovation literature is well-established, drawing insights from engineering, sociology, economics, psychology, and administration [26]. Within strategic management, research aims to enhance the consistency of investigations by developing theoretical models rooted in institutional theory, cognitive theories, transaction cost economics, market orientation, and the resource-based view [36]. Contemporary scholarship continues to explore how organizational design, leadership styles, and internal processes act as crucial antecedents to fostering robust innovation capabilities, particularly in light of rapid technological shifts and market turbulence [5].

The theoretical underpinning for studying innovation capability, particularly in dynamic contexts, largely draws from the resource-based view (RBV) Barney [8, 44, 60]. RBV posits that an organization's innovative output is intrinsically linked to its unique resources and processes, and competitive differences in innovation stem from the development of valuable, inimitable, and non-substitutable resources [8]. Consequently, competitive advantage often lies not merely in new products, but in the organizational capacity to consistently develop them [44]. Teece and Pisano [57] built upon this by developing the concept of dynamic innovation capacity, conceptualizing it as a subset of competencies enabling firms to create new products and processes and swiftly adapt to market shifts.

The dynamic capabilities framework remains highly pertinent for organizational innovation studies for three key reasons: (1) its holistic approach beyond mere technology; (2) its seamless integration of process and product innovation; and (3) its acknowledgment of asset heterogeneity, precluding a generic formula for innovation capacity [36]. **Recent research further underscores the vital role of dynamic capabilities in sensing opportunities, seizing them through innovation, and reconfiguring resources, which is essential for sustained competitive advantage, especially in highly turbulent or digitally transforming environments** [6, 55].

Empirical research on innovation has continuously sought more effective measurement of organizational innovation capacity. Various indicators have been developed to capture its diverse dimensions, including strategic, behavioral, process, product, and market innovation, reflecting an innovative orientation [14, 30, 39, 61]. **The increasing complexity of innovation, driven by factors like digitalization and open innovation paradigms, necessitates refined measurement approaches that capture the multifaceted nature of contemporary innovation capabilities more accurately** [4].

2.1 Innovation Ecosystems

Innovation ecosystems are conceptualized as dynamic networks of interconnected organizations and diverse stakeholders that coalesce around a common technological platform, focal firm, or shared vision, orchestrating efforts to collaboratively create and capture value through innovation [7, p. 205]. This framework emphasizes the critical importance of systemic collaboration and interdependence among various actors for successful innovation outcomes. **Recent scholarship highlights that these ecosystems are not merely static networks but evolving configurations requiring strategic orchestration and adaptability to environmental shifts** [31].

A central characteristic of contemporary innovation ecosystems is the active involvement of diverse stakeholders, particularly through user-driven innovation, where users assume varied roles from conceptualization to product and service deployment [33]. This participatory approach fosters heightened interactivity and deep collaboration, reinforcing the interdependent nature of ecosystem participants.

The increasing complexity of these interactions, especially within digitally-enabled ecosystems, has led to sophisticated mechanisms for co-creation and open innovation, integrating external knowledge sources more deeply into the innovation process [10]. Metaphorically likened to biological systems, the term "ecosystem" aptly describes these groups of heterogeneous actors working in cooperation and interdependence, necessitating nuanced governance to foster collective value creation [31, 33].

Universities and research institutions consistently play a pivotal role within innovation ecosystems, acting as key knowledge generators and catalysts for local and regional development. Their engagement involves fostering robust cooperation with the productive sector, facilitating intellectual property commercialization, and fostering the emergence of technology-based ventures [33].

This evolving role of academia underscores its strategic importance in bridging fundamental research with market applications, driving both technological transfer and entrepreneurial activity within the ecosystem [63].

Beyond simply a network, innovation ecosystems are increasingly viewed as a critical infrastructure designed to foster innovation, facilitating dynamic interactions between diverse stakeholders, including suppliers, demanders, and external actors [43]. **The advent of digital platforms has profoundly transformed this infrastructure into dynamic digital innovation ecosystems, where modular architectures and network effects redefine participation and value capture, enabling**

rapid scaling and fluid reconfiguration of resources [41]. These digital ecosystems facilitate new forms of interaction, where the user is often invited to participate as a co-creator of innovation, generating value flows that can be described both conceptually and through real-world examples, including those stemming from societal changes and the impacts of these innovations.

Governmental policies and regulatory frameworks play a significant role in nurturing and shaping innovation ecosystems. For instance, in Brazil, the establishment of legal frameworks such as the Innovation Law (Law No. 10,973) and the Good Law (Law No. 11,196) from 2004 to 2005 provided legal certainty and tax incentives for R&D cooperation between firms and scientific and technological institutions [25]. Such supportive regulatory environments are instrumental in strengthening national innovation ecosystems. Consequently, understanding the dynamics of innovation ecosystems is paramount for all stakeholders, including firms of all sizes, research institutions, and governmental/non-governmental organizations, in formulating effective innovation strategies.

While early attention was primarily on policymakers and sectoral articulators, there is a growing recognition among companies of the strategic imperative to engage in external collaborative innovation to enhance strategic planning and access diverse knowledge sources. **This increasing strategic complexity further necessitates sophisticated approaches to ecosystem governance and active participation from all actors to foster resilient and competitive innovation landscapes in a continually evolving global context** [23].

3 Methodology

This study employed a systematic literature review (SLR) methodology **to comprehensively map and synthesize the existing knowledge at the intersection of Innovation Ecosystems and Innovation Capacity**. Our approach was rigorously guided by the **Preferred Reporting Items for Systematic Reviews and Meta-Analyses (PRISMA) framework** [40, 42] **to ensure transparency, replicability, and minimize bias throughout the review process**.

The primary objectives of this research were threefold: (1) to analyze the theoretical frameworks developed concerning Innovation Ecosystems and Innovation Capacity; (2) to identify the main research topics and prevalent themes within this specific intersection; and (3) to delineate a robust agenda for future investigations. These objectives directly informed our research questions, structured as follows:

1. How has the academic literature on Innovation Ecosystems and Innovation Capacity developed between 2010 and 2023?
2. What are the main research themes and conceptual areas explored regarding Innovation Ecosystems and Innovation Capacity?
3. What are the key gaps and promising avenues for future research concerning Innovation Ecosystems and Innovation Capacity?

The article identification process rigorously followed the PRISMA guidelines, as visually summarized in Fig. 1, encompassing identification, screening, eligibility, and inclusion phases.

Regarding the search for articles, the Scopus database established the following guiding criteria: keywords (Innovation Ecosystem and innovation capacity), search fields in the title, abstract and keywords, between the years 2010 and 2023. The development of the research analysis continued by thematic area and defined through a co-occurrence analysis of the keywords using the R software. To carry out the analysis, the questions presented in Table 1 were defined as collection criteria.

The integration of the Scopus database through the "R" system made it possible to identify the main characteristics of the research constructs, all of which were associated with research that referred to the object "Innovation Ecosystems". In this sense, the research results were presented by explaining the findings in the analysis, placing each construct with its respective research objective.

4 Analysis of Results and Discussions

4.1 Innovation Ecosystems and Innovation Capacity: Bibliometric Trends

This section details the results derived from the systematic literature review and provides an initial discussion on the development of the field of study. The initial identification phase of the systematic literature review, focused on the intersection between "Innovation Ecosystems" and "Innovation Capacity," and covering publications from 2010 to 2023 in the Scopus database, resulted in the retrieval of 125 articles.

The subsequent bibliometric analysis, performed using R software, allowed us to answer the first research question (RQ1): "How has the literature on Innovation Ecosystems and Innovation Capacity developed between 2010 and 2023?" This preliminary phase of the analysis focused on quantifying key bibliometric indicators, including annual publication output, the most influential journals, the most prolific authors, and the research institutions with the highest contribution.

The longitudinal analysis of academic output reveals a remarkable growth trajectory. It is observed that the number of publications began an upward trend from 2017, and from 2021 onwards, the field experienced a significant acceleration in the volume of scientific production, as illustrated in detail in Fig. 2.

Regarding the **most prolific international journals** publishing articles on these study topics, the **leading contributors were identified** as the "Journal of Business Research" with eight articles, and "Technological Forecasting and Social Change" with five articles, as depicted in Fig. 3. The "Journal of Business Research" is **categorized as a Q1 journal** in the Scimago Journal & Country Rank, boasting an H-Index of 236 and an SJR of 13.44. Similarly, "Technological Forecasting and

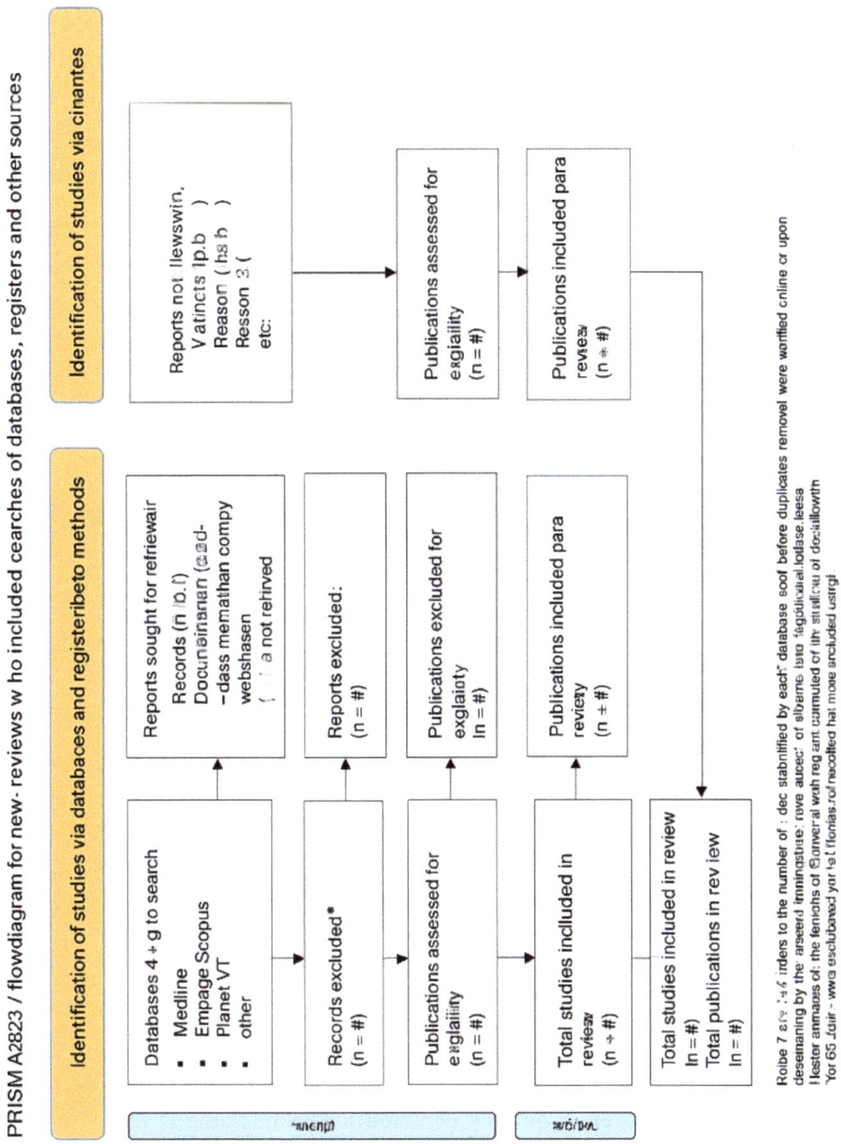

Fig. 1 Systematic data analysis flows using PRISMA

Table 1 Collection and analysis criteria

Steps	Actions
(1) Articles search	(a) Articles of an empirical and theoretical nature; (b) Peer-reviewed journals; (c) Work focused on innovation ecosystems; (d) Data analysis and bibliographic reviews, both primary and secondary
(2) Launching articles in the software	Distributing by criteria: year, authors, reference, theme, journal, impact factor, research approach, co-authorship, co-citations, concept, research variables, keywords
(3) Quantitative analysis	Citation analysis, network analysis performed in "R" Software
(4) Qualitative analysis	To identify the evolution of research constructs as well as gaps (full reading of the 20 recent articles on each topic)

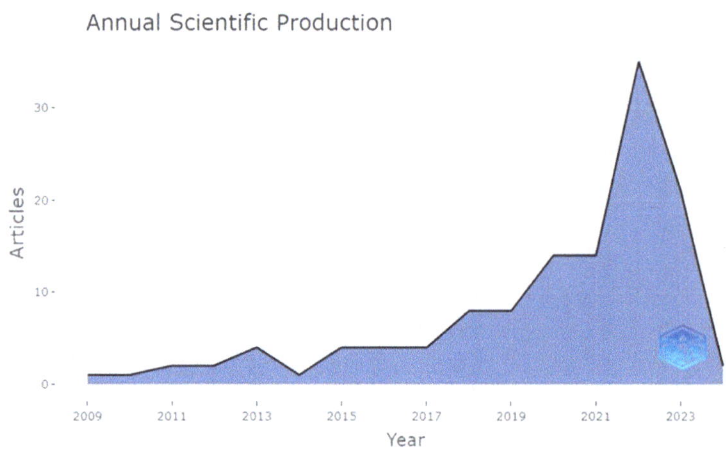

Fig. 2 Distribution of innovation ecosystem and innovation capacity articles by year

Social Change", with its five articles, is also **classified within the Q1 quartile** by Scimago, presenting an H-Index of 155 and an SJR of 2.64. **This prominence in highly-ranked, internationally recognized journals underscores the substantial scholarly interest and growing significance of research at the intersection of innovation ecosystems and innovation capacity**.

An examination of the most cited authors reveals a **relatively dispersed authorship landscape**, with no single author or research group overwhelmingly dominating the field. **However, a notable contribution** in this area is the study by Rohrbeck et al. [47], entitled "Opening up for competitive advantage–How Deutsche

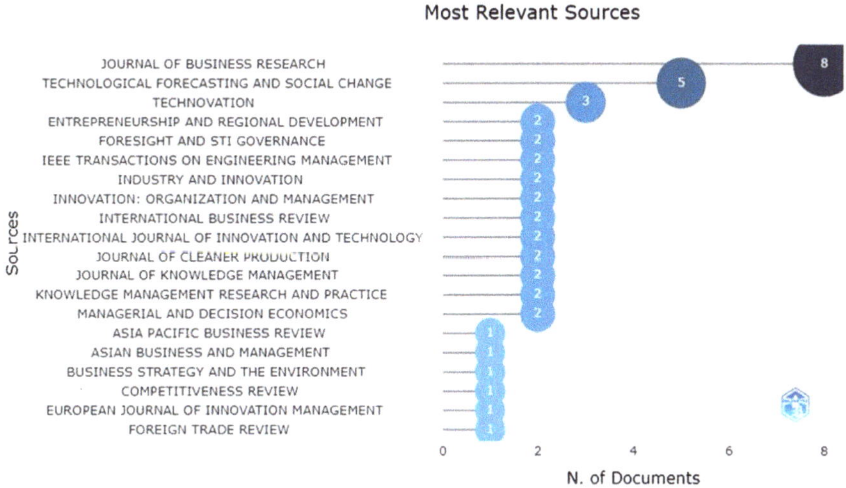

Fig. 3 International journals with studies on innovation ecosystem and innovation capacity

Telekom creates an open innovation ecosystem." **This pattern of distributed influence suggests that a comprehensive understanding of the literature necessitates consulting the diverse body of work from the prominent authors highlighted in Fig.** 4, **as their collective contributions are foundational to the field's development**.

Regarding the **institutional contributions** to research on Innovation Ecosystems and Innovation Capacity, the analysis indicates a **relatively dispersed landscape**

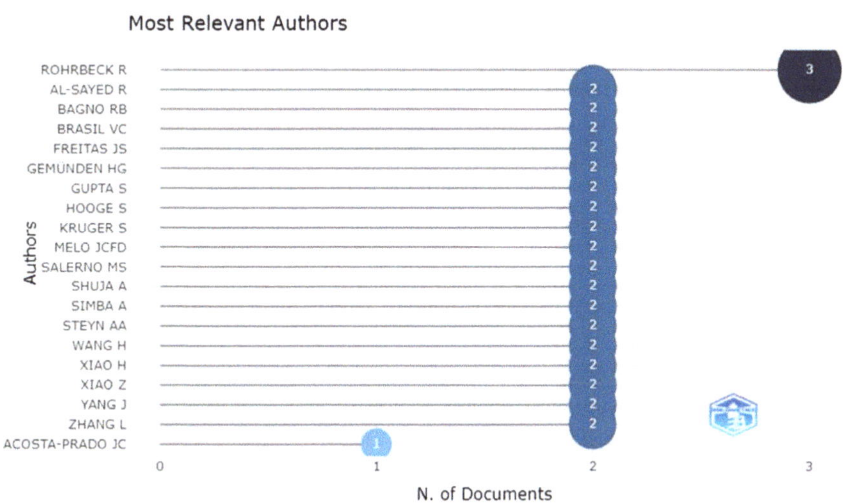

Fig. 4 Relevant authors in innovation ecosystem and innovation capacity research

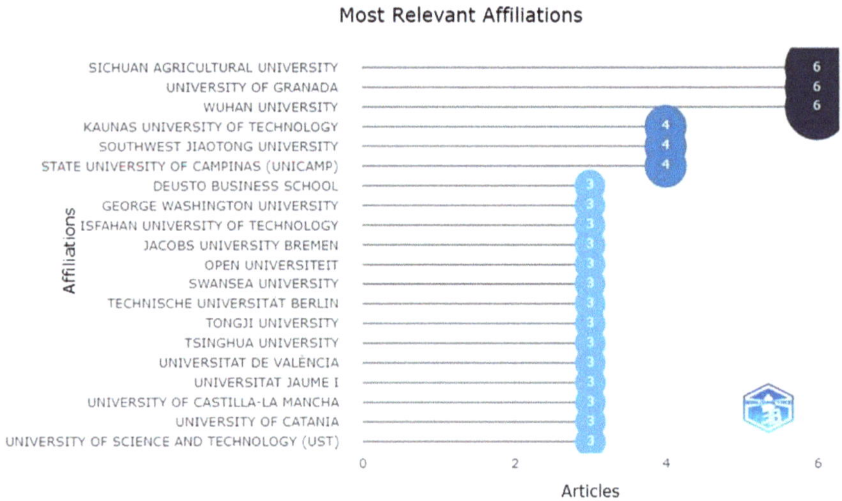

Fig. 5 Relevant universities in innovation ecosystem research and innovation capacity

among universities. Specifically, Sichuan Agricultural University, the University of Granada, and Wuhan University emerge as **the most prolific institutions**, each contributing six articles, as depicted in Fig. 5. This **absence of a single dominant institution** underscores the broad and global engagement in this interdisciplinary research area.

Among the most productive countries within the thematic area, China stands out, followed by Germany. Brazil is in tenth place in the production ranking, as shown in Fig. 6.

To further elucidate the thematic landscape and address the **second research question (RQ2): "What are the main research topics concerning Innovation Ecosystems and Innovation Capacity?"**, a deeper analysis was conducted. This involved several advanced bibliometric techniques, including a word cloud analysis, an examination of annual production by thematic topic, a cluster analysis, and the identification of prominent authors within the derived clusters.

The word cloud visualization revealed prominent concepts such as "innovation," "open innovation," "innovation capacity," "dynamic capability," and "knowledge sharing." These terms signify the most frequently occurring keywords, providing an initial mapping of the field's conceptual core, as visually represented in Figs. 7 and 8.

Figure 9 depicts the **frequency distribution of keywords across the analyzed studies**. This quantitative overview **further reinforces the thematic prominence initially identified through the word cloud analysis** presented in Figs. 7 and 8.

The integration of data from the Scopus database, processed through the "R" system, enabled the identification of the **core conceptual constructs** relevant to the intersection of Innovation Ecosystems and Innovation Capacity. Based on the aggregated insights derived from analyses such as word clouds, annual production

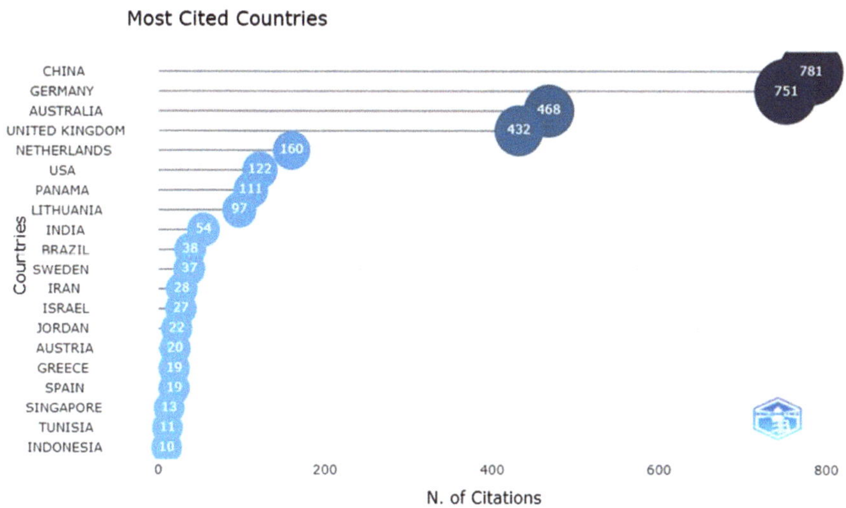

Fig. 6 Countries most cited in the innovation ecosystem and innovation capacity survey

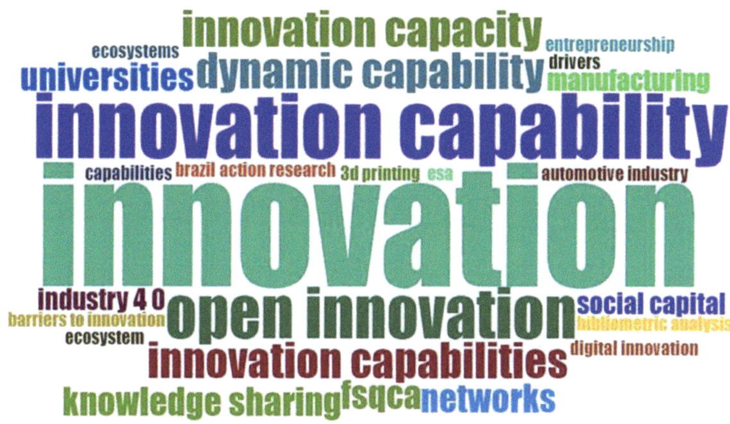

Fig. 7 Innovation ecosystem and innovation capacity—word cloud

by topic, and cluster analysis (as depicted in Figs. 7, 8, and 9), the **most salient themes** emerging from the literature are:

- Innovation
- Innovation Capacity
- Open Innovation
- Dynamic Capability
- Innovation Ecosystem
- Innovation Management
- Innovation Performance.

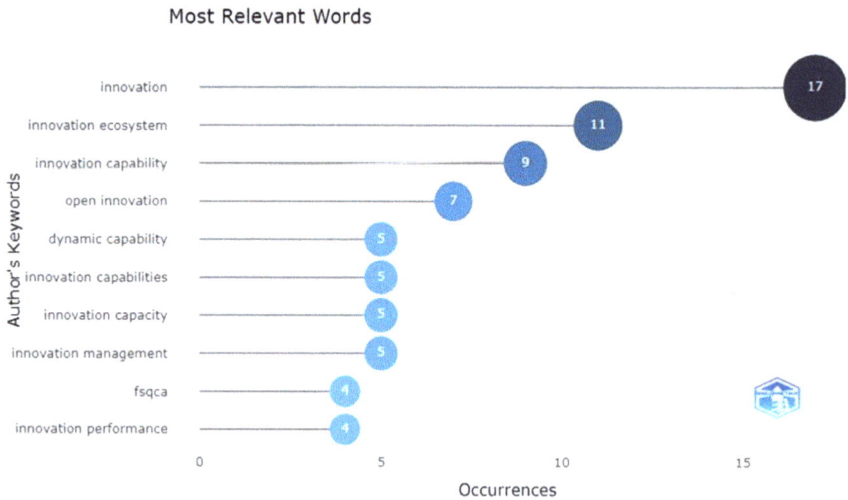

Fig. 8 Innovation ecosystem and innovation capacity—most relevant words

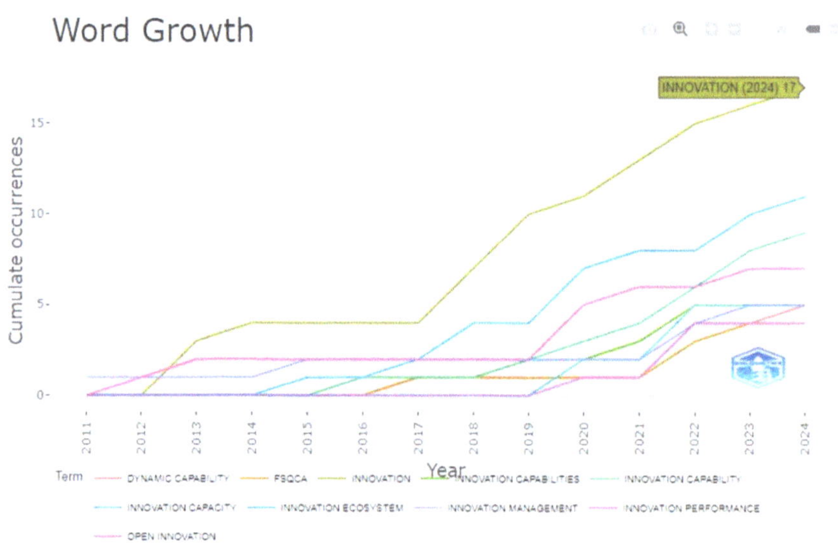

Fig. 9 Annual scientific production of innovation ecosystem and innovation capacity

4.2 Construction of the Thematic Map

Employing the visualization technique **pioneered by Callon et al.** [12] and implemented through **VOSviewer and R-Studio**, a thematic map was constructed. This map graphically represents the intellectual structure of the field, segmenting themes

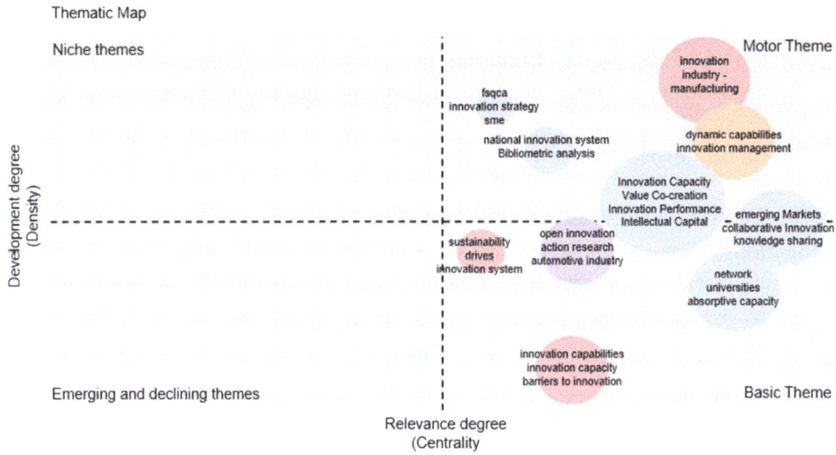

Fig. 10 Thematic areas of innovation ecosystem and innovation capacity research

into four quadrants based on their **centrality** (degree of interaction with other themes, indicating relevance) and **density** (strength of internal connections, indicating development). The size of each thematic cluster within the map is proportional to the number of keyword occurrences it comprises, thereby reflecting its associated article count. Clusters are labeled according to their most predominant keywords. For each quadrant, a detailed discussion will follow, prioritizing clusters with the highest number of related articles.

The 125 articles retrieved from the Scopus database were systematically processed: data was exported and subsequently analyzed in R-Studio for extraction and preparation of inputs for VOSviewer. Figure 10 illustrates the resulting distribution of studies across these thematic quadrants, forming the research clusters identified in the analysis.

Analysis of the Thematic Map**

The thematic map, constructed based on centrality and density metrics, delineated four distinct quadrants, each representing different stages of thematic development and relevance within the literature. The distribution of clusters per quadrant is presented and discussed below.

Quadrant I: Driving Themes

This quadrant comprises concepts characterized by high centrality and high density, signifying well-developed and fundamental areas that are integral to **structuring the** research field on Innovation Ecosystems and Innovation Capacity. Within this quadrant, five distinct thematic clusters were identified:

- The first cluster encompasses the themes "Innovation" and "Manufacturing Industry."

- The second cluster integrates "FSQCA," "Innovation Strategy," and "SMEs" (Small and Medium-sized Enterprises).
- The third cluster includes the themes "National Innovation System" and "Bibliometric Analysis."
- The fourth cluster comprises "Dynamic Capabilities" and "Innovation Management."
- The fifth and final cluster is composed of "Innovation Capacity," "Value Co-creation," "Innovation Performance," and "Intellectual Capital."
- An additional cluster, positioned at the boundary between the driving and basic themes, was also identified, encompassing "Emerging Markets," "Collaborative Innovation," and "Knowledge Sharing."

Quadrant II: Basic Themes

This quadrant includes themes exhibiting **high centrality but low density**, indicating their fundamental importance to the field of Innovation Ecosystems and Innovation Capacity, yet suggesting they are still nascent or less comprehensively developed in the current literature. This quadrant consists of four primary thematic clusters:

- The first cluster comprises the themes "Network," "Universities," and "Absorptive Capacity."
- The second cluster covers "Sustainability," "Drivers," and "Innovation System."
- The third cluster integrates "Innovation Capacity" and "Barriers to Innovation."
- The fourth and final cluster is formed by the themes "Open Innovation," "Action Research," and "Automotive Industry."

Quadrant III: Niche Themes

The Niche Themes quadrant, typically characterized by low centrality and high density (i.e., well-developed but peripheral themes), did not reveal any represented clusters within this analysis. This absence indicates a lack of highly specialized but isolated research areas at the intersection of Innovation Ecosystems and Innovation Capacity.

Quadrant IV: Emerging and Declining Themes

Similarly, the Emerging and Declining Themes quadrant, defined by **low centrality and low density (indicating themes with limited development and weak connections), also presented no discernible clusters. This suggests an absence of both nascent, weakly connected themes and areas that have significantly diminished in research focus during the analyzed period for these constructs.

Research Network

The co-authorship network analysis (Fig. 11) revealed the emergence of four distinct research networks within the broader field of Innovation Capacity and Innovation Ecosystems. These networks, although relatively nascent in their formation, demonstrate discernible interconnections, indicating collaborative bridges between research communities rather than complete isolation.

Analysis of Research Networks

The co-authorship network analysis (Fig. 11) revealed four distinct, albeit somewhat interconnected, research networks. Each network is anchored by seminal works that collectively shape the theoretical and empirical discourse within the broader field of Innovation Capacity and Innovation Ecosystems.

Blue Network The blue network is notably characterized by foundational contributions that link knowledge management with innovation performance. A seminal study by Cohen and Levinthal [16], "Absorptive capacity: A new perspective on learning and innovation," established the crucial link between a firm's capacity to recognize, assimilate, transform, and apply external knowledge, deeming it fundamental for enhancing innovative capacity performance. Further underpinning this network's theoretical lineage is Teece [53], whose work, grounded in the Resource-Based View (RBV), posits that competitive advantage derives from proprietary resources. This perspective, a natural extension of the RBV, notably broadens the definition of resources to encompass intangible assets, particularly knowledge-based resources. Lawson and Samson [36], in their paper "Developing innovation capability in organizations: a dynamic capabilities approach," synthesized insights from diverse domains to conceptualize innovation management as a distinct organizational capability. They argued that high-performing organizations strategically invest in and nurture this capability, leading to effective innovation processes, new product/service/process innovations, and superior business performance. Grounded in dynamic capabilities theory, they delineated innovation capability into seven constituent elements: vision and strategy, competency base, organizational intelligence, creativity and idea management, organizational structures and systems, culture and climate, and technology management.

Red Network The red network is prominently defined by the work of Adner and Kapoor [2], specifically their study "Value creation in innovation ecosystems: how the structure of technological interdependence affects firm performance in new technology generations." This research meticulously investigated value creation within innovation ecosystems, exploring how challenges encountered by external innovators impact firm performance. Adner and Kapoor [2] hypothesized that the impact of external innovation challenges is contingent not only on their magnitude but critically on their location within the ecosystem relative to the focal firm. Their findings revealed a fundamental asymmetry: increased upstream innovation challenges in components tend to augment the benefits for technology leaders, whereas heightened downstream innovation challenges in complements conversely erode these benefits.

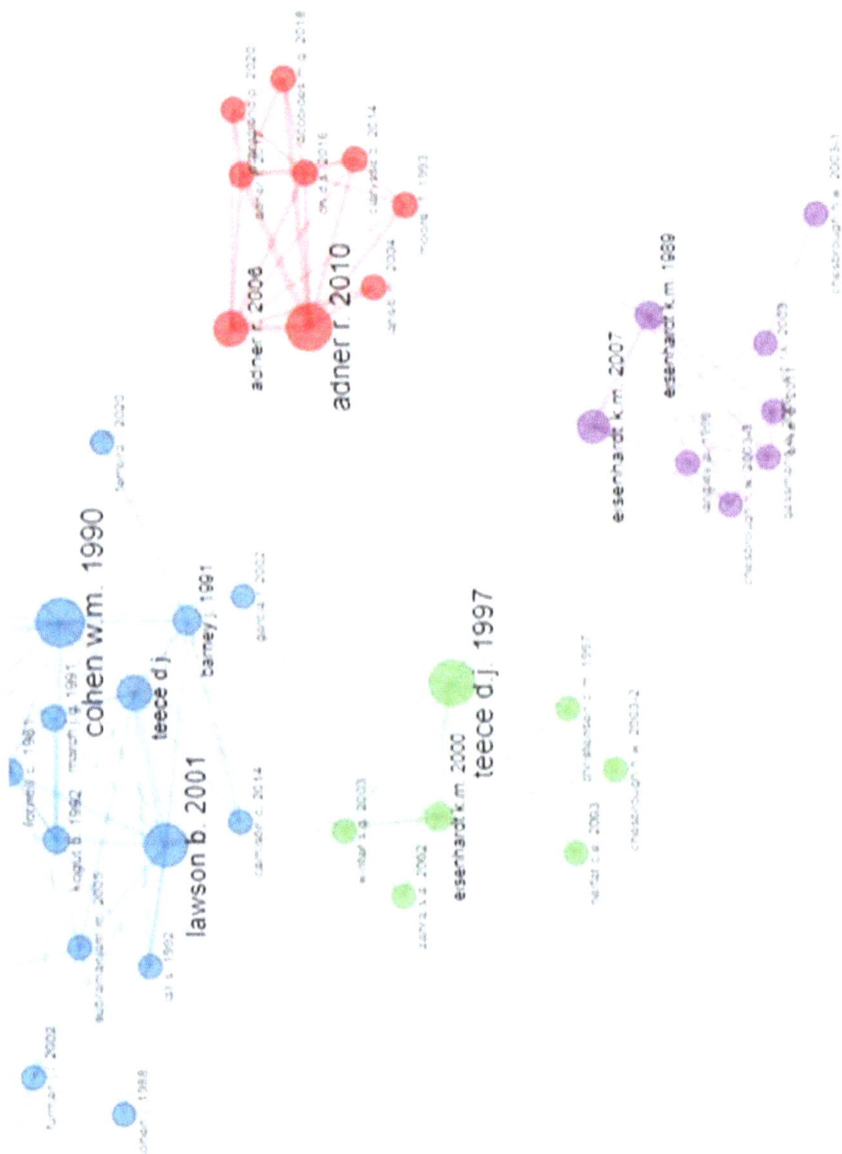

Fig. 11 Highlighted words among the clusters derived from the innovative ecosystem and innovation capacity research

Green Network The green network is anchored by the influential work of Teece et al. [56], "Dynamic capabilities and strategic management." This seminal paper not only positions absorptive capacity as an integral component of dynamic capabilities but also rigorously analyzes the mechanisms for wealth creation and capture by private firms navigating environments characterized by rapid technological change. Their framework provides a robust lens for understanding how firms can sustain competitive advantage by effectively adapting to and shaping turbulent market conditions.

Purple Network The purple network features two significant contributions from Eisenhardt. First, Eisenhardt [20], in "Making fast strategic decisions in high-velocity environments," inductively investigated how executive teams formulate rapid decisions within the high-velocity microcomputer industry. This study of eight microcomputer firms led to several propositions, notably that fast decision-makers utilize more information and generate more alternatives than their slower counterparts, employing a two-tiered advisory process. Furthermore, effective conflict resolution and the seamless integration of strategic decisions with tactical plans were identified as critical determinants of decision pace.

Ultimately, decisions made following this behavioral pattern were linked to superior performance. Second, Eisenhardt and Graebner [19], in "Theory building from cases: Opportunities and challenges," provided a comprehensive discussion on constructing theory from cases, particularly through multiple case studies. They articulated that this strategy involves leveraging one or more cases to develop theoretical constructs, propositions, or middle-range theories grounded in empirical evidence. The inherent logic of replication in this approach implies that each case functions as a distinct experiment, evaluated on its analytical merits. While acknowledging the frequent use of case studies, they also highlighted associated challenges, advocating for precise formulation and rigorous research design to mitigate potential limitations.

5 Final Considerations

This systematic literature review successfully addressed its primary objectives by meticulously analyzing the evolving academic landscape at the intersection of Innovation Ecosystems and Innovation Capacity from 2010 to 2023. Through bibliometric and thematic analyses, we effectively charted the development trends, identified core research themes, and delineated key conceptual clusters. Our findings reveal a dynamic and increasingly active field, marked by significant scholarly interest, particularly since 2021. This section synthesizes the main findings, highlights the study's theoretical and practical contributions, acknowledges its limitations, and proposes a comprehensive agenda for future research.

5.1 Summary of Key Findings

In response to Research Question 1 (RQ1): "How has the literature on Innovation Ecosystems and Innovation Capacity developed between 2010 and 2023?, our bibliometric analysis of 125 articles demonstrated a clear upward trend in publications, with a notable acceleration in scientific output since 2021. This indicates a burgeoning interest in understanding the interplay between these two critical constructs. The most influential journals, prolific authors, and contributing institutions were identified, providing a foundational overview of the field's intellectual landscape.

Addressing Research Question 2 (RQ2): "What are the main research topics concerning Innovation Ecosystems and Innovation Capacity?", our thematic analysis, including word clouds and cluster analysis, revealed core concepts such as "innovation," "open innovation," "innovation capacity," "dynamic capability," and "knowledge sharing" as central to the discourse. The thematic map further elucidated the intellectual structure, identifying "driving themes" (e.g., innovation, manufacturing industry, innovation strategy, SMEs, dynamic capabilities, innovation management, value co-creation, innovation performance, intellectual capital), which are well-developed and fundamental. "Basic themes" (e.g., network, universities, absorptive capacity, sustainability, barriers to innovation, open innovation) were also identified, characterized by high centrality but lower density, suggesting emerging areas with significant potential. Notably, no clusters were identified within the "niche" or "emerging/declining" quadrants, indicating a current focus on fundamental and emerging areas rather than highly specialized or waning topics within this specific intersection.

5.2 Contributions

This study offers several **theoretical contributions**. First, by systematically mapping the intellectual structure of the combined Innovation Ecosystems and Innovation Capacity literature, it provides a comprehensive and updated overview of a field previously fragmented, offering a consolidated conceptual map. Second, the identification and characterization of driving and basic themes clarify the current state of knowledge, highlighting mature areas that underpin the field and nascent research avenues ripe for further exploration. Third, by synthesizing the most influential articles within distinct research networks, this review offers a structured understanding of how core concepts are interconnected and have evolved, thereby providing a robust framework for future conceptual development and theory building.

From a practical perspective, this review offers valuable insights for managers and policymakers. It underscores the critical importance of understanding both internal innovation capabilities and the external dynamics of innovation ecosystems for fostering sustainable competitive advantage. The identified core themes and research networks can guide organizations in strategically aligning their innovation efforts, fostering collaboration, and leveraging external knowledge to enhance their innovative output. For policymakers, the review highlights areas where support mechanisms (e.g., legal frameworks, funding for specific clusters) can be most effectively implemented to nurture vibrant innovation ecosystems and strengthen regional innovation capacity.

5.3 *Limitations*

Despite its rigorous methodological approach, this study acknowledges several limitations inherent to systematic literature reviews and bibliometric analyses. First, restricting the search to the Scopus database, while comprehensive for many disciplines, might have excluded relevant publications from other specialized databases. Second, the defined keyword string, while carefully crafted for precision, might not have captured every nuance of related concepts. Third, the timeframe (2010–2023) inherently limits the historical scope of the analysis, precluding insights from earlier foundational works not explicitly captured by the search string. Finally, while bibliometric analyses provide a powerful macroscopic overview, they do not replace in-depth qualitative content analysis of every included article, which would offer deeper contextual understanding of specific theoretical arguments and empirical findings.

5.4 *Future Research Agenda*

Building upon the identified gaps and limitations, and directly addressing Research Question 3 (RQ3): "What is the research agenda concerning Innovation Ecosystems and Innovation Capacity?", we propose a comprehensive agenda for future research:

Methodological and Contextual Expansion:

Future research should expand database searches to include Web of Science, Google Scholar, or regional databases to enhance comprehensiveness and mitigate single-database bias. Longitudinal studies employing qualitative or mixed-methods approaches, particularly in Small and Medium-sized Enterprises (SMEs) and specific industrial sectors (e.g., food industries, as suggested by Taneo et al. [52, p.13], are recommended to capture process innovations and their dynamics over time.

Studies should also consider incorporating multiple data sources (e.g., combining bibliometric with survey or case study data) for a more holistic understanding [35, p. 9].

Conceptual and Theoretical Deepening:

Further investigation is warranted into the interaction between Intellectual Capital, innovation ambidexterity, firm value creation, and employee competencies, knowledge, and best practices [51].

The roles of organizational culture, leadership style, and engagement with external stakeholders are crucial for effective learning and adaptation in crisis management contexts [35, p. 9].

Integrating frugal innovation into capability frameworks, and developing scales to measure a company's frugal innovation level [48], could offer new perspectives, especially in resource-constrained environments.

Further studies could explore other types of capital (beyond intellectual capital) and assess potential mediation analyses involving human, relational, and structural capital [46].

Context-Specific and Policy Implications:

Empirically, research could focus on context-specific analyses, particularly exploring time-varying factors influencing Regional Innovation Capacity and analyzing entrepreneurship policy configurations in regions with low innovation capacity [28]. Replicating research in other countries is needed to identify common patterns and differences across business ecosystems [46].

The transferability of findings from specific crises (e.g., the COVID-19 pandemic context) to other crisis situations requires explicit investigation, with implications for policymakers and practitioners [34, 35].

Specific Actors and Intermediaries:

A particularly promising avenue involves examining the function of innovation intermediaries, especially in contexts of high uncertainty or volatility, such as emerging market multinationals [18].

These studies would benefit from quantitative and longitudinal designs to determine the true effectiveness of these intermediaries over time, and could focus on studies of Open Innovation in a single host country to control biases in institutional contexts while concentrating on a specific industry [18, p. 13].

Advanced Analytical Techniques:

Finally, employing panel studies to examine trends over time and conducting moderation analyses (e.g., by firm size or sector) would significantly deepen the insights into the relationships examined in this study [46]. In conclusion, the intersection of Innovation Ecosystems and Innovation Capacity remains a dynamic and vital field of inquiry. By addressing the identified gaps and rigorously pursuing these future research directions, scholars can further enhance our theoretical understanding and provide more actionable insights for practitioners seeking to foster resilient and high-performing innovation systems.

References

1. Adner, R. (2006). Match your innovation strategy to your innovation ecosystem. *Harvard Business Review, 84*(4), 98–107.
2. Adner, R., & Kapoor, R. (2010). Value creation in innovation ecosystems: How the structure of technological interdependence affects firm performance in new technology generations. *Strategic Management Journal, 31*(3), 306–333. https://doi.org/10.1002/smj.821
3. Adner, R., & Helfat, C. (2002). Corporate effects and dynamic managerial capabilities. *Strategic Management Journal, 23*(12), 1011 1025.
4. Akhtar, S., Arslan, M., & Ahmad, S. (2022). Innovation capability measurement: A systematic review and research agenda. *Journal of Cleaner Production, 340*, 130635.
5. Alawiah, E. T., & Tukiran, M. (2024). Digital leadership and digital transformation: Systematic literature review. *International Journal of Social and Management Studies, 5*(3), 27–34. https://doi.org/10.5555/ijosmas.v5i3.410
6. Ambidi, A., & Yusuf, S. (2021). Dynamic capabilities and innovation performance: The mediating role of organizational learning. *Journal of Management & Organization, 27*(5), 896–915.
7. Autio, E., & Thomas, L. D. W. (2014). Innovation ecosystems: Implications for innovation management? In M. Dodgson, D. M. Gann, & N. Phillips (Eds.), *The Oxford handbook of innovation management*. Oxford University Press. https://doi.org/10.1093/oxfordhb/9780199694945.013.012
8. Barney, J. (1991). Firm resources and sustained competitive advantage. *Journal of Management, 17*(1), 99–120. https://doi.org/10.1177/014920639101700108
9. Barney, J. (2001). Resource-based theories of competitive advantage: A ten-year retrospective on the resource-based view. *Journal of Management, 27*(6), 643–650.
10. Bogers, M., Burcharth, A. L. G., & Chesbrough, H. (2021). Open innovation in Brazil: Exploring opportunities and challenges. *International Journal of Professional Business Review, 6*(1), e213. https://doi.org/10.26668/businessreview/2021.v6i1.213
11. Breznik, L. D., & Hisrich, R. D. (2014). Dynamic capabilities vs. innovation capability: Are they related? *Journal of Small Business and Enterprise Development, 21*(3), 368–384.
12. Callon, M., Courtial, J. P., & Laville, F. (1991). Co-word analysis as a tool for describing the network of interactions between basic and technological research: The case of polymer chemistry. *Scientometrics, 22*(1), 155–205. https://doi.org/10.1007/BF02019280
13. Camisón, C., & Villar-López, A. (2014). Organizational innovation as an enabler of technological innovation capabilities and firm performance. *Journal of Business Research, 67*(12), 2891–2902.
14. Capon, N., Farley, J. U., Hulbert, J. M., & Lehmann, D. R. (1992). Profiles of product innovators among large US manufacturers. *Management Science, 38*(2), 157–169.
15. Chen, L., et al. (2019). An artificial intelligence-based data-driven approach for design ideation. *Journal of Visual Communication and Image Representation, 61*, 10–22.
16. Cohen, W. M., & Levinthal, D. A. (1990). Absorptive capacity: A new perspective on learning and innovation. *Administrative Science Quarterly, 35*(1), 128–152.
17. Cohen, W. M., & Levinthal, D. A. (1989). Innovation and learning: The two faces of R&D. *Economic Journal, 99*(397), 569–596.
18. Conroy, K. M., Jacobs, S., & Liu, Y. (2023). The dual knowledge role of open innovation intermediaries: Internal weaving and external filtering for MNE subsidiaries. *Technovation, 123*, 102721.
19. Eisenhardt, K. M., & Graebner, M. E. (2007). Theory building from cases: Opportunities and challenges. *Academy of Management Journal, 50*(1), 25–32. https://doi.org/10.5465/AMJ.2007.24160888
20. Eisenhardt, K. M. (1989). Agency theory: An assessment and review. *The Academy of Management Review, 14*(1), 57–74. https://doi.org/10.2307/258191
21. Eisenhardt, K., & Martin, J. (2000). Dynamic capabilities: What are they? *Strategic Management Journal, 21*(10), 1105–1121.

22. Ferreira, J., Cardim, S., & Coelho, A. (2021). Dynamic capabilities and mediating effects of innovation on the competitive advantage and firm's performance: The moderating role of organizational learning capability. *Journal of the Knowledge Economy, 12*, 620–644. https://doi.org/10.1007/s13132-020-00655-z
23. Ghazinoory, S., Sarkissian, A., Farhanchi, M., & Saghafi, F. (2020). Renewing a dysfunctional innovation ecosystem: The case of the Lalejin ceramics and pottery. *Technovation, 96*, 102122.
24. Gligor, D. M., Gligor, N., Holcomb, M. C., & Williams, L. L. (2021). Ambidexterity in supply chain management: Conceptualization, dimensions, and outcomes. *Journal of Business Logistics, 42*(4), 513–535.
25. Gomes, L. A. V., Facin, A. L. F., Salerno, M. S., & Ikenami, R. K. (2018) Unpacking the innovation ecosystem construct: Evolution gaps and trends. *Technological Forecasting and Social Change, 136*, 30-48. https://doi.org/10.1016/j.techfore.2016.11.009
26. Gopalakrishnan, S., & Damanpour, F. (1997). A review of innovation research in economics, sociology and technology management. *Omega, 25*(1), 15–28. https://doi.org/10.1016/S0305-0483(96)00043-6
27. Granstrand, O., & Holgersson, T. (2020). Innovation ecosystems: A systematic review and a conceptual framework. *Technovation, 90*, 102098. https://doi.org/10.1016/j.technovation.2019.102098
28. Huang, Y., Li, Q., Lin, X., & Deng, Z. (2022). How can the combination of entrepreneurship policies activate regional innovation capability? A comparative study of Chinese provinces based on fsQCA. *Journal of Innovation & Knowledge, 7*(3), 100227.
29. Hughes, M., Hughes, P., & Schewe, G. (2022). Organizational routines, managerial cognition, and innovation capability. *Journal of Product Innovation Management, 39*(2), 241–260.
30. Hurley, R. F., & Hult, T. M. (1998). Innovation, market orientation, and organizational learning: An integration and empirical examination. *Journal of Marketing, 62*(3), 42–54.
31. Jacobides, M. G., & Lianos, I. (2021). *Ecosystems and competition law in theory and practice*. SSRN. https://doi.org/10.2139/ssrn.3772366.
32. Khan, S. A. R., Yu, Z., Umar, M., & Ashraf, S. F. (2021). Towards green innovation capability: Evidence from emerging economies. *Business Strategy and the Environment, 30*(7), 3508–3519.
33. Koslosky, M. A. N., Speroni, R. D. M., & Gauthier, O. (2015). Ecossistemas de inovação–Uma revisão sistemática da literatura. *Revista Espacios, 36*(3).
34. Kruger, S., & Steyn, A. A. (2024). Developing breakthrough innovation capabilities in university ecosystems: A case study from South Africa. *Technological Forecasting and Social Change, 198*, 123002.
35. Kyrdoda, Y., Balzano, M., & Marzi, G. (2023). Learn to survive crises: The role of firm resilience, innovation capabilities, and environmental dynamism. *Technology in Society, 74*, 102285.
36. Lawson, B., & Samson, D. (2001). Developing innovation capability in organisations: A dynamic capabilities approach. *International Journal of Innovation Management, 5*(3), 377–400.
37. Li, Y., Chen, Y., Wang, J., Zhou, Y., & Wang, C. (2025). Digital platform capability and innovation ambidexterity: The mediating role of strategic flexibility. *Journal of Business Research, 186*(C), 114971.
38. Mendoza-Silva, A. (2021). Innovation capability: A systematic literature review. *European Journal of Innovation Management, 24*(3), 707–734. https://doi.org/10.1108/ejim-09-2019-0263
39. Miller, D., & Friesen, P. H. (1983). Strategy making and environment: The third link. *Strategic Management Journal, 4*(3), 221–235. https://doi.org/10.1002/smj.4250040306
40. Moher, D., Liberati, A., Tetzlaff, J., & Altman, D. G. (2009). Preferred reporting items for systematic reviews and meta-analyses: The PRISMA statement. *PLoS Medicine, 6*(7), e1000097. https://doi.org/10.1371/journal.pmed.1000097
41. Oberländer, A. M., Karnebogen, P., Rövekamp, P., Röglinger, M., & Leidner, D. (2025). Understanding the influence of digital ecosystems on digital transformation: The OCO (orientation,

cooperation, orchestration) theory. *Information Systems Journal, 35*(1), 368–413. https://doi.org/10.1111/isj.12539
42. Page, M. J., McKenzie, J. E., Bossuyt, P. M., Boutron, I., Hoffmann, T. C., Mulrow, C. D., Shamseer, L., Tetzlaff, J. M., Akl, E. A., Brennan, S. E., Chou, R., Glanville, J., Grimshaw, J. M., Hróbjartsson, A., Lalu, M. M., Li, T., Loder, E. W., Mayo-Wilson, E., McDonald, S., et al. (2021). The PRISMA 2020 statement: An updated guideline for reporting systematic reviews. *Systematic Reviews, 10*, 89. https://doi.org/10.1186/s13643-021-01626-4
43. Piantoni, G., Arena, M., & Azzone, G. (2023). Exploring how different innovation ecosystems create shared value: Insights from a multiple case study analysis. *European Journal of Innovation Management, 26*(7), 206–232. https://doi.org/10.1108/EJIM-09-2022-0495
44. Prahalad, C. K., & Hamel, G. (1990). The core competence of the corporation. *Harvard Business Review, 68*(3), 79–91.
45. Rialti, R., & Zollo, L. (2020). Digital transformation capability: A systematic literature review and research agenda. *Journal of Business Research, 118*, 64–75.
46. Rodrigues, A. O. A., Marques, C. S., & Ramadani, V. (2023). Artisan entrepreneurship, resilience, and sustainable development: The quintuple helix innovation model in the low-density and cross-border territories. *Journal of Enterprise Information Management.* https://doi.org/10.1108/JEIM-03-2023-0100
47. Rohrbeck, R., Hölzle, K., & Gemünden, H. G. (2009), Opening up for competitive advantage – How Deutsche Telekom creates an open innovation ecosystem. *R&D Management, 39*, 420–430. https://doi.org/10.1111/j.1467-9310.2009.00568.x
48. Rossetto, D. E., Leme, B. H. M. P., Tortorella, G. L., & Sawhney, R. (2023). Measuring frugal innovation capabilities: An initial scale proposition. *Technovation, 121*, 102674.
49. Schumpeter, J. A. (1934). *The theory of economic development: An inquiry into profits, capital, credits, interest, and the business cycle*. Transaction Publishers.
50. Sebastian, I. M., Antonelli, A. G., & Ross, J. W. (2023). Digital capabilities and the future of the firm. *MIS Quarterly, 47*(1), 1–24.
51. Taleb, M., & Pheniqi, Y. (2023). Building innovation capability in the Moroccan high-tech manufacturing industry: An intellectual capital perspective. *Journal of System and Management Sciences, 13*(1), 415–443.
52. Taneo, S. Y. M., Agustina, M., & Noya, S. (2023). The moderating influence of government policy on the innovation capability-competitiveness relationship in food SMEs. *Journal of Technology Management & Innovation, 18*(2), 6–17.
53. Teece, D. J. (1986). Profiting from technological innovation: Implications for integration, collaboration, licensing and public policy. *Research Policy, 15*(6), 285–305. https://doi.org/10.1016/0048-7333(86)90027-2
54. Teece, D. J. (1994). The dynamic capabilities of enterprises: An introduction. *Industrial and Corporate Change, 3*(3), 537–556.
55. Teece, D. J. (2014). A dynamic capabilities-based entrepreneurial theory of the multinational enterprise. *Journal of International Business Studies, 45*(1), 8–37.
56. Teece, D. J., Pisano, G., & Shuen, A. (1997). Dynamic capabilities and strategic management. *Strategic Management Journal, 18*(7), 509–533.
57. Teece, D. J., & Pisano, G. (1994). The dynamic capabilities of enterprises: An introduction. *Industrial and Corporate Change, 3*(3), 537–556.
58. Teece, D. J., Pisano, G., & Schuen, A. (1997). Dynamic capabilities and strategic management. *Strategic Management Journal, 18*(7), 509–533.
59. Teece, D. J. (2021). The dynamic capabilities framework and entrepreneurship: Strategy, technology, and organizations. *Journal of Business Venturing Insights, 15*, e00234.
60. Tidd, J., Bessant, J., & Pavitt, K. (1997). *Managing innovation: Integrating technological, market, and organizational change*. Wiley.
61. Wang, C. L., & Ahmed, P. K. (2007). Dynamic capabilities: A review and research agenda. *International Journal of Management Reviews, 9*(1), 31–51.

62. Wang, K. Y., Hermens, A., Huang, K.-I., & Chelliah, J. (2015). Entrepreneurial orientation and organizational learning on SME´s innovation. *International Journal of Organizational Innovation, 7*(2), 43–56.
63. Zarea, F., Douglas, E. J., Obschonka, M., Pletzer, L., & Schimbera, C. (2025). When the marketplace comes to the research laboratory: Technology transfer efficiency of innovation-driven publicly funded research centers. *Journal of Technology Transfer, 50*, 2556–2586. https://doi.org/10.1007/s10961-025-10188-9

Sustainability, Circular Economy and Smart Cities

Technology Parks in Unequal Territories: Evidence and a Framework for Inclusive Innovation

Luiz César de Oliveira and Luciano Tadeu Esteves Pansanato

Abstract *Background* The literature shows that science and technology parks tend to consolidate in already privileged urban centres, reinforcing spatial inequalities instead of mitigating them. *Objective* This study analyses the possibilities and limits of technology parks as instruments for regional development in unequal territories, and proposes a framework to guide their implementation in emerging regions. *Methods* A qualitative strategy combined interviews (n = 19) and on-site visits to three consolidated parks (Brazil, Portugal, Spain) with complementary documentary analysis (2018–2024). Fieldwork was conducted in 2018–2019; the 2024 mapping was used as contextual and descriptive material. *Results* The findings confirm that park consolidation depends on pre-existing structured ecosystems—strong universities and advanced infrastructure—reproducing territorial asymmetries. However, emerging evidence reveals academic, institutional and territorial strategies capable of reducing this bias. *Contribution* The article proposes a framework based on four pillars—emerging academic anchoring, multilevel cooperation, innovation infrastructure and social-territorial integration—expanding the inclusive innovation agenda applied to technology parks. *Implications* Decentralising investments and adopting inclusive models are crucial for parks to move from reproducing inequalities to acting as drivers of territorial cohesion and justice.

Keywords Technology parks · Triple helix · Regional inequality · Inclusive innovation · Decentralization policies

L. C. de Oliveira (✉) · L. T. E. Pansanato
Federal University of Technology – Paraná (UTFPR), Cornélio Procópio, Paraná, Brazil
e-mail: luizcesar@utfpr.edu.br

L. T. E. Pansanato
e-mail: luciano@utfpr.edu.br

L. C. de Oliveira
Faculty of Economics, University of Coimbra (FEUC), Coimbra, Portugal

L. Farinha et al. (eds.), *Sustainable Innovations in Mechanical Engineering and Management Sciences*, Lecture Notes on Multidisciplinary Industrial Engineering,
https://doi.org/10.1007/978-3-032-30372-1_9

1 Introduction

Technology parks have been recognized as strategic elements in regional innovation systems, playing a significant role in promoting economic and social development. More than simple instruments to boost national production or increase business profits, these parks represent environments conducive to interaction between universities, companies, and governments, fostering the transfer of knowledge and technology [1, 2].

The literature highlights that technology parks facilitate the creation of cooperation networks that promote innovation and entrepreneurship, contributing to regional competitiveness [3, 4]. These networks are characterized by dynamic interactions between the actors involved, resulting in mutual benefits and the strengthening of local social capital [5].

In the Brazilian context, it is observed that most technology parks are located in developed urban regions, close to consolidated universities, which raises questions about the effectiveness of these ventures in peripheral or less developed regions [6]. This geographical concentration can limit the potential of technology parks as instruments for equitable regional development.

The Triple Helix theory, which emphasizes collaboration between university, industry, and government, has been widely used to understand and promote regional innovation [7]. Recent studies suggest that adapting this model to include civil society (Quadruple Helix) can expand its impact, especially in contexts of sustainable development and social inclusion [8, 9]. Empirical evidence from interviews with park managers and academics, discussed later in the Results section, reinforces the importance of these interactions for strengthening local innovation networks and internationalization processes.

Despite advances in mature ecosystems, the literature still underexplores how technology parks may operate in unequal and less-favoured territories with lower scientific, institutional, and infrastructural density—a gap that remains unaddressed in a systematic manner. This article, derived from an ongoing doctoral project, aims to analyse the possibilities and limits of technology parks in peripheral contexts and to propose a framework oriented toward inclusive innovation in unequal territories. The theoretical contribution lies in shifting the debate on technology parks from an urban-efficiency axis to a territorial-equity axis, while the practical contribution consists of offering actionable guidelines for policymakers and managers seeking to reduce regional asymmetries.

The article is organised as follows: Sect. 2 revisits the literature on technology parks, territorial inequalities, and inclusive innovation; Sect. 3 presents the methodological approach; Sect. 4 synthesises the empirical results; Sect. 5 discusses the findings and introduces the proposed framework; and Sect. 6 concludes with theoretical implications, practical implications, and a future research agenda.

2 Literature Review

2.1 Fundamentals and Structuring of Technology Parks

Technology parks are inserted into innovation and entrepreneurship ecosystems, acting as managers and policymakers in the university–industry–government cooperation network. They are characterized as open systems of interaction with the environment, dynamic and constantly transforming, in which the actors involved must recognize and enhance existing networks to promote regional growth and development [7].

In this context, politics loses its hierarchical centrality, and the actors in the cooperative network simultaneously increase their independence and interdependence. The multiplicity of interests and the plurality of actors require the State to delegate functions and promote greater interaction between state agencies and social actors, aiming at articulation and cooperation. Burt emphasizes that strong networks in business opportunities are those that have structural gaps, and entrepreneurs are experts in building interpersonal bridges that cross these gaps [10, 11]. Clark [11] warns that, although innovation initiatives such as technology parks have transformative potential, they can also reinforce regional inequalities if they are not planned with an inclusive perspective and adapted to local specificities.

The structuring of technology parks is influenced by classic theories of regional development, such as Perroux's concept of the *growth pole*, which argues that economic growth manifests itself in poles of varying intensities that affect the economy as a whole through the transmission of positive effects. Myrdal introduces the idea of *circular and cumulative causation*, suggesting that market forces operate in the direction of inequality, requiring state intervention to contain these forces. Hirschman proposes that imbalances are key to economic dynamism, advocating for development plans that maintain tensions and asymmetries to mobilize scarce resources [12–14]. These theories contribute to the analysis of parks as central axes for knowledge transfer in locations that produce this knowledge, influencing local development policies implemented in various countries since the 1950s.

2.2 Innovation as Productive Connection and Social Development

Technological innovation is considered an essential tool to increase the productivity and competitiveness of organizations and to boost the economic and social development of regions and countries. Schumpeter introduced the concept of *creative destruction*, in which economic growth incessantly revolutionizes structures, destroying the old and creating the new. However, this process can discourage innovation when innovators are unable to capture all the profit from their discoveries, as imitation spreads rapidly. Easterly argues that, since the social profit from innovation exceeds individual profit, individuals tend not to innovate at a socially beneficial rate, suggesting strong incentives for research and development, technology adoption, and intellectual-property protection [15, 16].

The development of these technologies depends on public-policy goals and social options, considering the actors involved. Lahorgue [17] states that a society's innovation capacity depends on the innovation capacity of its various actors—university, company, and government—and that growth processes are intrinsic to the region, requiring alignment with local culture and micro-environments. In this context, the university is a key element directly related to learning and innovation, capable of articulating with the local culture and transferring that partnership into the technology park. Regional development is directly related to investments in universities, which often become founders of the parks. This incentive to higher education fosters growth and social inclusion by providing qualified labour to integrate the research and production processes of resident institutions [18]. Complementarily, Delgado and Murray [19] highlight that universities, especially in less developed regions, play a central role as catalysts of inclusive innovation by fostering collaboration with local actors and supporting equitable regional economic growth.

2.3 The Importance of Seminal Theories of Regional Development

The absence of social capital disqualifies a territory in the competitiveness process, and the technology park fits into this context through its role in managing the distribution of knowledge and technology for regional development. Revisiting seminal theories remains fundamental: classic contributions from Perroux, Myrdal, and Hirschman already anticipated how economic polarization and cumulative causation could shape territorial asymmetries [12–14].

The Triple Helix model, which emphasizes interaction and cooperation between university, industry, and government, is particularly relevant here. Etzkowitz and Leydesdorff argue that innovation occurs through overlapping institutional and functional spheres, generating new organizational forms [7]. However, critiques of the model indicate that it may exclude entrepreneurs or neglect broader societal dimensions. In response, scholars such as Carayannis and Campbell [8, 9] proposed the *Quadruple* and *Quintuple Helix* frameworks, integrating civil society, culture, and environmental sustainability as complementary dimensions to innovation systems. Importantly, these perspectives are understood here not as an evolution of the Triple Helix but as normative extensions that emerge from institutional and social pressures—precisely the kind of mechanisms explained by Institutional Theory, which interprets such pressures as expressions of legitimacy and adaptation [20–22].

From an institutional-economics standpoint, social norms, cultural expectations, and legitimacy constraints define the field in which organizations operate [20–22]. The rise of new helixes can therefore be read as institutional responses to contextual imperatives—environmental, civic, or communicational—without altering the core logic of university–industry–government interaction.

Recent regional-economics literature reinforces that innovation is not territorially neutral: it tends to amplify pre-existing advantages, reproducing asymmetries between dynamic centres and peripheral regions. Rodríguez-Pose demonstrates that poorly calibrated innovation policies may reinforce "geographies of discontent," where regions excluded from knowledge flows accumulate socio-economic frustration [23]. This mechanism aligns with the logic of knowledge spillovers: diffusion depends on institutional density and cognitive proximity, as shown by Audretsch and Feldman [24].

A growing body of work on *inclusive innovation*—for instance, Paunov and George et al.—argues that innovation policies must also operate in less-favoured territories, or else they risk deepening regional divergence [25, 26]. The central gap, therefore, does not lie in proving that parks thrive in consolidated centres—this is already documented—but in identifying the conditions under which they may act as instruments of territorial cohesion in unequal regions. Existing studies still lack operational guidance for implementing and governing parks outside mature ecosystems, remaining mostly at diagnostic or prescriptive levels [25–27].

This theoretical and practical void justifies the development of a framework oriented toward peripheral territories, anchored in the functional articulation of helixes and the institutional constraints that shape absorptive capacity, coordination, and legitimacy within such contexts.

3 Methodology

3.1 Research Approach

This study adopts a qualitative, exploratory and descriptive design aimed at understanding the role of technology parks in regional development. The empirical basis derives from fieldwork associated with a broader doctoral research project focused on institutional dynamics of science and technology parks; for the purpose of this article, the present analysis constitutes an autonomous analytical excerpt. Data collection was carried out in 2018–2019 through interviews and document collection, and the analytical procedures were resumed in 2023 after a pandemic-related interruption of academic activities. The article specifically examines, from the perspective of international experience and field observation in three countries, the challenges and potentials of inclusive innovation as a vector of territorial development mediated by technology parks. This excerpt presents theoretical and empirical consistency to stand as a self-contained contribution aligned with the scope of the journal.

3.2 Data Collection

Data collection involved two main sources:

- **Semi-structured interviews**: Interviews were conducted with stakeholders from technology parks, including managers, representatives of resident companies, academics, and public policy makers. The interviews were conducted based on a semi-structured script, allowing flexibility to explore emerging themes during the conversations.
- **Documentary analysis**: Institutional documents and reports published by entities such as the National Association of Innovative Ventures Promoters (ANPROTEC) and the University of Brasília (UnB) were analyzed, aiming to extract relevant information about the functioning and impact of technology parks.

Data collection involved two complementary sources. First, semi-structured interviews were conducted with stakeholders from technology parks, including managers, representatives of resident companies, academics and public policy actors. The interviews followed a semi-structured guide, ensuring comparability while allowing flexibility to explore emergent themes during the conversations. Second, documentary evidence was examined from institutional reports, public policy documents and official repositories from entities such as ANPROTEC and the University of Brasília, with the purpose of contextualising governance arrangements and institutional positioning of parks as instruments of regional development.

The empirical sample comprised 19 participants, including park managers, institutional representatives and external actors connected to innovation ecosystems. All interviews were conducted between 2018 and 2019 and form the primary dataset of this research, while documentary updates until 2024 were used exclusively to contextualise the empirical material without altering the original corpus. All participants signed informed consent forms and confidentiality is strictly preserved, ensuring compliance with ethical principles applicable to research in the social sciences.

Institutional documents from parks and universities, policy reports and official online sources were also examined to complement missing information and to support contextual interpretation. These sources fulfilled a descriptive function and were not coded in MAXQDA. Table 3, presented later in the article, synthesises the fieldwork effort (conducting, transcribing and analysing interviews) and is referenced here exclusively as a component of the research design.

The qualitative corpus was coded thematically using MAXQDA software. To ensure analytical reliability, a double-coding procedure was applied: the corpus was initially coded by the main researcher and independently validated by a second experienced researcher, converging by consensus. This procedure reduced interpretive bias and increased the consistency of evidence categorisation.

3.3 Ethical Procedures

The research followed ethical principles applicable to studies in the social sciences. All participants were informed about the objectives of the study and voluntarily consented to take part in the fieldwork. Anonymity and confidentiality were guaranteed throughout all stages of data handling. Although the sample consisted primarily of managers, academics and public sector representatives—actors who may express favourable positions towards technology parks—this characteristic is acknowledged as an inherent feature of the field rather than as a methodological flaw. Future studies could incorporate critical or peripheral perspectives, such as companies that exited the parks or local communities not integrated into innovation networks, to expand the spectrum of perceptions.

3.4 Data Analysis

Data analysis was conducted using content analysis as the central analytical strategy. This technique enables systematic categorisation of qualitative material and facilitates the identification of patterns, recurrences and analytical convergences relevant to the research objectives. The dataset was organised and coded using MAXQDA 12, which provided structured support for category creation, coding of textual segments and visualisation of relationships between codes. The analytical process followed the stages of preparation, organisation and reporting, comprising careful reading of transcripts and documents, construction of thematic categories through inductive–deductive logic, and interpretation of findings in articulation with the theoretical framework and research questions. Criteria of authenticity, credibility, representativeness and significance guided the analytical decisions to ensure methodological transparency and interpretive robustness.

3.5 Methodological Limitations

This study presents limitations inherent to qualitative research, such as the interpretive nature of the analysis and the non-generalizability of results to other contexts. These constraints are compensated by the depth and richness of the empirical material, which allow for analytical insight into the phenomenon under investigation. Additionally, although the primary data were collected in 2018–2019, the completion of the analytical procedures was postponed due to the interruption of academic activities during the COVID-19 pandemic, with the analysis resumed in 2023. Thus, while the empirical evidence reflects a pre-pandemic scenario, the findings remain theoretically and analytically relevant, particularly considering the structural transformations imposed by the pandemic on innovation ecosystems and on the governance of technology parks. This clarification is provided in accordance with transparency and integrity practices in scientific research.

4 From Theory to Practice: The Convergence Between Regional Development and the Triple Helix in the Park

The spatial localization of technology parks is not a neutral decision, but the outcome of structural forces associated with urban hierarchy, agglomeration economies, and the unequal spatial distribution of scientific and institutional density. Medium-sized cities have gained strategic relevance in regional systems of innovation, assuming functions previously concentrated in metropolitan cores and enabling new territorial arrangements for knowledge production and firm formation. These dynamics align with the literature that associates regional performance with the combination of human capital accumulation, institutional thickness and spatial frictions in mobility of knowledge and resources [3, 4, 12].

In this context, the Triple Helix configuration functions not only as a cooperation heuristic but as a selective mechanism that filters which territories achieve capacity for innovation-led development and which remain structurally peripheral. The literature shows that the intensity of collaboration and knowledge transfer inside innovation infrastructures tends to exceed that observed outside them, but this effect is contingent on governance, incentives and the presence of strong anchor institutions [5, 7, 24]. Government action is not auxiliary: it operates as a structural precondition that enhances or inhibits the competitiveness of firms through infrastructure, regulation and public financing instruments [7, 18, 19].

However, the mere co-location of actors is not sufficient to guarantee developmental outcomes. Rodríguez-Pose demonstrates that innovation policies poorly calibrated to territorial conditions may reinforce "geographies of discontent," amplifying pre-existing structural disadvantages rather than correcting them [23]. Likewise, Audretsch and Feldman show that knowledge spillovers are spatially and institutionally selective, benefitting territories with dense cognitive and organizational assets [24]. In parallel, Carayannis and Campbell argue that additional social and environmental helixes emerge as institutional pressures for legitimacy, but do not cancel the central functional role of university–industry–government articulation as the core driver of systemic change [8, 9, 22].

Thus, the convergence between regional development and the Triple Helix in technology parks does not follow automatically from their installation, but from the alignment between urban hierarchy, institutional capacity and state action capable of inducing coordination, mitigating structural asymmetries and sustaining inclusive innovation trajectories. When these conditions are not present, parks risk reproducing the very inequalities they are expected to mitigate, operating as enclaves of efficiency instead of instruments of territorial cohesion [12, 14, 23].

Table 1 Socioeconomic profile of the cities hosting the central case study parks

City	Country	Inhabitants	EUR/BRL	MHDI
Porto Alegre	Brazil	1.5 million	42,000	0.805
Porto	Portugal	214 thousand	35,000	0.841
Valencia	Spain	800 thousand	30,000	0.890

Source Data adapted from national statistics institutes of each country

5 Results

5.1 Anchoring in Knowledge Density and Pre-existing Academic Capital

The empirical evidence indicates that technology parks do not emerge as instruments to create academic density, but rather as structures that capitalize on territories where this density is already consolidated. As stated by one interviewee, "a park is born where there is already a strong university; it does not come to build that" (E7). Another respondent reinforced this logic: "without research density, a park becomes an empty building" (E12). This pattern was consistently observed in TECNOPUC, UPTEC and PCUV, all embedded in cities that already concentrate research institutions, postgraduate programs and a diversified productive base.

This finding is aligned with the literature showing that knowledge spillovers and innovation outputs are highly dependent on institutional concentration and cognitive proximity, and that innovation policies implemented only in consolidated regions tend to amplify cumulative advantages and territorial asymmetries. In other words, parks are not drivers of academic density but amplifiers of pre-existing knowledge ecosystems, which explains their prevalence and higher performance in large urban centers with mature scientific and technological infrastructures, which explains their prevalence and higher performance in large urban centers with mature scientific and technological infrastructures [23, 24].

Table 1 presents the hierarchy of the cities where the parks are located, considering the number of inhabitants, GDP per capita, and MHDI.

5.2 Multilevel Cooperation and Actor Orchestration

Evidence shows that parks operate as orchestrators of cooperation, connecting firms to academic competences and public instruments, albeit with variation in intensity and territorial spillovers. As one manager put it, "the park speeds up connections; it does not fund, but opens doors" (E8). Another respondent noted that some firms "seek physical space and security, but the differential is access to competencies and talent" (E11). These findings are consistent with the literature on regional innovation systems, in which **collaborative governance** and **organised proximity** are key to knowledge circulation. However, cooperation tends to remain within the park's perimeter and among the most capitalised actors, with limited reach over peripheral territories, which helps explain why dense regions capture more benefits, which helps explain why dense regions capture more benefits [28, 29].

These infrastructures increase the possibilities of interaction and cooperation, promoting the transfer of knowledge and technology for the creation of new products and innovative businesses, benefiting the regional ecosystem through the formation of qualified labor and better income distribution.

5.3 Innovation Infrastructure and Services as a Risk-Reduction Mechanism

Across the three cases, laboratory facilities, workspaces, security and specialised services (IP, technology transfer, incubation, acceleration, internationalisation) operate as **risk-reduction mechanisms** for knowledge-intensive firms. As one respondent summarised, "we moved to the park because of the services and the bridges to the university" (E2). Another emphasised the market angle: "the park shortens the path to competencies and anchor clients" (E9). These operational elements explain the stronger attractiveness of parks in dense regions where scientific infrastructure and human capital are readily available, consistent with density-dependent spillovers consistent, with density-dependent spillovers [24]. The characteristics of the three case-study technology parks are summarized in Table 2.

Table 2 Summary of studied technology parks

Park	City	Country	Year of foundation	Associated university
TECNOPUC	Porto Alegre	Brazil	2003	PUCRS
UPTEC	Porto	Portugal	2007	University of Porto
PCUV	Valencia	Spain	2009	University of Valencia

Source Data provided by the parks themselves

Table 3 Summary of field research

Activity	Quantity
Interviews conducted	19
Average time per interview	46 min
Average transcription time	231 min
Average analysis time	1270 min

Source Research data

Table 4 Technology parks in operation in Brazil (2019)

City	Inhabitants	GDP per capita (R$)	MHDI	Associated university
São Paulo	12.2 million	55,000	0.805	USP
Campinas	1.2 million	45,000	0.805	UNICAMP
Florianópolis	500 thousand	40,000	0.847	UFSC
São José dos Campos	700 thousand	42,000	0.807	ITA
Campo Bom	65 thousand	28,000	0.835	Feevale

Source Data adapted from IBGE and associated universities

The field research involved technical visits and semi-structured interviews with 19 experts representing the parks, partner governments, universities, and resident companies. The interviews were transcribed and analyzed using MAXQDA software, allowing the categorization of 409 references into four main categories and 32 subcategories.

The analysis revealed that technology parks function as entrepreneurial and innovative ecosystems, connecting institutions capable of providing specialized and technological resources. This strategic configuration allows companies to reduce costs and risks in the innovation process, becoming more globally competitive. Interviews highlight the importance of collaboration between universities and companies:

- EN02: "The parks share the vision, objectives, values, this issue of interaction between academia, knowledge, universities, government, companies, and society."
- EN09: "Whenever a company looks for the park, we inform them what the university has to offer. And vice-versa, when the university identifies a company that could benefit from the park, we also alert the park to get in touch."

Furthermore, parks are seen by the government as promoters of regional economic and social development, generating jobs and income, and creating an environment conducive to learning and innovation. National and international associations, such as ANPROTEC, APTE, TECPARQUES, and IASP, play an important role in promoting these ecosystems and in formulating public policies favorable to entrepreneurship and innovation.

Table 4 presents an analysis of the parks operating in Brazil in 2019, highlighting characteristics of the regions where they are installed.

In addition to HDI and GDP per capita indicators, it would be relevant to include direct economic impact metrics, such as the number of jobs created, the average revenue of resident companies, and the survival rate of startups after 5 years. ANPROTEC (2023) data indicate that parks in metropolitan regions, such as São Paulo and Campinas, generate on average 30% more high-skilled jobs than parks in smaller cities. This disparity reinforces the need for specific policies to reduce regional asymmetries.

Table 5 Consolidated—technology parks (Brazil, Portugal, and Spain)

Technology park	Partner university	Year of foundation	City HDI (approx.)	City population (approx.)	Predominant economic sector
"TECNOPUC (Porto Alegre, BR)"	PUCRS	2003	0.805	1,500,000	Technology and services
"UPTEC (Porto, PT—metropolitan)"	University of Porto	2007	0.841	2,000,000	Technology and creative industries
"PCUV (Valencia, ES)"	University of Valencia	2009	0.893	2,400,000	Life sciences and ICT
São José dos Campos Technology Park	ITA/ UNIFESP	1994	0.807	720,000	Aerospace and Defense
Porto digital (Recife)	UFPE	2000	0.772	1,600,000	Information technology
Belo Horizonte Technology Park (BH-TEC)	UFMG	2006	0.81	2,700,000	Biotechnology and ICT
Itaipu Technology Park (PTI—Foz Do Iguaçu)	UNILA/ Unioeste	2003	0.755	260,000	Energy and engineering
PCT Guamá (Belém)	UFPA/UFRA	2002	0.751	1,500,000	Technology and education
Sapiens Park (Florianópolis)	UFSC	2003	0.847	508,000	ICT and research
São Carlos Technology Park	USP/UFSCar	2001	0.805	250,000	Engineering and ICT
Unicamp Technology Park (Campinas)	Unicamp	2003	0.805	1,200,000	Education and innovation
UFRJ Technology Park (Rio de Janeiro)	UFRJ	2003	0.799	6,700,000	Oil and gas
Parque Tecnológico de Sorocaba	UNESP	2010	0.8	680,000	Manufacturing and ICT
Parque Tecnológico de Londrina	UEL	2005	0.789	575,000	ICT and services
Parque Tecnológico de Brasília (BioTIC)	UnB	2011	0.824	3,000,000	Public services and government
Parque Tecnológico de Joinville	UDESC	2013	0.809	600,000	Manufacturing and ICT
Taguspark (Oeiras)	Universidade de Lisboa	1992	0.9	175,000	ICT and telecommunications
UPTEC (Porto)	Universidade do Porto	2007	0.841	2,000,000	Technology and creative industries

(continued)

Table 5 (continued)

Technology park	Partner university	Year of foundation	City HDI (approx.)	City population (approx.)	Predominant economic sector
Biocant Park (Cantanhede)	Universidade de Coimbra / IPC	2005	0.86	37,000	Biotechnology
Madan Parque (Almada)	Universidade Nova de Lisboa	2000	0.9	175,000	ICT and renewable energies
Avepark (Guimarães)	Universidade do Minho	2004	0.86	157,000	Engineering and ICT
Tecmaia (Maia)	Universidade do Porto	2002	0.87	135,000	Engineering and ICT
Parque de Ciência e Inovação PCI (Aveiro)	Universidade de Aveiro	2008	0.86	78,000	ICT and Eletrônica
Madeira Tecnopolo (Funchal)	Universidade da Madeira	2004	0.87	111,000	Tourism and ICT
Sines Tecnopolo (Sines)	Universidades do Algarve, Évora, IPBeja e IPS	2007	0.86	56,000	Renewable energy
Óbidos Parque (Óbidos)	Diversas Instituições de Ensino superior	2009	0.85	11,000	Design and ICT
Parque Científico de Barcelona (PCB)	Universidade de Barcelona	1997	0.9	1,600,000	Life sciences
Ciudad Politécnica de la Innovación (CPI - Valência)	Universidade Politécnica de Valência	2000	0.893	800,000	Engineering and ICT
Parque Científico da Universidade de Valência (PCUV)	Universidade de Valência	2009	0.893	2,400,000	Life sciences
Parque Tecnológico de Andaluzia (PTA - Málaga)	Universidade de Málaga	1992	0.803	580,000	ICT and renewable energy
Parque Científico de Alicante (PCA)	Universidade de Alicante	2005	0.87	330,000	Life sciences
Parque Científico de Madri (PCM)	Universidade Autônoma de Madri	2001	0.9	3,200,000	Research and ICT
Parque Científico de Salamanca (PCUSAL)	Universidade de Salamanca	2005	0.85	144,000	Life sciences

(continued)

Table 5 (continued)

Technology park	Partner university	Year of foundation	City HDI (approx.)	City population (approx.)	Predominant economic sector
Parque Tecnológico de La Rioja (PTR)	Universidade de La Rioja	1999	0.87	152,000	Life sciences
Parque Científico de Múrcia (PCMUR)	Universidade de Múrcia	2001	0.84	453,000	Life sciences
INtech Tenerife	Universidade de La Laguna	2006	0.83	166,000	Technology and tourism

Source prepared by the author

It is observed that most parks are located in large urban centers or regional hubs with a high concentration of higher education and research institutions. This concentration favors the formation of qualified labor and the transfer of knowledge and technology. The spatial analysis of the implementation of technology parks in Brazil, Portugal, and Spain reveals significant patterns that corroborate the trend of concentration of these initiatives in large urban centers and regional hubs. Studies indicate that the geographical proximity between universities and companies is a determining factor for the location of technology parks, favoring the formation of cooperation networks and the transfer of knowledge [30]. In addition, the presence of consolidated infrastructure and qualified labor in these regions enhances the success and consolidation of the parks [28].

As discussed, the empirical data and analyzed maps reveal a clear trend in the location of technology parks in consolidated urban centers and regional hubs. To visually synthesize this concentrating logic, Fig. 1 presents the distribution of parks in relation to the most developed areas, reinforcing the association between pre-existing infrastructure, university presence, and the consolidation of innovation ecosystems. As shown in the figure below:

As observed in Fig. 1, the concentration of parks in central and developed regions reinforces the need for public policies that encourage decentralization and the creation of parks in emerging locations, capable of stimulating inclusive regional development.

Next, Maps 2, 3 and 4 illustrate the geographical distribution of technology parks in the three countries studied, highlighting the correlation between their location and the presence of higher education institutions and research centers. This visual representation reinforces the understanding that the consolidation and performance of technology parks are intimately linked to pre-existing territorial and institutional factors. The following are the Maps and Technology Parks—Brazil, Portugal, and Spain (2024 base):

The joint analysis of the maps shows that technology parks in Brazil, Portugal, and Spain are predominantly located in regions with a high concentration of universities and research centers. This geographical concentration aligns with the Triple

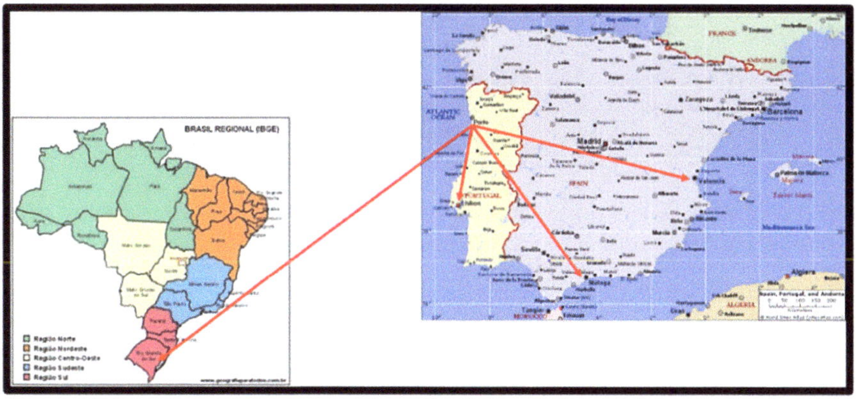

Fig. 1 Field research scenario—interviews and observations. *Source* prepared by the author

Helix model, which emphasizes the interaction between university, company, and government as a driver of regional innovation [7].

However, this concentrating logic can exacerbate regional inequalities by limiting the access of peripheral regions to the benefits provided by technology parks. In this sense, it is fundamental to rethink the implementation strategies of these parks, considering the need to promote decentralization and the development of less favored areas [31]. Therefore, the data presented not only illustrate the current situation but also point to the importance of public policies that encourage the creation of technology parks in emerging regions, aiming to reduce regional inequalities and promote more inclusive and territorially distributed innovation.

5.4 Social-Territorial Integration and Limits to Spillovers

The comparative analysis indicates that when embedded in consolidated centres, parks **reinforce local trajectories** rather than induce new ones in peripheral territories. The geographical distribution shown in Figs. 2, 3 and 4 reveals that Brazil, Portugal and Spain concentrate parks in areas with high university and productive density, which favours immediate performance but **limits spillovers** into low-density regions. This pattern aligns with the argument on **cumulative territorial inequalities** (Rodríguez-Pose [23]) and with the literature on "learning spaces" and territorial learning conditions, limits spillovers into low-density regions [23, 31].

In addition to the privileged location, the Consolidated Table of Technology Parks (Consolidated Table 5) presents relevant data that strengthen the comparative analysis. The data on the year of foundation, HDI, and predominant economic sectors of the host cities reinforce the idea that the creation of parks is not random, but strategic and, frequently, linked to universities of excellence. This connection between universities and parks aligns with the Triple Helix model [1], in which cooperation between

Fig. 2 Location of the 15 main technology parks in Brazil and their partner universities. *Source* prepared by the author based on publicly available data from MEC, ANPROTEC and MCTI (2024). (1) São José dos Campos (ITA/UNIFESP), (2)Recife (Porto Digital/UFPE), (3)Porto Alegre (TECNOPUC/PUCRS), (4) Curitiba (PUCPR/Parque Científico), (5) Foz do Iguaçu (PTI/ UNILA/ Unioeste), (6) Belém (PCT Guamá/ UFPA/UFRA), (7) Florianópolis (Sapiens Parque/UFSC), (8) Belo Horizonte (BH-TEC/UFMG), (9) São Carlos (USP/UFSCar), (10) Campinas (Parque Unicamp), (11) Rio de Janeiro (Parque UFRJ), (12) Sorocaba (Parque UNESP), (13) Londrina (Parque UEL), (14) Brasília (BioTIC/UnB), (15) Joinville (Parque UDESC)

university, company, and government is a pillar for regional and global economic development. Consolidated Table 5 follows:

Taken together, the data reinforce that the presence of top universities and specialised services is a **necessary** condition for park consolidation, but **insufficient** to generate broad territorial inclusion without specific policies for diffusion.

It is worth noting that, by expanding the sample beyond the three central cases, recurring patterns are observed. In Brazil, for example, the strong presence in cities such as São José dos Campos, Campinas, and Florianópolis reveals the search for regions already recognized as technological and industrial hubs. In Portugal, parks such as Taguspark and Biocant Park follow the same logic, locating in areas with a strong scientific and academic tradition. In Spain, the network of parks reflects

Fig. 3 Location of the 10 main technology parks in Portugal and their partner universities. *Source* prepared by the author based on publicly available data from TECPARQUES and Portuguese education statistics (2024). (1) Taguspark (Oeiras)—University of Lisbon (Instituto Superior Técnico), (2) UPTEC (Porto)—University of Porto, (3) Biocant Park (Cantanhede)—University of Coimbra and Polytechnic Institute of Coimbra, (4) Madan Parque (Almada)—NOVA University Lisbon, (5) Avepark (Guimarães)—University of Minho, (6) Tecmaia (Maia)—University of Porto, (7) Science and Innovation Park—PCI (Aveiro)—University of Aveiro, (8) Madeira Tecnopolo (Funchal)—University of Madeira, (9) Sines Tecnopolo (Sines)—Universities of Algarve, Évora, IPBeja, and IPS, (10) Óbidos Science and Technology Park (Óbidos)—Various Higher Education Institutions

Fig. 4 Location of the main technology parks in Spain and their partner universities. *Source* prepared by the author based on publicly available data from APTE (2024). (1) Barcelona Science Park (PCB)—University of Barcelona, (2) Polytechnic City of Innovation (CPI)—Polytechnic University of Valencia, (3) University of Valencia Science Park (PCUV)—University of Valencia, (4) Andalusia Technology Park (PTA)—University of Málaga, (5) Alicante Science Park (PCA)—University of Alicante, (6) Madrid Science Park (PCM)—Autonomous University of Madrid, (7) Salamanca Science Park (PCUSAL)—University of Salamanca, (8) La Rioja Technology Park (PTR)—University of La Rioja, (9) Murcia Science Park (PCMUR)—University of Murcia, (10) INtech Tenerife—University of La Laguna

the diversity of the economic fabric and the decentralization of innovation hubs to regions such as Valencia, Málaga, and Tenerife.

Therefore, by integrating the qualitative data from the interviews with the empirical and geographical evidence presented in the tables and maps, the study advances the understanding of the importance of parks as instruments of regional development. They not only reinforce the role of universities as agents of the innovation ecosystem, but also create environments conducive to institutional cooperation and the emergence of new local and global economic dynamics.

However, the comparative analysis highlights a critical point: the most successful technology parks tend to emerge in already privileged locations, where consolidated universities and innovative environments pre-exist. As Shearmur and Doloreux [30] state, "the proximity of robust academic institutions is one of the most determining factors for park performance" (p. 120). This finding reinforces the idea that such ventures end up reproducing and enhancing existing regional inequalities, instead of combating them.

Thus, for the Triple Helix model to fully fulfill its role as a driver of balanced regional development, it is necessary to rethink the policies for installing technology parks. As Audretsch and Feldman [24] argue, true innovation capable of generating social transformations occurs when there is "diffusion of knowledge in peripheral regions and not just concentration in traditional centers of economic and academic power" (p. 56).

In this sense, decentralization policies and incentives for the creation of parks in less favored areas can promote the inclusion of peripheral regions in the global innovation ecosystem. As Vallance, Tewdwr-Jones, and McGuinness [31] conclude, "only by integrating low-density regions with less academic tradition will it be possible to consolidate a fairer and more sustainable national innovation system" (p. 214). In short, the expanded analysis of the parks reinforces that, although they are recognized as vectors of development, there is still a long way to go for them to effectively contribute to the reduction of regional inequalities and to the promotion of truly inclusive and collaborative development, in line with the principles of the Triple Helix.

5.5 Discussion and Proposition of a Framework for the Decentralization of Technology Parks

Findings suggest that the current siting logic of parks **maximises urban efficiency** while minimising **territorial equity**, reproducing cumulative trajectories (Rodríguez-Pose [23]) and density-dependent spillovers (Audretsch and Feldman [24]). This does not imply abandoning the parks' agenda, but rather reframing it: policies and governance must create **conditions of possibility** for effective action in unequal territories, otherwise the concentration bias will persist.

However, for the innovation ecosystem to fulfill its function as an inducer of inclusive socioeconomic development—a principle that underpins the concept of the Triple Helix [1]—it is necessary to overcome the logic of reproducing regional inequalities. The decentralization of park implementation policy is, therefore, not only a desirable but an urgent strategy. For Vallance, Tewdwr-Jones, and McGuinness [31], the transformation of the territory through innovation requires "the activation of peripheries and the integration of regions that have historically been at the margin of the national innovation system" (p. 214).

Based on the empirical analysis and the theoretical foundation discussed, a framework is proposed that synthesizes the main elements necessary to guide the implementation of technology parks in emerging regions. Figure 5 illustrates this proposal, organized into four interdependent pillars that seek to align innovation, territory, and inclusion.

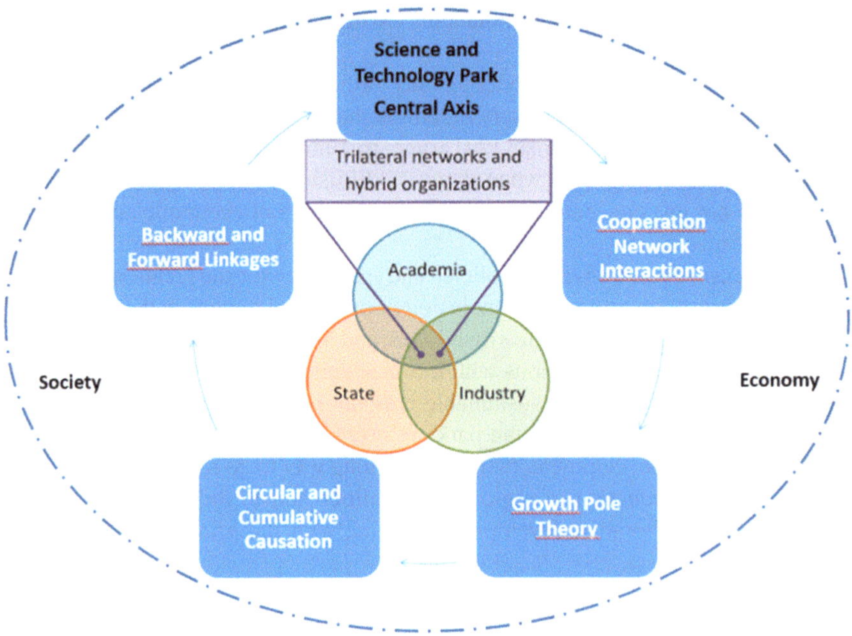

Fig. 5 Analysis of parks as a "central axis" of interactions in a knowledge transfer and cooperation network. *Source* prepared by the author, based on the Triple Helix conceptual model proposed by Etzkowitz and Leydesdorff [7]

The framework presented in Fig. 5 reinforces the need for integrated policies that simultaneously consider academic articulation, institutional support, infrastructure, and social inclusion, creating conditions for technology parks to effectively contribute to the reduction of territorial inequalities.

Framework Oriented to Inclusive Innovation in Unequal Territories
Building on the findings (Sects. 5.1–5.4) and literature (Cooke [28]; Audretsch & Feldman [24]; Vallance et al. [31]), a four-pillar framework is proposed: (i) **Emerging academic anchoring** (design of minimum capacities in applied research and training aligned with territorial vocation); (ii) **Multilevel cooperation** (governance mechanisms integrating local-state-national instruments and transnational networks); (iii) **Innovation infrastructure and services** (integrated offer of labs, IP, TT, incubation, acceleration and market access); (iv) **Social-territorial integration** (capacity-building and inclusion of local providers/users). Each pillar addresses empirical barriers observed in this study and is directly linked to the reported findings. Figure 6 illustrates the proposed framework, which articulates the four guiding pillars for the implementation of technology parks in emerging regions, aligned with the Triple Helix logic and aimed at promoting inclusive and sustainable regional development. The figure follows.

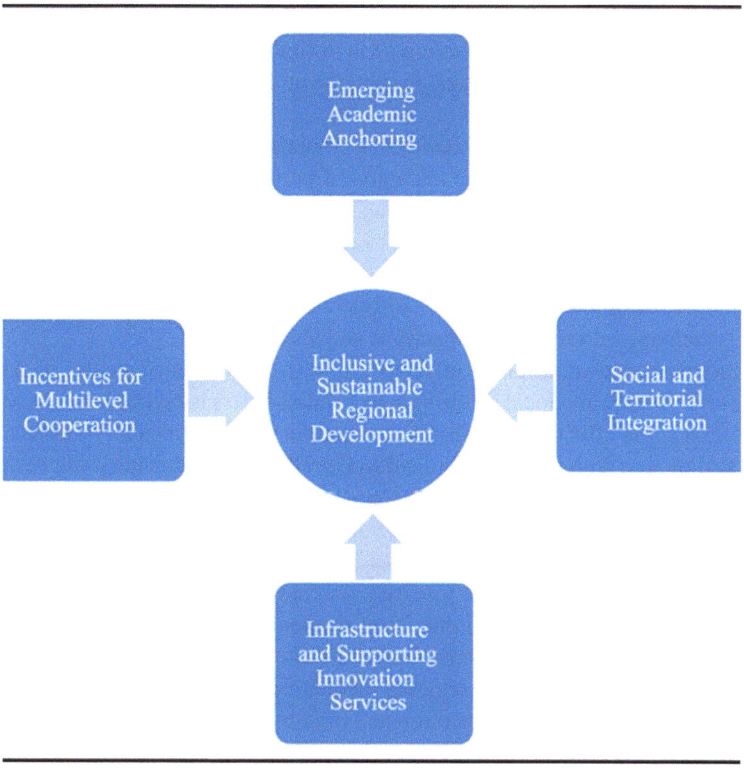

Fig. 6 Framework for the implementation of technology parks in emerging regions. *Source* Prepared by the author

This model thus seeks to break with the trend of concentrating innovation investments and build the foundations for a more equitable and transformative public policy for technology parks, contributing so that less favored regions also become protagonists in the knowledge economy.

5.6 Implications for Research and Policy

Academic implications: This study reframes the parks' debate from **urban efficiency** to **territorial equity**, bridging Triple Helix with regional inequality literature, and offering **operational guidance** (framework) for low-density contexts.

The results obtained in this study and synthesized in the theoretical and empirical analysis presented so far have relevant implications both for the formulation of public policies and for the conduct of future research on technology parks and regional innovation. Firstly, the finding that the most successful parks tend to be located in already privileged regions, with consolidated universities and pre-existing innovation ecosystems, reinforces the need for a shift in public strategies. As Shearmur and Doloreux [30] argue, the concentration of investments in already developed centers can exacerbate regional inequalities, moving away from the transformative purpose of technology parks.

From the perspective of public policy, it is imperative to consider mechanisms that stimulate decentralization and the inclusion of peripheral regions in the innovation map. As Vallance, Tewdwr-Jones, and McGuinness [31] propose, policies aimed at activating these regions are fundamental to ensure that innovation plays a social role beyond the economic. In this sense, the creation of national and regional programs that combine subsidies, tax incentives, infrastructure investments, and the promotion of local academic and research capacity is suggested.

Furthermore, the importance of redesigning the governance model of parks to include civil society and local actors as active stakeholders is highlighted. As Morgan [32] observes, the effectiveness of innovation policies is intrinsically linked to their ability to dialogue with the social and cultural realities of the territory where they are implemented. Thus, the articulation between governments, universities, and companies must be complemented by an approach that values social capital and promotes territorial cohesion.

Regarding the implications for research, the findings of this study suggest several opportunities for theoretical and empirical deepening. Firstly, quantitative investigations are recommended to robustly measure the impact of technology parks on the reduction or reproduction of regional inequalities. Longitudinal studies can also contribute to understanding the medium and long-term dynamics associated with the implementation of parks in different territorial contexts.

Additionally, the international comparative analysis proved to be a fruitful approach that deserves to be expanded in future research. The examination of cases in countries with different levels of development and institutional models can illuminate paths and alternative strategies for the management and governance of parks in emerging regions. As Audretsch and Feldman [24] suggest, innovation is a contextual process, and its trajectories are influenced by the specificities of regional and national systems.

In summary, this study reinforces the importance of technology parks as instruments of public policy for regional development but warns of the risks of their reproduction in ways that accentuate the concentration of resources and opportunities. Overcoming this challenge requires both a critical review of support policies and the advancement of academic research towards more inclusive models aligned with the objectives of sustainable territorial development.

Practical implications; For parks to operate as instruments of territorial cohesion, funding should be tied to **diffusion targets** (extension programmes, innovative public procurement, local supplier networks) and **social integration** (training and inclusion of local workers and entrepreneurs).

6 Conclusion

This study demonstrates that technology parks, while lauded as drivers of innovation, often function as amplifiers of pre-existing regional advantages rather than as instruments of territorial cohesion. Our international comparative analysis of parks in Brazil, Portugal, and Spain reveals a consistent pattern of concentration in large urban hubs endowed with robust universities, high HDI, and dense innovation ecosystems. This concentrating logic, a transnational phenomenon, risks reinforcing the very regional inequalities that innovation policy aims to overcome.

The findings necessitate a critical reframing of the technology park model from a paradigm of **urban efficiency** to one of **territorial equity**. In response, this article's primary contribution is a four-pillar framework for implementing parks in emerging regions, translating diagnostic insights into operational guidance. The pillars—**Emerging Academic Anchoring, Multilevel Cooperation, Innovation Infrastructure, and Social-Territorial Integration**—provide a strategic blueprint for leveraging the Triple Helix to foster inclusive development beyond consolidated centres.

Furthermore, we argue that the principles of inclusive innovation are best served not by adding helixes, but by deepening the social and territorial embeddedness of the existing three. This involves actively integrating civil society and local communities into the innovation process, ensuring that park-led development aligns with local needs and builds social legitimacy, as evidenced by initiatives in parks like Foz do Iguaçu.

Finally, this research opens avenues for future inquiry, particularly in quantifying the impact of decentralised parks, analysing the role of social capital, and evaluating the competitiveness of resident companies in peripheral contexts. By bridging Triple Helix theory with the literature on regional inequality, this study provides both a critical lens and a practical tool for policymakers and managers committed to harnessing innovation for a more equitable and inclusive territorial development.

This article is part of a broader doctoral research project on knowledge and technology transfer in technology parks in Brazil, Portugal and Spain, conducted at the Faculty of Economics of the University of Coimbra, from which this excerpt derives as an autonomous and publishable contribution.

Acknowledgements/Funding The author acknowledges financial support from the Coordination for the Improvement of Higher Education Personnel (CAPES), Brazil, (CAPES)—Finance Code 9556-13-3, under the PhD scholarship programme associated with the University of Coimbra.

References

1. Etzkowitz, H., & Klofsten, M. (2005). The innovating region: Toward a theory of knowledge-based regional development. *R&D Management, 35*(3), 243–255. https://doi.org/10.1111/j.1467-9310.2005.00387.x
2. Amaral, M., Ferreira, J. J., & Ratten, V. (2023). Technology parks and entrepreneurship: A systematic literature review. *The Journal of Technology Transfer, 48*(2), 567–589. https://doi.org/10.1007/s10961-022-09955-6
3. Cadorin, E., Johansson, M., & Klofsten, M. (2017). Science parks and talent attraction: Evidence from Swedish parks. *The Journal of Technology Transfer, 42*(2), 263–285.
4. Kolehmainen, J., Irvine, J., Stewart, L., Karacsonyi, Z., Szabó, T., Alarinta, J., & Norberg, A. (2016). Quadruple helix, innovation and knowledge-based development: Lessons from remote, rural and less-favoured regions. *Journal of the Knowledge Economy, 7*(1), 23–42. https://doi.org/10.1007/s13132-015-0289-9
5. McAdam, M., Miller, K., & McAdam, R. (2012). The development of university technology transfer stakeholder relationships at a regional level: The role of the technology transfer office. *Entrepreneurship and Regional Development, 24*(3–4), 301–324. https://doi.org/10.1080/08985626.2011.617823
6. Ratinho, T., & Henriques, E. (2010). The role of science parks and business incubators in converging countries: Evidence from Portugal. *Technovation, 30*(4), 278–290. https://doi.org/10.1016/j.technovation.2009.09.002
7. Etzkowitz, H., & Leydesdorff, L. (2000). The dynamics of innovation: From National Systems and "mode 2" to a triple helix of university–industry–government relations. *Research Policy, 29*(2), 109–123. https://doi.org/10.1016/S0048-7333(99)00055-4
8. Carayannis, E. G., & Campbell, D. F. J. (2009). 'Mode 3' and 'quadruple helix': Toward a 21st century fractal innovation ecosystem. *International Journal of Technology Management, 46*(3–4), 201–234. https://doi.org/10.1504/IJTM.2009.023374
9. Höglund, L., & Linton, G. (2018). Smart specialization in regional innovation systems: The role of science parks. *European Planning Studies, 26*(12), 2319–2334. https://doi.org/10.1080/09654313.2018.1530156
10. Burt, R. S. (2000). The network structure of social capital. *Research in Organizational Behavior, 22*, 345–423.
11. Clark, J. (2020). *Uneven innovation: The work of smart cities*. Columbia University Press.
12. Perroux, F. (1977). O conceito de polos de crescimento. In J. Schwartzman (Ed.), *Organizer Economia regional: textos escolhidos* (pp. 35–52). CEDEPLAR.
13. Myrdal, G. (1957). *Economic theory and under-developed regions*. Duckworth.
14. Hirschman, A. O. (1958). *The strategy of economic development*. Yale University Press.
15. Schumpeter, J. A. (1961). *Capitalismo, socialismo e democracia*. Fundo de Cultura.
16. Easterly, W. (2001). *The elusive quest for growth: Economists' adventures and misadventures in the tropics*. MIT Press.
17. Lahorgue, M. S. (2004). Tecnologia e desenvolvimento regional: a experiência da Universidade Federal do Rio Grande do Sul. *Revista de Administracao Publica, 38*(1), 151–171.
18. Brasil. Lei n. 13.005, de 25 de junho de 2014. Aprova o Plano Nacional de Educação (PNE) 2014–2024. Diário Oficial da União. 2014 Jun 26.
19. Delgado, M., & Murray, F. (2023). Universities and inclusive innovation for regional growth. *Issues in Science and Technology, 39*(4), 34–39. Available from: https://issues.org/universities-inclusive-innovation-regional-growth-delgado-murray/
20. Bardin, L. (2011). *Análise de conteúdo* (Vol. 70). Edições.
21. Kuckartz, U. (2014). *Qualitative text analysis: A guide to methods, practice and using software*. Sage.
22. Elo, S., & Kyngäs, H. (2008). The qualitative content analysis process. *Journal of Advanced Nursing, 62*(1), 107–115. https://doi.org/10.1111/j.1365-2648.2007.04569.x

23. Rodríguez-Pose, A. (2018). The revenge of the places that don't matter (and what to do about it). *Cambridge Journal of Regions, Economy and Society, 11*(1), 189–209. https://doi.org/10.1093/cjres/rsx024
24. Audretsch, D. B., & Feldman, M. P. (2004). Knowledge spillovers and the geography of innovation. In J. V. Henderson & J. F. Thisse (Eds.), *Handbook of regional and urban economics* (Vol. 4, pp. 2713–2739). Elsevier.
25. Rodríguez-Pose, A. (2013). Do institutions matter for regional development? *Regional Studies, 47*(7), 1034–1047. https://doi.org/10.1080/00343404.2012.748978
26. Paunov, C. (2013). Innovation and inclusive development: A discussion of the main policy issues. *OECD Science, Technology and Industry Working Papers, 2013*(1). https://doi.org/10.1787/5k4dd1rvsnjj-en
27. George, G., McGahan, A. M., & Prabhu, J. (2022). Innovation for inclusive growth: Towards a theoretical framework and a research agenda. *Journal of Management Studies, 59*(2), 377–405. https://doi.org/10.1111/joms.12764
28. Cooke, P. (2001). Regional innovation systems, clusters, and the knowledge economy. *Industrial and Corporate Change, 10*(4), 945–974. https://doi.org/10.1093/icc/10.4.945
29. Etzkowitz, H. (2008). *The triple helix: University–industry–government innovation in action.* Routledge.
30. Shearmur, R., & Doloreux, D. (2013). Innovation and knowledge-intensive business services: Contribution to innovation in manufacturing firms—Evidence from Canada. *Regional Studies, 47*(10), 1613–1631. https://doi.org/10.1080/00343404.2011.628651
31. Vallance, P., Tewdwr-Jones, M., & McGuinness, D. (2018). Rethinking the regional innovation strategy: The role of learning spaces. *European Planning Studies, 26*(10), 1835–1855. https://doi.org/10.1080/09654313.2018.1504890
32. Morgan, K. (2004). The exaggerated death of geography: Learning, proximity and territorial innovation systems. *Journal of Economic Geography, 4*(1), 3–21. https://doi.org/10.1093/jeg/4.1.3

Digital Pathways Towards Efficient Urban Mobility in Smart Cities

Duarte Fernandes, Jorge Julião, and Marcelo Gaspar

Abstract This study explores how digital technologies can improve the efficiency of urban mobility systems in smart cities. It highlights three key enablers: Artificial Intelligence (AI), the Internet of Things (IoT), and Mobility-as-a-Service (MaaS), while assessing their impact on five important areas: travel time, safety, accessibility, sustainability, and cost-effectiveness. Using a qualitative approach, the research gathers insights from interviews with urban planning, engineering, technology, and public policy experts. It examines the technical functions of these tools alongside the broader conditions necessary for their successful implementation in cities. Findings suggest that digital technologies enhance real-time decision-making, optimize routing, and create fairer transportation networks. However, their success depends on the presence of adequate infrastructure, good governance, and suitable policies. Major challenges include disjointed physical systems, poor data governance, strict regulations, and unequal access to digital services. Experts emphasise co-designing solutions with users, improving physical infrastructure, and managing data transparently and ethically. For digital mobility to thrive, cities need to adopt modular technologies, create flexible policy frameworks, and develop responsive and fair systems. This study highlights that smart mobility involves both technical and socio-technical challenges, with its success reliant on how well cities merge innovation with inclusive governance to create efficient, sustainable, and equitable transportation systems.

Keywords Smart cities · Urban mobility · Smart mobility · Digital transformation · Mobility-as-a-service

---

D. Fernandes (✉)
Católica Porto Business School, Universidade Católica Portuguesa, Porto, Portugal
e-mail: s-daafernandes@ucp.pt

J. Julião
Católica Porto Business School, Research Centre in Management and Economics, Universidade Católica Portuguesa, Porto, Portugal

M. Gaspar
School of Technology and Management, Polytechnic Institute of Leiria, Leiria, Portugal

L. Farinha et al. (eds.), *Sustainable Innovations in Mechanical Engineering and Management Sciences*, Lecture Notes on Multidisciplinary Industrial Engineering,
https://doi.org/10.1007/978-3-032-30372-1_10

1 Introduction

Urbanization remains one of the most transformative and complex phenomena of the twenty-first century. According to the United Nations [1], the global urban population has grown from 751 million in 1950 to over 4.2 billion in seven decades and is projected to encompass 68% of the world's population by 2050 [2]. This shift imposes multifaceted challenges on city governments, particularly in resource management, spanning energy, waste, housing, and most critically, urban mobility. Addressing these issues demands not only capacity expansion but a fundamental transformation in how urban systems are planned and managed [3].

Although cities occupy merely 2% of the Earth's land, they account for over 60% of energy consumption, 70% of greenhouse gas emissions, and generate an equally large share of global waste, yet paradoxically contribute around 70% of the world's GDP [4]. This contradiction underscores the imperative of embedding sustainability into urban development frameworks. In particular, urban mobility is emerging as a vital pillar for achieving sustainable growth. It influences not only environmental outcomes but also urban livability, economic performance, and social equity [5].

In recent years, digital innovation has profoundly reshaped the urban mobility landscape. The convergence of Internet of Things (IoT) infrastructure, Artificial Intelligence (AI), and big data analytics has enabled smarter, more adaptive transportation systems. These technologies facilitate real-time monitoring of traffic flows, predictive modeling of congestion patterns, and optimization of multimodal transport networks [6]. Intelligent Transport Systems (ITS), integrated mobility-as-a-service (MaaS) platforms, and sensor-driven infrastructure are now central to the mobility strategies of smart cities.

Global urban centers such as Amsterdam, Seoul, and Singapore exemplify how technology-driven urbanism can mitigate congestion, reduce emissions, and enhance commuter experiences. Barcelona's smart traffic lights, for instance, adjust dynamically based on pedestrian and vehicle flows, while New York City leverages AI to optimize its bus routes and prioritize emergency response [7]. These digital pathways are not merely conveniences but strategic tools to address broader sustainability goals.

As cities evolve into complex cyber-physical systems, a key question arises: how can digital transformation be harnessed equitably and efficiently to foster inclusive, sustainable urban mobility? This study investigates the intersection of digital technology and mobility in smart cities, with a focus on how emerging innovations can alleviate longstanding transportation issues. It critically examines international best practices, evaluates the socio-technical challenges of implementation, and outlines policy considerations for integrating these systems. By unpacking the potential and pitfalls of digital mobility strategies, this work aims to provide urban planners, policymakers, and technologists with actionable insights into crafting livable, connected, and future-ready cities.

2 Literature Review

2.1 Conceptual Foundations of Smart Cities

The notion of smart cities initially emerged from the integration of digital technologies, particularly ICT systems, into urban infrastructure, aiming to enhance service efficiency and urban living standards [8]. Early models predominantly emphasized technological solutions to urban challenges, often overlooking social, political, and environmental dimensions [9]. Critics like Hollands [10] highlighted the need for more inclusive and multidimensional frameworks that extend beyond digitization to encompass governance, human capital, and sustainability.

Modern perspectives on smart cities advocate a systems-based view, where cities operate as interconnected networks of infrastructure, data, people, and institutions [11]. These cities employ real-time analytics, IoT, AI, and big data to adapt dynamically to changing urban needs. George and George [12] emphasize that the effectiveness of smart technologies lies not only in automation but also in their integration with social values, equity, and participatory governance.

A defining feature of contemporary smart city models is the "system of systems" approach, wherein transportation, energy, health, and urban planning are digitally synchronized. This cross-sectoral interoperability supports urban resilience and real-time responsiveness. Bibri [13] argues that such integration is critical for cities to remain adaptive, efficient, and inclusive, particularly in the face of rapid urbanization and environmental pressures.

However, challenges persist, especially regarding data governance, surveillance risks, and the dominance of private tech firms. Cardullo and Kitchin [14] warn against techno-solutionism and emphasize the importance of democratic oversight and citizen agency. As such, smart cities must be designed not only to optimize services but also to safeguard ethical standards, promote equity, and foster sustainable urban development.

2.2 Digital Infrastructure and Emerging Technologies

Digital infrastructure is the cornerstone of smart city development, enabling real-time responsiveness, optimized resource management, and citizen-centric services. Core components such as the Internet of Things (IoT), Artificial Intelligence (AI), edge computing, and 5G networks allow for the continuous collection, transmission, and analysis of urban data [15]. These technologies form the operational backbone for urban applications, including traffic flow optimization, dynamic public transport scheduling, air quality monitoring, and energy grid management.

Recent advances have moved beyond passive data collection towards predictive and adaptive capabilities, most notably through Digital Twin technology. Digital

Twins replicate urban mobility systems in virtual environments to test policies, optimize transport operations, and predict the impact of infrastructure changes [16]. AI-powered Digital Twins can simulate multi-modal traffic scenarios, assisting planners in identifying congestion patterns before they materialize.

Furthermore, edge computing has emerged as a key enabler of low-latency decision-making, pushing data processing closer to the point of data generation [17]. This is particularly vital for time-sensitive urban mobility functions such as autonomous vehicle coordination or emergency service routing. When integrated with 5G networks, edge computing ensures faster communication and greater reliability in high-density urban areas.

However, critical challenges persist. Infrastructure disparities between global North and South cities limit the scalability of these solutions. Additionally, the deployment of surveillance-oriented technologies raises privacy and governance concerns, prompting debates about data ethics and democratic accountability [11]. The reliance on proprietary platforms by private tech firms can further fragment digital ecosystems, undermining interoperability and increasing long-term costs for municipalities [14].

2.3 *Urban Mobility and Smart Transformation*

Urban mobility lies at the heart of smart city initiatives, functioning as both a challenge and an opportunity for sustainable development. Traditionally centered on infrastructure expansion, contemporary mobility strategies now emphasize digital innovation, intermodal connectivity, and user-centric design [18]. The convergence of ICT, AI, and data analytics has enabled more responsive and adaptive transport systems, reshaping how cities manage congestion, emissions, and accessibility.

One transformative development is Mobility-as-a-Service (MaaS), a platform-based approach that integrates public transit, bike-sharing, ride-hailing, and micromobility into a unified interface. MaaS enhances travel efficiency and reduces car dependency, but its adoption varies globally, often limited by data silos and regulatory fragmentation [19]. Furthermore, research indicates that while MaaS can support sustainable mobility goals, it may also exacerbate inequality if affordability and digital literacy are not addressed [20].

Another key innovation is the application of AI and machine learning in dynamic traffic control and predictive demand modeling. These tools allow cities to optimize traffic flow, manage parking, and reduce transit waiting times. However, overreliance on algorithmic systems introduces governance risks, such as biases in routing algorithms that disadvantage low-income neighborhoods [21].

The shift toward shared and automated mobility, including e-scooters, autonomous shuttles, and electric vehicles, raises environmental and spatial questions. While these modes reduce carbon emissions, they also require regulatory oversight to avoid urban clutter, safety risks, and rebound effects in travel demand [17].

2.4 Barriers to Implementation

Despite the promising vision of smart urban mobility, real-world implementation remains hindered by complex and multifaceted barriers. A persistent challenge is fragmented governance, where smart mobility projects involve multiple stakeholders with overlapping responsibilities and misaligned agendas. This institutional complexity often leads to inefficiencies, a lack of policy continuity, and slow adoption [22].

Another significant barrier is the digital divide, particularly evident in cities of the Global South. Inadequate access to digital infrastructure and skills prevents equitable participation in smart systems and exacerbates socio-spatial inequalities [23]. Moreover, public mistrust of surveillance technologies and data handling practices poses ethical risks, often resulting in public resistance and reduced legitimacy of smart initiatives [24].

From a technological standpoint, the lack of interoperability between legacy infrastructure and advanced digital platforms continues to restrict integration. As Lindland [25] points out, even advanced cities struggle with real-time data harmonization due to incompatible systems, cybersecurity threats, and proprietary restrictions that fragment innovation ecosystems.

Financial constraints further exacerbate these issues. Many municipalities lack the long-term funding mechanisms needed to maintain and upgrade smart mobility infrastructure, relying heavily on inconsistent public-private partnerships [26]. This can lead to cost overruns and dependence on third-party platforms that undermine local control and resilience.

2.5 Dimensions of Urban Mobility Efficiency

Urban mobility efficiency is a core performance metric in smart city planning, reflecting how well a city's transport ecosystem meets environmental, economic, and social goals. In the context of digital transformation, these dimensions have evolved from traditional measures like speed and cost to include sustainability, equity, and adaptive capacity.

Travel Time Optimization remains a foundational metric. Through real-time traffic management, dynamic routing, and AI-based signal control, cities aim to reduce congestion and commuting time [16]. Roslan et al. [27] demonstrate that such systems can reduce delays by 20–30% when integrated into multimodal environments.

Multimodal Safety is increasingly vital with the proliferation of micro-mobility and shared services. Sensor-based collision avoidance systems and AI-driven monitoring are employed to mitigate pedestrian and cyclist risks. Cheong [28] stresses the need for robust safety analytics, while Abdjuljabbar et al. [29] emphasize inclusive safety for vulnerable populations.

Economic and Environmental Efficiency reflects the dual goals of reducing costs and emissions. Smart mobility reduces fuel use and improves operational efficiency, especially through electrification and shared vehicle use [17]. Sumitkumar and Al-Sumaiti [30] further argue that platform-based fare integration improves cost control and reduces redundancy in transit systems. Accessibility and Social Inclusion underscore the equity dimension of efficiency. Adejuwon [23] notes that without adaptive interfaces and affordable pricing, digital mobility can reinforce social exclusion.

Behavioral and Cultural Responsiveness refers to how well systems align with user preferences, habits, and trust. Wolniak and Turon [20] identify cultural dissonance in scooter-sharing schemes in Eastern Europe. Lindland [25] argues that ignoring behavioral norms can limit adoption even in tech-ready environments.

3 Methodology

This study employs a qualitative research approach to explore the research question: "How can the integration of digital advancements in smart cities' urban mobility systems enhance their efficiency?" The selected methodology highlights the exploratory aspect of the research, aiming to comprehend intricate, multifaceted phenomena that quantitative methods alone cannot fully capture. Considering the quickly changing landscape of urban mobility and digital innovation, a qualitative framework facilitates the gathering of expert insights that are contextually rich, rooted in real-world practice, and shaped by professional experience. This study seeks to reveal essential insights into the transformative capabilities of digital technologies in urban mobility systems by conducting semi-structured interviews with experts in urban planning, architecture, smart city development, and technology integration.

3.1 Research Design and Methods

To address the research question and its associated objectives, this study adopts a qualitative research design grounded in an interpretivist paradigm. This approach is particularly appropriate for examining complex sociotechnical systems such as urban mobility in smart cities, where context-specific insights, stakeholder perspectives, and experiential knowledge are central to understanding the phenomena under study [31].

Semi-structured interviews were employed as the primary method of data collection, allowing for a balance between consistency across interviews and the flexibility to explore emergent themes. This technique is well-suited for capturing in-depth narratives and nuanced expert opinions, particularly within exploratory research contexts [32]. The interview protocol was closely aligned with five research objectives: (1) assessing the impact of digital tools on travel time reduction; (2) evaluating contributions to user safety; (3) understanding improvements in accessibility; (4)

examining the role of technology in promoting sustainability; and (5) exploring the cost-efficiency of digital integration. Each objective was translated into specific questions designed to draw out expert insights grounded in real-world implementation and policy-making (Table 1).

A purposive sampling strategy was applied to recruit participants with relevant expertise. Professionals were identified through established networks, targeted LinkedIn searches, and speaker lists from smart city conferences. Inclusion criteria focused on individuals with demonstrable experience in urban planning, smart city development, or digital infrastructure implementation. Purposive sampling is widely recognized for its effectiveness in qualitative research, especially when the goal is to gather rich, contextually informed data from a specialized group [33].

Ten expert interviews were conducted between March and April 2024. Interviews were audio-recorded with participant consent, transcribed verbatim, and anonymized for analysis. The number of interviews was guided by the principle of thematic saturation, wherein additional data ceased to yield novel insights, a concept well-documented in qualitative research literature [34].

By integrating a robust qualitative design with a strategically developed interview protocol, the study was able to generate high-quality data aligned with its exploratory goals. This methodological foundation supports a rich, interpretive analysis of how digital advancements are shaping the future of urban mobility in smart cities.

Table 1 Range of questions utilised in interviews

Research Objective	Summary of Questions Asked
1. Assess the impact of digital advancements on travel time reduction	Asked about the effects of real-time traffic systems and route optimization on travel time, and challenges in implementation
2. Evaluate the role of digital technologies in enhancing user safety	Explored effective safety-enhancing technologies and the factors influencing their integration into mobility systems
3. Investigate how digital platforms improve accessibility	Investigated examples and barriers related to improving accessibility for marginalized groups through digital solutions
4. Examine the contribution of digital innovations to sustainability	Inquired about the role of current and emerging technologies in promoting environmental sustainability in urban transport
5. Explore the cost-efficiency of digital technology integration	Focused on financial implications of digital integration, including cost-benefit analysis and real-world savings examples

3.2 Data Analysis

The interview data were analyzed using a thematic analysis approach, a widely adopted method in qualitative research for identifying, analyzing, and interpreting patterns within textual data [35]. This technique was chosen for its flexibility and its capacity to generate insights grounded in participants' lived experiences and professional expertise.

Following transcription, all interview responses were systematically reviewed and coded. An initial round of open coding was conducted to capture key concepts and emergent themes relevant to the research objectives. Codes were then grouped into broader thematic categories aligned with the study's five research themes. This iterative process enabled a structured yet inductive exploration of the data.

To enhance analytical transparency, a comprehensive matrix was developed, comparing responses across participants. This allowed for the identification of converging insights as well as sector-specific divergences. Convergences were treated as indicators of shared priorities or trends, while divergences were explored to understand contextual variability.

The analysis followed established qualitative practices such as constant comparison, memo writing, and reflective interpretation to preserve rigor and minimize bias [31]. The resulting themes form the empirical basis for the findings presented in the subsequent chapters.

4 Results and Data Analysis

This section presents the results of the empirical research based on ten semi-structured expert interviews, analyzed through thematic analysis. The aim is to examine how digital transformation contributes to the efficiency of urban mobility systems within smart cities. The findings are structured around five research objectives, which also serve as analytical categories. The interviews included professionals from fields such as engineering, architecture, urban innovation, data governance, and public transport planning, ensuring a multidimensional understanding of the subject. Table 2 summarizes the participants' academic and professional backgrounds.

Thematic analysis of the interview data revealed five core areas aligned with the research objectives: (1) digital tools and infrastructure, (2) stakeholder perspectives, (3) barriers to implementation, (4) user-centered impacts, and (5) strategic recommendations. Each theme reflects not only distinct dimensions of digital transformation in urban mobility systems but also areas of overlap and interdependence.

Table 2 Participants characterization

Interviewer	Background	Professional activity
1	Architecture	Urban planner/public space designer
2	Engineer	Urban Systems engineer
3	Management/economics	Mobility consultant
4	Engineering	City infrastructure advisor
5	Engineering	Public transport planner
6	Management/economics	Urban innovation strategist
7	Technology	Mobility start-up founder
8	Engineering	Smart City specialist
9	Technology	Data governance expert
10	Architecture	Accessibility and inclusion advocate

4.1 *Digital Tools and Infrastructure*

Participants highlighted the increasing adoption of digital solutions aimed at improving the efficiency, responsiveness, and adaptability of urban mobility systems. Key technologies mentioned included:

- Real-time transit tracking to monitor traffic conditions and improve bus and train punctuality.
- Smart traffic management systems that use AI and sensor networks to optimize traffic flow.
- Mobility-as-a-Service (MaaS) platforms, integrating different modes of transportation (e.g., bus, bike, metro, ride-hailing) into a single user interface.
- Digital ticketing systems, enabling contactless payments and flexible pricing.

These technologies were viewed as essential tools for transitioning toward intelligent, demand-responsive transport networks. However, multiple interviewees noted that technical adoption remains uneven across cities due to limitations in legacy infrastructure and a lack of system interoperability.

Several experts also raised concerns about the fragmentation of platforms and data environments, which often results in isolated pilot programs rather than coherent system-wide solutions. Without standardized protocols and unified governance structures, the integration of these tools remains partial and inconsistent.

4.2 *Stakeholder Perspectives*

Interviewees brought forward a wide range of views shaped by their professional and institutional roles. A key finding was the contrast in priorities:

- Engineering and technology professionals focused on the technical efficiency, system integration, and reliability of digital tools.
- Policy advisors, innovation strategists, and urban planners emphasized public value, inclusiveness, long-term planning, and regulatory reform.
- Advocates and accessibility experts centered their attention on usability, equity, and access for marginalized groups.

Despite differences in focus, there was agreement on the need for stronger collaboration across institutional and disciplinary boundaries. Several interviewees described current mobility planning processes as fragmented, with overlapping responsibilities and insufficient knowledge exchange between departments or sectors. This gap was seen as a major bottleneck to innovation and a source of inefficiency in digital mobility strategies.

4.3 Barriers to Implementation

The research identified a combination of technical, organizational, regulatory, and financial barriers to the effective implementation of digital mobility systems:

- Technical limitations: Legacy infrastructure often lacks the flexibility to accommodate digital upgrades, especially in older or spatially constrained urban environments.
- Organizational fragmentation: Decision-making authority is typically distributed across multiple municipal departments, hindering coordinated action.
- Regulatory inertia: Many cities operate under outdated legal frameworks that do not accommodate fast-evolving digital tools or data practices.
- Data governance concerns: Issues related to privacy, cybersecurity, and ethical data use were highlighted as underregulated and potentially damaging to public trust.
- Financial constraints: The cost of deploying, maintaining, and scaling digital systems, particularly in smaller municipalities, was noted as a critical challenge.

These barriers were seen as interrelated. For example, a lack of institutional coordination often exacerbates technical and financial inefficiencies, while weak regulatory structures make it difficult to enforce or monitor data protection measures.

4.4 User-Centered Impacts

Interviewees emphasized that while digital tools offer significant potential for improved user experience, they also introduce new forms of exclusion and inequality. Key themes included:

- Accessibility gaps: Several experts noted that digital mobility platforms often overlook users with physical disabilities, low digital literacy, or limited access to smartphones and mobile data.
- Usability and personalization: In more digitally mature environments, users benefit from customized journey planning, flexible payment methods, and real-time updates. However, these advantages are not universally available.
- Data transparency and trust: Concerns were raised about the opaque collection and use of user data. Interviewees warned that a lack of transparency could lead to declining public confidence in digital services.
- Surveillance risks: Some participants questioned the long-term implications of continuous data collection, particularly in contexts where users are unaware of how their data is being stored, shared, or monetized.

Overall, the findings suggest that mobility efficiency cannot be disentangled from user equity. A system that excludes vulnerable populations—even unintentionally—undermines its effectiveness and legitimacy.

4.5 *Strategic Recommendations*

Based on their experience, participants offered targeted strategies for improving the design, implementation, and governance of digital mobility systems:

- Pilot-based approaches: Begin with small-scale, modular interventions to test feasibility and build institutional capacity before wider rollouts.
- Interoperability and open standards: Encourage platform-neutral, scalable technologies to ensure that systems can communicate across agencies and providers.
- Inclusive design: Prioritize universal access and usability through co-design processes, accessibility audits, and feedback mechanisms that engage diverse user groups.
- Ethical data frameworks: Establish clear rules for data collection, consent, anonymization, and user control, supported by robust regulatory oversight.
- Institutional coordination: Create cross-sector task forces or innovation hubs to align transport, planning, IT, and social policy under a shared strategy.

These recommendations reflect a shift in thinking from purely technical innovation toward a more systemic, inclusive, and policy-integrated approach to digital transformation in urban mobility.

5 Discussion

This study examined the role of digital technologies in transforming urban mobility systems (UMS) in smart cities (SCs) across five dimensions: travel time efficiency, safety, accessibility, sustainability, and cost-efficiency. The findings from expert interviews confirm the transformative potential of digital innovation but also expose systemic dependencies, particularly infrastructure readiness, data governance, and public policy frameworks, that shape their success.

5.1 Travel Time Efficiency: Enablers and Structural Limits

Digital solutions such as real-time traffic information, adaptive signal control, and route optimization systems are recognized for their contributions to reducing congestion and travel times [36]. Participants reinforced these findings, emphasizing that such improvements are conditional on high-quality infrastructure and network design. As one expert aptly noted, "digital systems cannot replace a good transportation network design," underscoring that technology augments—but does not substitute—physical mobility structures.

Moreover, the quality and availability of data were flagged as essential to the functionality of these systems. In alignment with Kitchin [11], who emphasizes the foundational role of data in real-time urbanism, respondents described inadequate or siloed datasets as a key barrier. Interoperability, timeliness, and data standardization thus emerge as critical enablers of travel time improvements.

5.2 Safety: A Double-Edged Sword of Innovation

Advanced technologies—ranging from AI-enhanced surveillance to predictive traffic analytics—are increasingly leveraged to improve urban transport safety. Bolobonov et al. [37] provide strong evidence of digital tools reducing incidents through proactive traffic management. Similarly, interviewees cited smart lampposts, accident detection systems, and ice warnings as impactful examples.

However, as Ismagilova et al. [38] argue, smart city technologies also pose risks related to surveillance, privacy, and ethical data use. This paradox was echoed in the interviews: while digital monitoring increases safety, the lack of public oversight and transparent data governance threatens trust and long-term adoption. The trade-off between passive data collection and individual privacy rights is a critical policy issue that must be resolved.

5.3 *Accessibility: Digital Inclusion as Infrastructure*

Digital solutions were widely recognized as facilitating accessibility, especially for marginalized groups. Examples such as real-time bus tracking or GPS navigation for the visually impaired align with findings by [39], who argue that mobility digital ecosystems can enhance inclusivity when barriers to access are addressed.

Still, participants repeatedly emphasized that accessibility remains uneven due to infrastructural and informational deficits. Litman [40] highlights the need for equity-based planning in transportation, warning that benefits often disproportionately accrue to already-privileged populations. In this study, the notion that "physical structure first" remains central reflects the interdependence of digital and material accessibility. Without universally accessible physical spaces, digital services risk excluding those with limited digital literacy or access to devices.

5.4 *Sustainability: Conditional Gains and Strategic Frictions*

Digital mobility solutions can support environmental goals through reduced emissions and more efficient transit use. Caragliu [8] discusses how ICTs can promote sustainability by optimizing energy use and enabling modal shifts. Interviewees echoed this view, particularly in the context of bike-sharing and electric vehicle infrastructure.

However, the net environmental benefit of these technologies is conditional. As Banister [41] argues, sustainability gains depend not only on technological deployment but on behavioral shifts and urban design choices. Participants raised concerns that shared mobility services may displace walking or public transit rather than private car use, a phenomenon also noted by Ewing et al. [42]. This points to the need for strategic integration of digital platforms with regulatory policies that actively discourage high-emission modes and incentivize truly green alternatives.

5.5 *Cost-Efficiency: Underexplored but Promising*

Although experts cited examples of cost-saving technologies, such as waste management tracking systems in Cascais or adaptive traffic lights in Lisbon, the financial performance of these innovations remains under-documented. As Al-Barhamtoshy et al. [43] suggest, there is a growing need to develop methods for assessing return on investment (ROI) in smart urban systems. The limited responses in this study mirror a broader lack of longitudinal, empirical data on financial sustainability in digital mobility.

Moreover, Litman [44] stress the importance of incorporating distributional impacts into transport planning. Even if it is cost-efficient in aggregate, digital solutions may benefit certain user groups or neighborhoods disproportionately, raising questions of equity and fairness. Future research should address both microeconomic efficiency and macro-level societal impacts to guide evidence-based investments.

6 Conclusions

This study investigated the transformative role of digital technologies in urban mobility systems (UMS) within the context of smart cities (SCs), exploring five interrelated dimensions: travel time efficiency, safety, accessibility, sustainability, and cost-efficiency. Drawing on expert insights and supported by academic literature, the findings reveal both the potential and limitations of digital interventions in shaping the mobility landscape.

A key conclusion is that digital technologies are not standalone solutions but enablers whose success hinges on systemic readiness. Across all categories, infrastructure quality, data governance, policy frameworks, and social equity emerged as critical enablers or constraints. While digital tools can enhance operational efficiency, safety, and user experience, they remain contingent upon well-developed physical networks and inclusive urban design.

In terms of travel time, the integration of real-time information systems and smart traffic management can significantly improve flow and reduce delays—but only when supported by coherent transport infrastructure. Similarly, safety gains through AI-driven monitoring and predictive analytics depend on public trust and regulatory clarity around data use and privacy.

Accessibility enhancements, especially for marginalized populations, show promise but remain limited by digital divides and the uneven deployment of supportive infrastructure. Here, the need for holistic and inclusive design strategies becomes especially salient.

Regarding sustainability, digital solutions such as shared mobility and smart public transit have the potential to reduce emissions and energy use. However, their environmental benefits are conditional and context-specific; success depends on whether they replace high-emission modes like private cars rather than displace low-carbon alternatives like walking or cycling.

Cost-efficiency, while indicated in several use cases, remains the least substantiated dimension. The lack of empirical, longitudinal financial evaluations suggests a research and policy gap. Moving forward, cities must develop more robust frameworks to quantify the return on investment of digital mobility systems, not just in economic terms but also in social and environmental value.

Ultimately, the findings reinforce the notion that smart mobility is a socio-technical challenge, not merely a technological one. The promise of digital innovation must be matched by adaptive governance, inclusive planning, and sustained investment in infrastructure. Future smart cities will not be defined solely by their technological capabilities, but by their capacity to integrate those capabilities within just, accessible, and resilient urban systems.

6.1 Implications for Theory and Practice

Theoretical implications stem from the study's framing of smart mobility as a socio-technical construct. It challenges techno-centric narratives in the smart city discourse and affirms that digital mobility outcomes are co-produced by institutional, infrastructural, and societal systems. The findings reinforce emerging theoretical frameworks that integrate infrastructure studies, urban planning, and digital governance, such as the "real-time city" and "inclusive smart urbanism". Future theory-building should consider not just the presence of technology but the relational dynamics between digital tools, physical environments, and social inclusion.

Practical implications are equally significant. Policymakers and urban mobility planners must treat digital innovation not as a substitute for infrastructure or regulation but as part of a broader transformation process. This includes:

- Investing in high-quality, inclusive infrastructure to support technology deployment;
- Establishing ethical and transparent data governance frameworks;
- Designing digital tools that are co-created with, and accessible to, all user groups;
- Developing clear, long-term cost-efficiency evaluation models to guide resource allocation.

Moreover, practitioners should consider deploying pilot projects that integrate digital solutions with traditional transport assets, measuring not only usage metrics but also societal outcomes such as equity, environmental impact, and user satisfaction. This approach could foster adaptive learning and help cities scale interventions based on localized needs and capacities.

Acknowledgments The authors sincerely thank all interview participants for generously sharing their time, insights, and experiences, which were invaluable for the development of this research.

This manuscript includes content developed with assistance from artificial intelligence (AI). AI-supported tools, such as OpenAI's ChatGPT and Grammarly, were utilised to aid various stages of the writing process, including text editing, idea refinement, grammar correction, sentence restructuring, and overall clarity enhancement. The authors have ensured that the use of AI tools complies with ethical research and authorship standards and does not compromise the originality or integrity of the work. All content was critically reviewed and edited by the authors, who remain solely responsible for the final version of the manuscript.

References

1. United Nations. (2019). *World urbanization prospects: T: The 2018 revision (ST/ESA/SER.A/420).*
2. Mekuriaw, T., & Gokcekus, H. (2019). The impact of urban expansion on physical environment in Debre Markos town, Ethiopia, *11*(5). https://doi.org/10.7176/CER
3. McCormick, K., Anderberg, S., Coenen, L., & Neij, L. (Jul. 2013). Advancing sustainable urban transformation. *Journal of Cleaner Production, 50*, 1–11. https://doi.org/10.1016/j.jclepro.2013.01.003
4. United Nations. (2016). *World cities report 2016: Urbanization and development—emerging futures (HS/038/16E).*
5. Azarbarzin, N., & Hoseinisiagoli, M. (2024). Modeling of urban Livability indicators with an emphasis on sustainable urban transportation: The case study of Ahvaz City. *Geographical Urban Planning Research Quarterly, 12*(2), 25–43. https://doi.org/10.22059/jurbangeo.2024.374294.1921
6. Zemmouchi-Ghomari, L. (Apr. 2025). Artificial intelligence in intelligent transportation systems. *Journal of Intelligent Manufacturing and Special Equipment, 6*(1), 26–42. https://doi.org/10.1108/JIMSE-11-2024-0035
7. Müller-Eie, D., & Kosmidis, I. (Dec. 2023). Sustainable mobility in smart cities: A document study of mobility initiatives of mid-sized Nordic smart cities. *European Transport Research Review, 15*(1). https://doi.org/10.1186/s12544-023-00610-4
8. Caragliu, A., del Bo, C., & Nijkamp, P. (Apr. 2011). Smart cities in Europe. *Journal of Urban Technology, 18*(2), 65–82. https://doi.org/10.1080/10630732.2011.601117
9. Nam, T., & Pardo, T. A. (2013). Smart city as urban innovation: Focusing on management, policy, and context. In *ICEGOV´11: Proceedings of the 5th international conference on theory and practice of electronic governance* (pp. 185–194). ACM Digital Library. https://doi.org/10.1145/2072069.2072100
10. Hollands, R. G. (2008). Will the real smart city please stand up? Intelligent, progressive or entrepreneurial? *City, 12*(3), 303–320. https://doi.org/10.1080/13604810802479126
11. Kitchin, R. (2016). Getting smarter about smart cities: Improving data privacy and data security. *Dublin*. https://doi.org/10.1017/dap.2021.13
12. George, A. S., & Hovan George, A. S. (2024). Towards a super smart society 5.0: Opportunities and challenges of integrating emerging Technologies for Social Innovation. *Partners Universal International Research Journal*. https://doi.org/10.5281/zenodo.11522048
13. Bibri, S. E., Huang, J., Jagatheesaperumal, S. K., & Krogstie, J. (Jul. 01, 2024). The synergistic interplay of artificial intelligence and digital twin in environmentally planning sustainable smart cities: A comprehensive systematic review. *Editorial Board, Research of Environmental Sciences*. https://doi.org/10.1016/j.ese.2024.100433
14. Cardullo, P., & Kitchin, R. (2019). Being a 'citizen' in the smart city: Up and down the scaffold of smart citizen participation in Dublin, Ireland. *Geo Journal, 84*(1), 1–13. https://doi.org/10.1007/s10708-018-9845-8
15. Albino, V., Umberto, B., & Dangelico, R. M. (Jan. 2015). Smart cities: Definitions, dimensions, performance, and initiatives. *Journal of Urban Technology, 22*(1), 3–21. https://doi.org/10.1080/10630732.2014.942092
16. Wajid, M. A., Wajid, M. S., Zafar, A., & Terashima-Marin, H. (2025). Digital twin technology for multimodal-based smart mobility using hybrid co-ABC optimization based deep CNN. *Cluster Computing, 28*(3), 194. https://doi.org/10.1007/s10586-024-04903-8
17. Fatorachian, H., Kazemi, H., & Pawar, K. (2025). Enhancing Smart City logistics through IoT-enabled predictive analytics: A digital twin and cybernetic feedback approach. *Smart Cities, 8*(2). https://doi.org/10.3390/smartcities8020056
18. Rodrigue, J.-P. (2020). *The geography of transport systems* (5th ed.). Routledge.
19. Kamargianni, M., Li, W., Matyas, M., & Schäfer, A. (2016). A critical review of new mobility services for urban transport. *Transportation Research Procedia, 14*, 3294–3303. https://doi.org/10.1016/j.trpro.2016.05.277

20. Wolniak, R., & Turoń, K. (2025). The problems of scooter-sharing in smart cities based on the example of the Silesian region in Poland. *Smart Cities, 8*(1). https://doi.org/10.3390/smartcities8010016
21. Feroz Khan, A. B., & Ivan, P. (Dec. 2023). Integrating machine learning and deep learning in smart cities for enhanced traffic congestion management: An empirical review. *Journal of Urban Development and Management, 2*(4), 211–221. https://doi.org/10.56578/judm020404
22. Akgün, E. Z., Gerli, P., Mora, L., & McTigue, C. (2024). Breaking barriers for breaking ground: A categorisation of public sector challenges to smart city project implementation. *Public Policy and Administration.* https://doi.org/10.1177/09520767241263233
23. Adejuwon, K. D. (2025). Smart cities and digital governance in Africa: Transforming urbanization for a sustainable future. *Policy and Governance Research (JPAPGR), 3*(1), 96–108.
24. van Zoonen, L. (2016). Privacy concerns in smart cities. *Government Information Quarterly, 33*(3), 472–480. https://doi.org/10.1016/j.giq.2016.06.004
25. Lindland, K. M. F. (2025). A pragmatist approach to the Smart City concept and practice. In *A Nordic smart Sustainable City* (Vol. 5, pp. 91–106).
26. Kutkov, O., Zolotov, A., Akimova, L., & Akimov, O. (2025). Digital transformation of social governance: Economic challenges and opportunities of smart cities. *Economics, Finance and Management Review, 1*(21). https://doi.org/10.36690/2674
27. Roslan, S. N. M., Gohain, K., Mustafa, A. M. A. A., Ismail, M. M., & Kumaran, V. V. (Mar. 2025). Designing affordable urban ecosystems: A quantitative model to enhance the quality of life for the urban poor in Malaysia through employment, housing, and digital access. *Challenges in Sustainability, 13*(1), 18–34. https://doi.org/10.56578/cis130102
28. Cheong, T. S., Li, T., Shi, X., & Yu, J. (Nov. 2024). Editorial: New directions of digital economy, energy transition, and climate change in the post-COVID-19 era: Application of machine learning and other advanced analytical techniques. *Frontiers in Environmental Science, 12.* https://doi.org/10.3389/fenvs.2024.1514467
29. Abduljabbar, R. L., Liyanage, S., & Dia, H. (2021). The role of micro-mobility in shaping sustainable cities: A systematic literature review. *Transportation Research Part D: Transport and Environment, 92*, 102734. https://doi.org/10.1016/j.trd.2021.102734
30. Sumitkumar, R., & Al-Sumaiti, A. S. (2024). Shared autonomous electric vehicle: Towards social economy of energy and mobility from power-transportation nexus perspective. *Renewable and Sustainable Energy Reviews, 197*, 114381. https://doi.org/10.1016/j.rser.2024.114381
31. Creswell, J. W., & Poth, C. N. (2018). *Qualitative inquiry and research design: Choosing among five approaches* (4th ed.). Sage.
32. Brinkmann, S., & Kvale, S. (2015). *Inter views: Learning the craft of qualitative research interviewing* (3rd ed.). Sage.
33. Patton, M. Q. (2015). *Qualitative Research & Evaluation Methods Integrating Theory and practice* (4th ed.). Sage.
34. Guest, G., Bunce, A., & Johnson, L. (2006). How many interviews are enough?: An experiment with data saturation and variability. *Field Methods, 18*(1), 59–82. https://doi.org/10.1177/1525822X05279903
35. Merriam, S. B., & Tisdell, E. J. (2016). *Qualitative research: A guide to design and implementation* (4th ed.). Wiley.
36. Batty, M., et al. (2012). Smart cities of the future. *The European Physical Journal Special Topics, 214*(1), 481–518. https://doi.org/10.1140/epjst/e2012-01703-3
37. Bolobonov, D., Frolov, A., Borremans, A., & Schuur, P. (2021). Managing public transport safety using digital technologies. *Transportation Research Procedia, 54*, 862–870. https://doi.org/10.1016/j.trpro.2021.02.140
38. Ismagilova, E., Hughes, L., Rana, N. P., & Dwivedi, Y. K. (2022). Security, privacy and risks within smart cities: Literature review and development of a Smart City interaction framework. *Information Systems Frontiers, 24*(2), 393–414. https://doi.org/10.1007/s10796-020-10044-1
39. Loos, E., Sourbati, M., & Behrendt, F. (2020). The role of mobility digital ecosystems for age-friendly urban public transport: A narrative literature review. *International Journal of Environmental Research and Public Health, 17*(20). https://doi.org/10.3390/ijerph17207465

40. Litman, T. A. (2023). *Evaluating accessibility for transportation planning: Measuring People's ability to reach desired services and activities.*
41. Banister, D. (2008). The sustainable mobility paradigm. *Transport policy, 15*(2), 73–80. https://doi.org/10.1016/j.tranpol.2007.10.005
42. Ewing, R., Bartholomew, K., Winkelman, S., Walters, J., & Chen, D. (2009). Growing cooler: The evidence on urban development and climate change. *Renewable Resources Journal, 25*(4), 6–13.
43. Al-Barhamtoshy, H. M., Abdullah, K. K., Dauda, M. K., & Himdi, T. F. (2023). Detecting available parking spaces in smart cities. In *2023 1st international conference on advanced innovations in smart cities (ICAISC)* (pp. 1–5). https://doi.org/10.1109/ICAISC56366.2023.10085386
44. Litman, T. (2022). Evaluating transportation equity: Guidance for incorporating distributional impacts in transport planning. *ITE Journal, 92*(4) Available: www.ite.org

EcoXperience Circular Business Models for Agri-Food Waste Valorization: A Case Study

Barbara Lamolinara, Mário Sérgio Teixeira, Cristina Galamba Marreiros, and Vítor Hugo dos Santos Ferreira

Abstract The circular economy is fundamental to sustainable development, emphasizing resource efficiency, waste reduction, and the regeneration of natural systems. Circular business models provide an effective framework for operationalizing circular principles by creating value from waste and extending product life cycles. However, their adoption in the agri-food sector remains underexplored, despite the sector's significant environmental impact and potential for waste reuse. This study examines EcoXperience, a Portuguese micro-enterprise that valorizes waste cooking oil by transforming it into eco-friendly detergents through a patented process. Using a qualitative single-case study approach, data were collected through a semi-structured interview with the company's CEO, complemented by document analysis and field observation to ensure triangulation. The findings show that EcoX integrates multiple circular business model archetypes—circular supplies, resource recovery, product life extension, sharing, and product-as-a-service—demonstrating how small firms can successfully operationalize comprehensive circular strategies. The study contributes to circular economy research by extending the application of circular business models to micro-enterprises in agri-food systems and offers practical insights for entrepreneurs, managers, and policymakers seeking to accelerate the transition toward sustainable and regenerative business practices.

B. Lamolinara · M. S. Teixeira
Centre for Transdisciplinary Development Studies (CETRAD), University of Trás-os-Montes e Alto Douro, Vila Real, Portugal
e-mail: mariosergio@utad.pt

B. Lamolinara · C. Galamba Marreiros
Center for Advanced Studies in Management and Economics (CEFAGE), Department of Management, School of Social Sciences, University of Évora, Évora, Portugal
e-mail: cristina@uevora.pt

B. Lamolinara (✉) · V. H. dos Santos Ferreira
Centre for Rapid and Sustainable Product Development (CDRSP), Polytechnic of Leiria, Marinha Grande, Portugal
e-mail: barbara.lamolinara@ipleiria.pt

V. H. dos Santos Ferreira
e-mail: vitor.ferreira@ipleiria.pt

L. Farinha et al. (eds.), *Sustainable Innovations in Mechanical Engineering and Management Sciences*, Lecture Notes on Multidisciplinary Industrial Engineering,
https://doi.org/10.1007/978-3-032-30372-1_11

Keywords Circular economy · Circular business model · Case study · EcoX · Waste valorization

1 Introduction

The circular economy (CE) is considered fundamental to building a sustainable society and shaping the future of business practices [1]. It is based on a dematerialization of the industrial economy, with a focus on resource efficiency, waste prevention and waste valorization [2]. In the context of the circular economy, the concept of circular business models (CBMs) [3, 4], rooted in circular economy principles and business model innovation, has garnered increasing attention in both academic and business domains as a strategic framework for advancing sustainability and circularity [5]. CBMs focus on slowing, narrowing, intensifying and/or dematerialising resource loops, reducing environmental impact and delivering superior customer value [6]. They are based on closed-loop value chains, industrial ecology and product-service systems, allowing material flow tracking to create value from waste [7]. The research by Donner et al. (2020) [7] expanded on the five archetypes of CBMs originally proposed by Bocken et al. 2016 [6], identifying their concrete application across agri-food supply chain: (1) circular supplies/circular design; (2) resource recovery; (3) product life extension (remanufacture, repair/upgrade and resell); (4) share; (5) product as a service. CBM is not only about closing the loop generating value from waste, but also the process of extending the life of the product, minimising the waste through eco-design, incentivising the reduction of consumption and the transition from the "ownership" of a product paradigm to the "sharing" of products and services. However, the recognition of these five types of CBM is still at the beginning, while waste valorization is the most used and recognized strategy to implement circular practices in a company [8], especially in agribusiness. Agribusiness is the field that is receiving more attention, as a crucial player in sustainable development and as a sector which has a considerable environmental impact [7]. The organic nature of agri-food products provides different possibilities for waste reuse and valorization [9, 10]. These facts explain the growing interest in CBMs and waste valorization practices in the agri-food industry sector, described by different studies [11–15]. Characteristics, success and risk factors [16], and typology of CBMs in agro-waste [7] have been discussed. However, just a few case studies of circular business models for agro-waste valorization have been reported in the literature [5, 17–20]. The lack of enough business case studies for CBMs in agri-food may represent a barrier to the adoption of circular practices in the sector [21, 22], especially in small and medium enterprises which are yet not familiar with the concept [23]. Therefore, there is a need for further circular case study descriptions in agri-food waste-to-resource practices, focused on the business aspects and strategies. Even if each case study has its proper peculiarities, it can inspire and incentivise the application of circular practices and circular business solutions in traditional linear economy companies, contributing to sustainability.

For this reason, the authors aim to contribute to the literature on circular strategies and waste-to-resource practices awareness by providing a successful example of agri-food waste valorization, recycling and management, describing all its different circular strategies. This work addresses a gap in the existing research, which has focused mainly on technological aspects of waste valorization, by emphasizing the managerial and strategic dimensions of circular adoption. The main objective of the paper is to describe and analyze how EcoXperience Lda (EcoX), specialized in transforming food oil waste into eco-friendly cleaning products, applies circular strategies within its business model. EcoX is a Portuguese micro-enterprise founded in 2016 and part of the Portuguese MSTN group.

The authors focus on the company's business model rather than on their technological/chemical solution. Besides describing the EcoX circular strategy for cooking oil upcycling, the authors address the following research questions:

- Is EcoX's Circular strategy based just on food oil waste valorization?
- Does EcoX include all five CBM types in its strategy?
- Can the different CBM types be successfully interconnected in a single company?

The methodology used to describe the business circularity implemented by EcoX and respond to these research questions is the qualitative approach of the single study case analysis [24] since it is appropriate to understand circular valorization practices in the agribusiness sector and has been already used in similar studies [25, 26].

From a theoretical perspective, this study advances knowledge on successful CBM strategies for agri-food waste valorization, while from a managerial perspective it provides inspiration and guidance for companies seeking to integrate CE and sustainability into their business models.

The article is organized as follows: Sect. 2 presents the conceptual framing behind the EcoX business; Sect. 3 explains the methodology used for the research; Sect. 4 describes the main results obtained by the case study analysis and related discussion; Sect. 5 presents the conclusion of this research, including the main limitations of the article and future research ideas. The interview protocol is presented in Annex 1.

2 The Theoretical Background of the Ecox Solution

Waste cooking oil is one of the abundant wastes in the food supply chain [27]. The increasing consumption of edible oils worldwide generates as a consequence a higher amount of waste cooking oil, which is not toxic per se, but it is a pollutant when handled and discharged unproperly [28]. Waste cooking oil from households is generally disposed of to drainage and soil, causing environmental damage to water, soil pollution and considerable gas emission [29]. This waste forms a layer on the water in lakes and rivers, which hinders the entry of light and oxygen into the aquatic environment. Collection treatment and disposal are crucial for waste cooking oil valorization, which has a recycling potential due to its chemical characteristics [28]. The most common way of recycling it is by turning it into a source of energy and bio-based products, such as the production of biodiesel [27, 30], biolubrificants and interesterification [31].

However, the company EcoX, find out an innovative and circular way to valorize waste cooking oil into eco-friendly detergents. The production of environmental-friendly detergents is an important issue around the world since the common chemical detergent are hazardous to water pollutants which are dangerous for humans, other organisms and ecosystems [32]. Eco-friendly laundry detergents are becoming more popular, as people are becoming more aware of the environmental impacts of common laundry detergents [33] and consequently more likely to purchase them [34]. Considering all these premises, the successful solution proposed by EcoX is based on a patented transformation process of food oil into ecological detergent, which allows the reduction of virgin materials and prevents water pollution, giving new life to waste that has serious environmental impacts and allowing for ecological and sustainable cleaning with biodegradable products.

3 Methodology

This section describes the methodological approach chosen for this study, i.e. the qualitative case study analysis [24]. Success case of circular practices and agro-waste valorization is selected as study cases, to describe and conceptualise the circular business models applied by the company. The case study can be used to capture the complexity of a case, including temporal changes, as well as explore its contextual conditions [24]. This qualitative approach has been chosen as a research methodology for this work because it allows to describe of success cases that can serve as an example for other companies and other contexts [24]. It allows one to describe and observe a phenomenon, in this case, circular practices, in its natural environment and without generalization, preserving the uniqueness of every company's practice and of every business model. The researcher can integrate into the company's team and procedures, being directly involved in the process to be described [24]. Moreover, the case study has been adopted widely by other recent studies in this field [25, 26, 35, 36]. Although the use of a single case study has been criticized by Kirchherr and van Santen [37] due to external validity challenge, other authors believe that the methodology is needed when knowledge and case studies are scarce about the topic of investigation [24, 26, 38]. The authors selected the Portuguese company EcoX because of different reasons: the new upcycling solution adopted for food oil reuse (unexplored in literature and the market before); the circular business models adopted; the circular and sustainable mission declared; the positive response of the Portuguese market to its products and therefore to its business model; the awareness of managers and founders about circular practice and CBMs concepts; the willingness to participate in this study; the geographical location of the company (same working country of the authors). Ethical considerations were addressed through informed consent, data anonymization, and adherence to confidentiality agreements. The CEO reviewed the main findings for factual accuracy, and permission was obtained to reproduce data provided by the company.

3.1 Data Collection and Interview Protocol

After the selection of the proper company to carry out this research, and the company's agreement to participate in the study, the authors and company managers have been in direct and constant contact to plan the data collection procedure. Starting from the literature review and the development of a conceptual framework and theory on qualitative research method [24], a design data collection protocol is created and a data collection tool is designed. For data collection, authors performed a market observation in a local store selling EcoX products, a semi-structured interview with the CEO and co-founder of the company and a document collection and analysis of material provided by the company (internal performance report, a sustainability report for the year 2022, power points presentations, the company website and press published articles). These data collection approaches will be used together to guarantee triangulation and reliability of results. The triangulation will assure that the qualitative research is not subjective, but that the results are confirmed by different approaches to data collection [24]. Also, internal validation must be taken into consideration for the quality of the research. In fact, in the case of studies, it is important to collect reliable data available for the research [39]. Thus, the authors performed a qualitative semi-structured interview with the EcoX co-founder and CEO. Semi-structured interviews allow a certain amount of freedom in comparison with a structured one since, during the interview, the researcher can get a lot of unexpected information and knowledge which stimulate new questions [40]. In this case, the interviews are not completely structured before, but they are open to questions and topics of the conversation that have not been prepared before [40–42]. In this study, the authors prepare an interview protocol to guide the discussion and remind fundamental topics. The review protocol, reported in Annex I, has been constructed based on guidelines of previous studies [26, 40] in the field and following essential points for the review protocol construction [40] to address author's research questions and study purpose:

- Brief company characterization
- Products and services offered
- Description of sustainable/circular strategy
- Description of circular practices (waste valorization, circular flow, input of waste from other companies)
- Description of other circular practices, if they exist: sharing, repair, remanufacturing, life cycle extension, product as a service
- Sustainability report
- Future implementation steps for circularity
- Some information about community support, subsidies, and tax incentives used for circularity

The interview with EcoX CEO, occurred in Portuguese language via videoconference, on the "google meet" platform, in January 2023. A critical study [43] about videoconference interviews assessed that they can be a useful supplement or replacement for traditional face-to-face interviews. The interview was recorded with the company's authorization to facilitate content transcription for analysis purposes. Besides recording, the researcher also took brief notes about every new information that represents an important contribution to the research. It is essential, during data collection, to properly report the information acquired [40].

The triangulation of the data collection (interview, document analysis and local shop observation) led to the results discussed in the following (Sect. 4).

4 Results and Discussion

By applying the methodological framework outlined in the previous (Sect. 3), the authors analyzed the data collected regarding the circular business models of EcoX [44]. EcoX is a microenterprise (with fewer than 10 employees) dedicated to producing cleaning and hygiene detergents aimed at both domestic and professional markets. The company boasts over 150 professional clients and more than 180 specialized retail partners in Portugal, having conducted over 600 awareness campaigns. In 2022 alone, EcoX valorized 10,000 L of used cooking oil, reduced CO_2 emissions by 340 kg, prevented the contamination of 5 billion liters of water, and sold 140 tons of detergent. Most of their products hold both ecolabel and v-label certifications. Their detergents are entirely biodegradable, vegan, and available for purchase online in recycled and refillable bottles or in bulk at over 200 partner stores, minimizing the use of single-use bottles (over 40,000 single-use 1 L bottles have been eliminated). EcoX has established local partnerships for distribution within professional sectors. Currently, their products are sold in Portugal and Spain, and the brand is beginning to expand into Europe and China. Each product comes with a technical data sheet detailing its environmental impact and the pollution mitigated through its use, aiding communication with customers regarding the genuine sustainable impact of their purchase. Furthermore, EcoX is dedicated to educational initiatives in schools to raise awareness among younger generations about sustainability and circularity. They also provide educational kits for children to create eco-friendly soap from waste oil.

The company has secured various forms of funding to develop its sustainable business and recruit human resources, including Portugal 2020 funding, R&D projects, and circularity support initiatives. However, while EcoX benefits from R&D tax incentives, there are currently no additional tax relief schemes or incentives available in Portugal for circular companies. EcoX was founded with a circular identity, establishing a CBM from the outset, which negated the need for a shift from a linear to a circular model. This case study showcases excellent examples of the various types of CBMs discussed earlier in this work. EcoX implements all five distinct types of CBMs [44]: circular supplies/circular design; resource recovery; product life extension (remanufacturing, repair, upgrading, and resale); sharing; and product as a service.

Table 1 summarizes the main circular practices implemented by EcoX, organized according to the five CBM adapted to the agri-food sector by [7].

Primarily, the company employs the "circular supply/circular design," "resource recovery," and "sharing" models by valorizing waste cooking oil to create a valuable commercial product, while subcontracting goods and services. To facilitate this circular production, EcoX has forged strong partnerships with suppliers who handle the collection and management of cooking oil waste. Subsequently, EcoX

Table 1 EcoX circular practices organized by circular business models types

CBMs	EcoX practices and examples
1. Circular supplies/circular design	•Zero-waste packaging policy based on eco-design and use of recycled materials •Bottles designed for durability and multifunctional reuse (refillable and washable) •Bulk sales system encouraging consumers to reuse containers (RecoX program) •Reuse of packaging materials such as boxes and paper tape in distribution
2. Resource recovery	•Transformation of used cooking oil into eco-friendly detergents •Partnerships with suppliers for waste oil collection and management •Production in MSTN group facilities powered by photovoltaic panels •Environmental outcomes (2022): 10,000 L of oil recovered, 140 tons of detergents sold, and strong reduction in CO_2 and water contamination
3. Product life extension	•Collection, cleaning, and reuse of dispensers and bottles for bulk sales (RecoX) •Design of refillable packaging and long-lasting molds used for multiple products •Reuse of operational materials and investment in used equipment to extend machinery lifespan
4. Sharing platforms	•Shared industrial processing and logistics services with MSTN group and external partners •Distribution network of over 150 professional clients and 180 retail partners •Educational campaigns and school activities promoting circularity and awareness
5. Product as a service (PaaS)	• EcoX spot initiative allowing business clients (e.g., hotels, restaurants) to signal environmental responsibility •Educational kits and training activities for younger generations •Integration of industry 4.0 tools to enhance production, quality, and reverse logistics efficiency

Source Authors' elaboration, 2023

converts this waste into eco-detergent using the industrial facilities of their partner MSTN group, which features 542 photovoltaic panels, enabling sustainable production of clean energy. For product distribution, EcoX is committed to minimizing the environmental impact of shipping by selecting a logistics provider that prioritizes sustainability in its operations, aiming to decrease fossil fuel consumption and CO_2 emissions. Thus, EcoX has structured its model around subcontracted logistics services for waste oil collection and product distribution. Additionally, they share industrial processing and transformation with their strategic business partner, allowing EcoX to operate with just two facilities: one operational location shared with their industrial partner and an administrative office situated within a start-up incubator.

EcoX's packaging strategy aligns with the "circular supply/circular design" and "product life extension" CBMs. The company has adopted a zero-waste packaging policy, with bulk sales being an essential component of EcoXperience's commercial approach. Their products are offered in their own eco-designed recycled plastic bottles and paper boxes or sold in bulk, with customers bringing their own containers to partner stores. EcoX retrieves used dispensers for bulk sales, cleans, sanitizes, and refills them before redistributing them to bulk-selling retailers. This packaging collection and refill program is named RecoX. The packaging produced by EcoX must be reusable, multifunctional, and include recycled materials. The bottles are designed to have a longer lifespan, allowing consumers to refill them. They are also multifunctional, as the same molds and production processes are used for various applications, such as spray bottles, multipurpose bottles, dishwashing and toilet detergent containers, and dosing caps for laundry. The bottles incorporate 50% recycled polyethylene and are crafted to endure. The packaging is designed to be easily washable, sanitized, and reusable, with refill capabilities. All paper boxes are reused for different shipping orders and feature a tracking system to indicate how many times they have been reused. A 100% eco-friendly tape made from Kraft paper and natural rubber adhesive is also utilized for packaging. Additionally, the company invests in used equipment, such as forklifts, to extend the lifecycle of operational machinery.

EcoX also adopts "product as a service" circular business models. In addition to conducting over 600 awareness-raising initiatives and educational programs through 2022, they developed an initiative called EcoX Spot, aimed at assisting customers (particularly professional ones) in signaling their use of environmentally friendly detergents. Businesses like hotels and restaurants utilizing EcoX products can display a label indicating their commitment to environmental responsibility. In conclusion, EcoX serves as a successful case study demonstrating the effective application of CBMs in running a sustainable enterprise. Furthermore, the company has already begun implementing Industry 4.0 strategies in areas such as administration and accounting, production and quality control, customer management, and reverse logistics. Moving forward, EcoX plans to enhance and invest further in Industry 4.0 and digitalization, anticipating that their CBM will become increasingly efficient and refined in the near future.

5 Conclusions

EcoX represents a successful case study in which circular business models are adopted to run a sustainable and circular business which reuse cooking oil waste to produce eco-friendly cleaning detergents. The study responded to the authors' research questions presented at the beginning, confirming that the EcoX business model is not only based on the oil waste valorization practice, but also on the extension of product life, minimising waste through eco-design, incentivising reduction of consumption, and transitioning from an ownership to a sharing paradigm. Therefore, EcoX presents a mixture of all five CBM types in its strategy, demonstrating that is possible for a company to interconnect them in a clear, functional, and productive way. However, as this research relies on a single-case study, the results cannot be generalized to all agri-food companies or sectors. This methodological limitation, common in qualitative approaches, should be considered when interpreting the findings. Nevertheless, the in-depth nature of the case provides rich contextual insights that can inspire further research and practical applications in other contexts. The triangulation of data collected from interviews, documents, and observations contributes to strengthening the study's reliability and internal validity. For the future, EcoX plans to invest in Industry 4.0 and digitalisation, which are expected to enhance its circular business models toward smart circular business models. Exploring how Industry 4.0 technologies can enable the interconnection of the five CBM archetypes is recommended as a promising direction for future research.

Acknowledgements This research was supported by national funds through the Portuguese Foundation for Science and Technology (FCT) under the PhD grant 2021.07012.BD awarded to Barbara Lamolinara, co-funded by the European Social Fund (ESF) and national funds through the "Norte 2020" Programme and MCTES, and under the research unit projects UID/4011/2025 (CETRAD—Centro de Estudos Transdisciplinares para o Desenvolvimento; DOI: 10.54499/UID/PRR/04011/2025), UID/04007/2025 (CEFAGE—Centro de Estudos e Formação Avançada em Gestão e Economia; DOI: 10.54499/UID/04007/2025), UID/04044/2025 (CDRSP—Centre for Rapid and Sustainable Product Development; DOI: 10.54499/UID/04044/2025), and the Associate Laboratory ARISE (LA/P/0112/2020; DOI: 10.54499/LA/P/0112/2020).

References

1. Geissdoerfer, M., Savaget, P., Bocken, N. M. P., & Hultink, E. J. (2017). The circular economy—A new sustainability paradigm? *Journal of Cleaner Production, 143*, 757–768. https://doi.org/10.1016/j.jclepro.2016.12.048
2. Geissdoerfer, M., Pieroni, M. P. P., Pigosso, D. C. A., & Soufani, K. (2020). Circular business models: A review. *Journal of Cleaner Production, 277*. https://doi.org/10.1016/j.jclepro.2020.123741
3. Donner, M., & de Vries, H. (2021). How to innovate business models for a circular bioeconomy? *Business Strategy and the Environment*. https://doi.org/10.1002/bse.2725

4. Galvão, G. D. A., Homrich, A. S., Geissdoerfer, M., Evans, S., Ferrer, P. S. S., & Carvalho, M. M. (2020). Towards a value stream perspective of circular business models. *Resources, Conservation and Recycling, 162*. https://doi.org/10.1016/j.resconrec.2020.105060
5. Zucchella, A., & Previtali, P. (2019). Circular business models for sustainable development: A 'waste is food' restorative ecosystem. *Business Strategy and the Environment, 28*, 274–285. https://doi.org/10.1002/bse.2216
6. Bocken, N. M. P., de Pauw, I., Bakker, C., & van der Grinten, B. (2016). Product design and business model strategies for a circular economy. *Journal of Industrial and Production Engineering, 33*(5), 308–320. https://doi.org/10.1080/21681015.2016.1172124
7. Donner, M., Gohier, R., & de Vries, H. (2020). A new circular business model typology for creating value from agro-waste. *Science of the Total Environment, 716*. https://doi.org/10.1016/j.scitotenv.2020.137065
8. Wang, X., et al. (2022). Emerging waste valorization techniques to moderate the hazardous impacts, and their path towards sustainability. *Journal of Hazardous Materials, 423*. https://doi.org/10.1016/j.jhazmat.2021.127023
9. Garcia-Garcia, G., Stone, J., & Rahimifard, S. (2019). Opportunities for waste valorization in the food industry—A case study with four UK food manufacturers. *Journal of Cleaner Production, 211*, 1339–1356. https://doi.org/10.1016/j.jclepro.2018.11.269
10. Maina, S., Kachrimanidou, V., & Koutinas, A. (2017). A roadmap towards a circular and sustainable bioeconomy through waste valorization. *Current Opinion in Green and Sustainable Chemistry, 8*, 18–23. https://doi.org/10.1016/j.cogsc.2017.07.007
11. Dagevos, H., & de Lauwere, C. (2021). Circular business models and circular agriculture: Perceptions and practices of Dutch farmers. *Sustainability, 13*(3), 1–15. https://doi.org/10.3390/su13031282
12. Donner, M., & Radić, I. (2021). Innovative circular business models in the olive oil sector for sustainable Mediterranean agrifood systems. *Sustainability, 13*(5). https://doi.org/10.3390/su13052588
13. Cadinu, L. A., Barra, P., Torre, F., Delogu, F., & Madau, F. A. (2020). Insect rearing: Potential, challenges, and circularity. *Sustainability, 12*. https://doi.org/10.3390/su12114567
14. Bocken, N. M. P., Short, S. W., Rana, P., & Evans, S. (2014). A literature and practice review to develop sustainable business model archetypes. *Journal of Cleaner Production, 65*, 42–56. https://doi.org/10.1016/j.jclepro.2013.11.039
15. Singh, S., et al. (2021). Thematic exploration of sectoral and cross-cutting challenges to circular economy implementation. *Clean Technologies and Environmental Policy, 23*, 915–936. https://doi.org/10.1007/s10098-020-02016-5
16. Donner, M., Verniquet, A., Broeze, J., Kayser, K., & de Vries, H. (2021). Critical success and risk factors for circular business models valorising agricultural waste and by-products. *Resources, Conservation and Recycling, 165*. https://doi.org/10.1016/j.resconrec.2020.105236
17. Pampuri, A., et al. (2021). Environmental impact of food preparations enriched with phenolic extracts from olive oil mill waste. *Food, 10*. https://doi.org/10.3390/foods10050980
18. Lagrasta, F. P., Pontrandolfo, P., & Scozzi, B. (2021). Circular economy business models for the Tanzanian coffee sector: A teaching case study. *Sustainability, 13*. https://doi.org/10.3390/su132413931
19. Martin, M., & Herlaar, S. (2021). Environmental and social performance of valorizing waste wool for sweater production. *Sustainable Production and Consumption, 25*, 425–438. https://doi.org/10.1016/j.spc.2020.11.023
20. Ryabchenko, O., et al. (2017). Sustainable business modeling of circular agriculture production: Case study of circular bioeconomy. *Journal of Security and Sustainability Issues, 7*, 301–309. https://doi.org/10.9770/jssi.2017.7.2(10)
21. Yamoah, F. A., Sivarajah, U., Mahroof, K., & Peña, I. G. (2022). Demystifying corporate inertia towards transition to circular economy: A management frame of reference. *International Journal of Production Economics, 244*. https://doi.org/10.1016/j.ijpe.2021.108388

22. Zucchella, A., Previtali, P., & Strange, R. (2021). Proactive and reactive views in the transition towards circular business models: A grounded study in the plastic packaging industry. *International Entrepreneurship and Management Journal*. https://doi.org/10.1007/s11365-021-00785-z
23. D'Amato, D., Veijonaho, S., & Toppinen, A. (2020). Towards sustainability? Forest-based circular bioeconomy business models in Finnish SMEs. *Forest Policy and Economics, 110*. https://doi.org/10.1016/j.forpol.2018.12.004
24. Hollweck, T. (2016). Robert K. Yin (2014). Case study research design and methods (5th ed.). Thousand oaks, CA: Sage, 282 pages. *The Canadian Journal of Program Evaluation*. https://doi.org/10.3138/cjpe.30.1.108
25. Cielo, I. D., Ribeiro, M. C. P., Rocha Júnior, W. F. D., Fragoso, R. M. D. S., & Schmidt, C. M. (2023). Generation of renewable energy (biogas) in the Western region of Paraná/Brazil—A multicase study from the viewpoint of contracts. *Sustainability, 15*. https://doi.org/10.3390/su15021458
26. Del Vecchio, P., Secundo, G., Mele, G., & Passiante, G. (2021). Sustainable entrepreneurship education for circular economy: Emerging perspectives in Europe. *International Journal of Entrepreneurial Behavior & Research*. https://doi.org/10.1108/IJEBR-03-2021-0210
27. Kulkarni, M. G., & Dalai, A. K. (2006). Waste cooking oil—An economical source for biodiesel: A review. *Industrial & Engineering Chemistry Research, 45*(9), 2901–2913. https://doi.org/10.1021/ie0510526
28. Lam, M. K., Lee, K. T., & Mohamed, A. R. (2010). Homogeneous, heterogeneous and enzymatic catalysis for transesterification of high free fatty acid oil (waste cooking oil) to biodiesel: A review. *Biotechnology Advances, 28*(4), 500–518. https://doi.org/10.1016/j.biotechadv.2010.03.002
29. Zaharia, C., & Suteu, D. (2012). Wastewater discharges from food industry as significant contributors to surface water pollution. *Environmental Engineering and Management Journal, 11*(12), 2237–2247. https://doi.org/10.30638/eemj.2012.278
30. Nascimento, L., et al. (2021). Turning waste cooking oils into biofuels - valorization technologies: A review. *Energies, 15*. https://doi.org/10.3390/en15010116
31. Khodadadi, M. R., Malpartida, I., Tsang, C. W., Lin, C. S. K., & Len, C. (2020). Recent advances on the catalytic conversion of waste cooking oil. *Molecular Catalysis, 494*. https://doi.org/10.1016/j.mcat.2020.111128
32. Azizullah, A., Khan, S., Rehman, S., Taimur, N., & Häder, D.-P. (2021). Detergents pollution in freshwater ecosystems. In *Anthropogenic pollution of aquatic ecosystems* (pp. 245–270). https://doi.org/10.1007/978-3-030-75602-4_12
33. Giagnorio, M., Amelio, A., Grüttner, H., & Tiraferri, A. (2017). Environmental impacts of detergents and benefits of their recovery in the laundering industry. *Journal of Cleaner Production, 154*, 593–601. https://doi.org/10.1016/j.jclepro.2017.04.012
34. Paul, J., Modi, A., & Patel, J. (2016). Predicting green product consumption using theory of planned behavior and reasoned action. *Journal of Retailing and Consumer Services, 29*, 123–134. https://doi.org/10.1016/j.jretconser.2015.11.006
35. Bao, Z. (2023). Developing circularity of construction waste for a sustainable built environment in emerging economies: New insights from China. *Developments in the Built Environment, 13*. https://doi.org/10.1016/j.dibe.2022.100107
36. Roy, E. D., Esham, M., Jayathilake, N., Otoo, M., Koliba, C., Wijethunga, I. B., & Fein-Cole, M. J. (2021). Compost quality and markets are pivotal for sustainability in circular food-nutrient systems: A case study of Sri Lanka. *Frontiers in Sustainable Food Systems., 5*. https://doi.org/10.3389/fsufs.2021.748391
37. Kirchherr, J., & van Santen, R. (2019). Research on the circular economy: A critique of the field. *Resources, Conservation and Recycling, 151*. https://doi.org/10.1016/j.resconrec.2019.104480
38. Eisenhardt, K. M. (1989). Building theories from case study research. *Academy of Management Review, 14*(4), 532–550. https://doi.org/10.5465/amr.1989.4308385

39. Miles, M. B., Huberman, A. M., & Saldaña, J. (2014). *Qualitative data analysis: A methods sourcebook* (3rd ed.). Sage.
40. King, N., Horrocks, C., & Brooks, J. (2019). *Interviews in qualitative research.* Sage Publications Ltd.
41. Dresch, A., Pacheco Lacerda, D., & Cauchick Miguel, P. A. (2015). A distinctive analysis of case study, action research and design science research. *Revista Brasileira de Gestao de Negocios*, 1116–1133. https://doi.org/10.7819/rbgn.v17i56.2069
42. Silverman, D. (2013). *Doing qualitative research: A practical handbook* (4th ed.). Sage Publications.
43. Khan, T., & MacEachen, E. (2022). An alternative method of interviewing: Critical reflections on videoconference interviews for qualitative data collection. *The International Journal of Qualitative Methods, 21*, 1–12. https://doi.org/10.1177/16094069221090063
44. Glinik, M., Lamolinara, B., Ropposch, C., dos Santos Ferreira, V. H., & Pacher, C. (2024). Circular business models: Overcoming barriers and unlocking potentials. In P. Mora & F. G. Acien Fernandez (Eds.), *Circular economy on energy and natural resources industries: New processes and applications to reduce, reuse and recycle materials and decrease greenhouse gases emissions* (pp. 269–279). Springer International Publishing. https://doi.org/10.1007/978-3-031-56284-6_15

Strategic Communication of Sustainability in the Agri-Food Sector: Challenges and Digital Opportunities

Adriano Andreghetto, Marisa R. Ferreira, and Markos Kourgiantakis

Abstract This study examines the intersection of sustainability and digital promotion of Geographical Indication (GI) agricultural products in Portugal. Through a case study of products certified under Protected Designation of Origin (PDO) and Protected Geographical Indication (PGI), the research investigates how values of sustainability and origin are communicated across digital platforms. Using data from the EU's GI View portal 27 producer groups were identified, of which 15 maintained active Facebook Pages between January 2023 and April 2024. Quantitative analysis assessed the extent and type of digital presence, while qualitative analysis of Facebook posts explored content and communication strategies. Findings reveal that digital presence is irregular, producer group names were prioritised over product names, leading to inconsistencies and reduced visibility of the GI designation as a protected brand asset. Content analysis showed a strong emphasis on local and regional identity but limited reference to sustainability, despite the intrinsic link between GIs and sustainable practices. These results highlight a missed opportunity: although GI products are uniquely positioned for authentic storytelling, their digital strategies often fail to align with consumer demand for transparency, sustainability, and purpose. The study underscores the need for strategic capacity building and policy support to strengthen the role of GIs in sustainable rural development.

Keywords Geographical indications · Digital communication · Sustainability

A. Andreghetto · M. R. Ferreira · M. Kourgiantakis
Department of Business Administration and Tourism, Hellenic Mediterranean University, Heraklion, Greece
e-mail: mferreira@estg.ipp.pt

M. Kourgiantakis
e-mail: mkourg@hmu.gr

A. Andreghetto (✉)
CIICESI, ESTG, Polytechnic of Porto, Felgueiras, Portugal
e-mail: aan@estg.ipp.pt

L. Farinha et al. (eds.), *Sustainable Innovations in Mechanical Engineering and Management Sciences*, Lecture Notes on Multidisciplinary Industrial Engineering,
https://doi.org/10.1007/978-3-032-30372-1_12

1 Introduction

Consumers are showing increasing awareness and preference for sustainable products and practices. Recent studies reveal that 65% of consumers prefer to purchase from environmentally responsible companies and are willing to pay, on average, a 9.7% premium for eco-friendly products [1, 2]. However, sustainability is interpreted and managed differently across stakeholders, including governments, companies, and producers [3] requiring coordination across agricultural production, processing, distribution, and consumption [2, 4].

Within this context, digital marketing has emerged as a key driver in promoting sustainable practices. Evidence shows that it contributes to sustainability by improving communication, widening market access, and supporting decision-making among stakeholders [5, 6]. Furthermore, digital marketing techniques can shape behaviour by sharing information, raising awareness, and encouraging the adoption of sustainable practices [3]. Many agri-food companies have already integrated websites and social media into their communication strategies [7] and many farmers and cooperatives also recognise that digital marketing tools, strengthen customer relationships and product promotion [8].

Through a case study, this research examines how sustainability is promoted on digital platforms, using as a sample, Portuguese agricultural products certified under the Geographical Indication (GI) scheme. The GI system was created to harmonise the protection of agricultural products and foodstuffs with specific regional origins across the European Union (EU). GIs provide intellectual property rights to products whose qualities are closely tied to their place of production. The system includes two main labels: Protected Designation of Origin (PDO) and Protected Geographical Indication (PGI) [9]. These designations not only guarantee authenticity and traceability but also contribute to rural development, cultural heritage preservation, environmental sustainability, and community empowerment [9, 10]. Within the GI framework, local producers form associations that take responsibility for the preservation, commercialization, and promotion of the product, supported by a legal recognition and protection system managed by the EU. These groups, organized by individual farmers, may also engage in processing raw products and/or establish partnerships with intermediaries, who become key actors in the value chain [10].

Preliminary findings show that digital presence is inconsistent across GI-certified agricultural products, and sustainability is largely absent from their communication. Structured, long-term marketing strategies are also missing. Despite the intrinsic association of GI products with sustainability, cultural heritage, and community identity, these values are rarely reflected in digital promotion. This reveals a critical gap: while GI products have strong narratives rooted in sustainability that could be leveraged for storytelling, producer groups fail to translate these attributes into coherent and consistent digital strategies—particularly on social media. As a result, there is a disconnect between the potential of GIs to foster sustainable development and the way they are currently presented in the digital landscape.

2 Theoretical Framework

2.1 *Growing Consumer Interest in Sustainable Products*

Consumers are demonstrating a heightened awareness and preference for sustainable products and practices. Recent studies indicate that 65% of consumers prefer to buy from environmentally responsible companies and are willing to pay a 9.7% sustainability premium for eco-friendly products [1, 2]. This growing trend underscores the importance of encouraging responsible consumption worldwide [3]. Furthermore, approximately 70% of consumers are prepared to pay more for products that positively impact the environment [11].

This interest extends to various aspects of sustainable consumption, including the selection of products with sustainable packaging, reducing plastic use, and purchasing from companies that embrace sustainable production [3]. Consumers with strong green consumption values are more receptive to green communication and are more likely to advocate for green products within their social circles [1]. Their increasing sensitivity to environmental issues translates into a stronger response to promotion campaigns, emphasising the significance of enhancing public environmental awareness as an integral part of digital promotion strategies [3].

2.2 *Marketing and Digital Communication as a Key Channel*

Marketing strategies play a crucial role at the final stage of the food supply chain, where consumer preferences meet the marketing practices of agricultural producers. Principles such as product differentiation and market segmentation are essential for increasing product value and responding to changing consumer expectations [12–14]. These expectations include a renewed interest in traditional products, the importance of proximity retailing, a preference for high-quality goods produced locally [15–17], sustainable practices [3, 18] and a stronger demand for transparent communication about added value [19, 20].

To meet these emerging preferences, many agri-food companies have integrated websites and social media into their communication strategies, which has significantly reshaped their organisational models and marketing approaches [7]. The COVID-19 pandemic brought major challenges, but it also created new opportunities for entrepreneurship in the agri-food sector [21], including the strengthening of the online presence of organisations and businesses [22]. Today, a digital presence is essential for small and medium-sized enterprises (SMEs) in the agri-food industry [7].

Digital tools and technologies can be used strategically to secure a competitive advantage, which is increasingly vital for survival in a highly competitive market [23]. Those tools are key elements of a company's e-business strategy and serve as the main channels for communication and interaction with consumers [24]. Social networks, websites, and smartphone applications provide marketers with extensive opportunities to influence consumers worldwide, creating awareness and fostering support for sustainable practices [3]. They also enable farmers and agri-food businesses to reach wider markets, increase sales, and reinforce more sustainable production and distribution practices [6]. From a marketing standpoint, the inherently social nature of today's environment means that nearly every stage of the consumer decision-making process can be influenced by a somekind of digital platform [25].

Seen as the cheapest way to enter the digital world [26], social media is increasingly viewed as a tool for networking, collaboration, and co-learning, thereby supporting innovation in agricultural development [26, 27]. Today, social media platforms are used not only for personal communication but also as simple and fast channels for sales and business interaction [26]. When these platforms are aligned with the needs of agricultural producers, the intention to adopt social media for marketing and sales grows stronger, facilitating wider use in the sector [8].

Digital communication is also a key tool for promoting sustainability. Encouraging actions through interactive content and direct participation in sustainable initiatives has proven effective in driving change [11]. Research indicates that digital marketing contributes to sustainability by improving communication, widening market access, and supporting decision-making among stakeholders [5, 6]. In addition, digital marketing techniques can influence behaviour by sharing information, raising awareness, and encouraging the adoption of sustainable practices [3].

Digital technologies make it possible to deliver sustainability messages in ways that are more engaging and accessible for consumer [11]. Campaigns centred on health-conscious and environmentally responsible consumption are becoming more visible, encouraging people to adopt sustainable eating habits [28, 29]. At the same time, companies are embedding sustainability into broader marketing strategies by aligning communication with sustainable practices across the product life cycle [11]. Environmental awareness also plays a mediating role between digital marketing and sustainable consumption, showing that informed consumers are more likely to adopt sustainable behaviours [3]. Strategies such as ensuring message relevance, creating appealing content, and linking products to sustainability themes have proven particularly effective in shaping consumer choices [20, 30]. Visual content is often central in encouraging participation in sustainable initiatives, as it strengthens both emotional and cognitive engagement. Interactive content, such as quizzes and challenges, further promotes direct consumer participation, enhancing overall engagement [11].

2.3 *Sustainability in the Agri-Food Sector: Triple Helix Perspectives*

The concept of sustainability in the agri-food sector is multifaceted and understood differently across stakeholder groups. There is increasing recognition that sustainable innovation must encompass the entire food system, requiring coordination across the supply chain between agricultural production, processing, and consumption. Established firms are gradually embedding sustainability priorities into their vision, goals, and practices, often in response to regulations, shareholder demands, or consumer pressure [2]. For the complexity of a practice-based perspective to be fully embraced by actors in the food sector and relevant governmental institutions, more sophisticated interventions and strategies will be required. This will demand a triple helix involvement, and cooperation of multiple strategic stakeholders, both within the food supply system and in public policy and related sectors. Such an approach places significant responsibility for systemic change beyond the individual level [4].

Government and Policymakers Governments globally are engaged in regulatory efforts to improve nutrition and protect consumer health, exemplified by initiatives at the European Union level like the General Food Law Regulation [31]. The Sustainable Development Goals (SDGs), adopted by all United Nations Member States in 2015, represent a universal call to action to end poverty, protect the planet, and ensure prosperity by 2030, with food security being a central aim. Achieving these goals necessitates strong regulatory bodies and robust laws governing the production and distribution of sustainable food [31].

Governments are increasingly investing in Information and Communication Technologies (ICTs) for the agricultural sector to boost food production and sustainability, especially in developing countries and rural areas. Yet, the high cost of internet data and persistent connectivity issues remain major barriers [32]. In some cases, government policies may also hinder change, as political interests often support the status quo and resist transformation [2].

Policymakers highlight the need for cohesive frameworks to guide the use of ICTs in food system interventions [32]. Public policies have been important in supporting the adoption of information technologies, helping ICTs act as tools for social inclusion and for improving farmers' lives, rather than deepening inequalities [33].

Governments are also expected to take a more active role in supporting agricultural demonstration zones to promote practical training for small-scale farmers [34]. However, as not all farmers are digitally literate, inequalities in agriculture may widen in a context where educators, extension agents, and governments increasingly rely on the internet for communication and information dissemination [33]. Previous research has shown that training positively impacts farmers' entrepreneurial and organisational skills, particularly in business planning, branding, and marketing knowledge [35].

Agri-Food Firms and Industries Agri-food firms and industries play a critical role in sustainability transitions, often being assessed on how they contribute to, or hinder, progress [2]. For these organisations, sustainable marketing entails a holistic perspective that spans the entire product life cycle—from design to disposal—and reflects how companies manage their overall business processes [11]. The agri-food sector includes all businesses engaged in the production, processing, manufacturing, distribution, wholesale, retail, and consumption of food products [2].

Multinational companies are now developing digital platforms across the agri-food system [36]. Yet, they face persistent challenges, such as power asymmetries with farmers and resistance to change within entrenched systems [2]. However, the sector is frequently dominated by large agribusiness firms and major food retailers, who exert considerable influence [2, 32]. Despite broader transformations and the spread of digital technologies in the food supply chain, supermarkets continue to hold a dominant position [37]. Through digital supply chain tools, they are further consolidating their control over producers and commodity chain [36].

These power structures have important implications for sustainability. Mainstream actors may actively restrict the market visibility of sustainable products, while retailers are often criticised for the limited promotion of small-scale producers [4]. Such dynamics remain a significant barrier to the adoption and communication of sustainable practices. Importantly, the degree and nature of retailer involvement are key variables in shaping consumption trends and behavioural change [37]. Evidence from previous research suggests that responsibility for sustainable food consumption ultimately lies with retailers and policymakers, who must ensure that unsustainable products are not readily available [4].

Producers and Farmers Farmers are increasingly recognised as central actors in sustainability transitions at the farm level, given the significant environmental impact of agricultural production [2]. Their active involvement is essential for implementing sustainable practices and is driven by multiple factors, including the recognition that conventional systems are unsustainable, the wish to leave a legacy for future generations, and a proactive stance that anticipates regulatory requirements [2].

Nevertheless, farmers—especially small-scale producers—continue to face considerable challenges. These include limited financial resources, structural power imbalances with large agribusinesses and food retailers, and a lack of technical skills [2]. Technology can address only some of these barriers, and the anticipated positive impacts have not always materialised [38]. Strengthening the role of farmer organisations in knowledge governance is therefore seen as essential for improving access to research and information that support the adoption of sustainable farming practices [2].

The expansion of ICTs in rural areas of developing countries has important implications for economic development, as these tools can provide farmers with timely and low-cost information services [39]. In certain contexts, ICTs can enhance rural households' agricultural production, farm profitability, job opportunities, healthier practices, and risk management [30] by broadening access to information on good agricultural practices and markets [32]. Digital technologies have also improved transparency in agricultural markets, lowering the costs of identifying new markets and promoting the inclusion of rural producers in regional, national, and global trade [38]. Moreover, they have enabled the creation of farmer networks and communities that facilitate the sharing of knowledge, experiences, and best practices, strengthening community ties and fostering innovation and technology adoption [36, 39].

Despite these benefits, low digital literacy and high internet costs remain significant obstacles for many small-scale producers and other vulnerable groups, limiting their ability to use digital tools for sustainability [32, 40]. The extent to which households can benefit from ICTs depends largely on their ability to interpret and act on the information received [30].

Horizontal collaboration among producers has proven valuable for exchanging ideas on innovation, discussing performance, and achieving the scale and reliability of supply demanded by large buyers [41]. Smallholder farms, however, are less likely to use computers for business management [33]. Instead, they often rely on cooperatives and aggregators that employ digital tools to improve collection, transportation, and quality control [38]. For this reason, cooperatives are encouraged to adopt a "digital" differentiation strategy and to strengthen digital skills in order to adapt their business models and processes to current industry developments [42].

3 Case Study: Sustainability and Digital Marketing—Evidence from Portuguese Geographical Indications

This study explores the digital promotion of GI-certified agricultural products in Portugal, with a focus on the extent to which sustainability and origin-based values are reflected in online communication. The products under analysis are registered as PDO or PGI within the EU protection scheme and belong to the Food category—Class 1.6 (fruits, vegetables, and cereals, fresh or processed)—see Appendix A.

The main data source is the GI View portal, the official EU database managed by the European Union Intellectual Property Office. This platform provides detailed information for each registered product, including the official name, country of origin, and the producer group responsible. Desk research was conducted to assess each product based on three criteria: (1) whether the producer group name appears on the GI View portal, (2) whether the product has a digital presence (websites, blogs, social media, e-commerce), and (3) whether promotion uses the product name or the producer group name. "Product name" refers to the registered designation (e.g., Cereja do Fundão—Fundão Cherry), while "producer group name" refers to the association or cooperative managing the GI product.

To evaluate digital presence, the study first used producer group data available on the GI View portal. When this information was missing, further research was carried out through publications of the Official Journal of the European Union and, when needed, through Google searches and related websites, to confirm whether an active producer association existed.

A quantitative analysis was used to measure the proportion of products with digital presence and to classify this presence by platform and naming strategy. In addition, an analysis was conducted to examine how names were used in online promotion (product name vs. producer group name).

The qualitative analysis considered the producer group with a Facebook Page with any type of publication from January 2023 to April 2024. The dataset was built using Apfy, a web-crawling service that collected publicly available Facebook posts. Since these pages were public, it was possible to access and analyse the data without restrictions. The extracted dataset contained publication details, including consumer engagement levels and the types of content shared by producer groups. The qualitative analysis was carried out using Voyant Tools, a web-based text analysis platform. The content of posts was examined, and results were grouped by country to understand how producer associations used Facebook. The tools applied included: Cirrus (word cloud of most frequent terms), Count (absolute frequency of terms) and Relative (relative frequency calculated by occurrences per million terms).

This study focused on products for which a producer group was listed in the GI View portal (n = 32). Among the 33 registered products, only one did not provide information about its producer group. It is worth noting that some of the 32 products are managed by the same producer group, which led to their evaluation as unique entities. In total, 27 producer groups were considered for analysis.

The qualitative analysis included only those products with a digital presence through a Facebook Page (n = 17). Facebook was selected because it is the most widely used channel among producer groups, and in many cases, it represents their sole form of digital presence. Since some producer groups are responsible for more than one product, they were assessed as unique Facebook Pages. The final sample included only pages with at least one publication between January 2023 and May 2024 (n = 15). This limitation reflects both the short-lived nature of social media content and the fact that a lack of information sharing may negatively affect consumer trust Table 1.

4 Results

4.1 *GI Agricultural Products—Overall Digital Presence and Platform Usage*

The data revealed a diverse scenario regarding the online promotion of Portuguese agricultural products. Overall, 27% (n = 9) of the products analysed showed no digital presence. For this study, digital presence was defined as the use of at least one of the following channels: website, Facebook, Instagram, e-commerce, Twitter, or YouTube. None of the producer groups maintained a digital presence on LinkedIn, Pinterest, or TikTok.

Among the 23 producer groups with a digital presence, differences in their level of adoption were observed. Seven relied exclusively on a Facebook Page, another seven maintained a presence on only one channel (either website or Facebook), while four had a broader digital presence across more than two platforms. Facebook was the most widely adopted platform, representing 63% of the sample's digital presence, followed by Instagram at 19%. This result is consistent with previous findings, as Facebook remains the most widely used social media platform globally [43].

Table 1 Baseline characteristics of portuguese GI products and their digital presence

Baseline characteristic	n	%
Type of GI (n = 33)		
PDO	16	48
PGI	17	52
Digital presence (n = 33)		
Product name	3	9
Producers group name	20	61
No digital presence	9	27
Producer group—Not found	1	3
Channel (n = 27)[a]		
Website	11	41
Facebook	17	63
Instagram	5	19
e-commerce	2	7
Twitter	1	4
YouTube	1	4

[a] Some of the 32 products are managed by the same producer group, which led to their evaluation as unique entities. In total, 27 producer groups were considered for analysis

Many companies are drawn to social media marketing—particularly via Facebook—because it provides a faster and low-cost entry into the digital environment [26, 44] Although Portugal is one of the few markets where Instagram surpasses Facebook in terms of user numbers [43] this trend has not yet been fully reflected in the promotion of agricultural products, where producers remain more attached to Facebook.

In addition, the analysis considered the type of digital presence—whether platforms were named after the GI product or the producer group. The Portuguese sample shows that most pages are named after the producer group (64%). Some inconsistencies were also observed, with websites registered under the producer group's name but corresponding Facebook Pages using the GI product name. This lack of consistency in naming strategies across platforms may weaken brand recognition and consumer trust.

4.2 *Facebook Content Strategy*

Considering that Facebook accounts for the majority of the digital presence among GI producer groups and remains the most widely used social platform globally, this study focuses on Facebook as the primary channel to examine how "sustainability" and "product identity" are communicated online in the agri-food sector. Of the 17 producer groups with a digital presence, 15 had active Facebook Pages with posts between January 2023 and April 2024.

A preliminary global corpus of words was constructed by analysing the content (posts from Facebook Pages) of Portuguese agricultural products with GI. The corpus comprised 10,053 words.

The analysis of the Portuguese corpus revealed that terms directly related to geographical indications were rarely used in Facebook publications. For instance, "PGI" appeared only 16 times, while "Quality" was mentioned 36 times. Terms such as "PDO", "Geographical Indication", and "Origin" were not mentioned at all. In contrast, terms associated with territory had a strong presence: "Local" (n = 260), "Regional" (n = 46), "National" (n = 24), and "Traditional" (n = 22).

References to sustainability were also scarce. The term "Sustainability" appeared only 17 times, while other related terms were marginally mentioned, such as "Environmental" (n = 13) and "Biologic" (n = 18).

Due to the prevalence of Facebook Pages being named after producer groups, it is unsurprising that producer names dominated the corpus. "Aecsclo" (n = 282), the group responsible for *Marmelada Branca de Odivelas*, was among the most frequently mentioned. Other notable terms included "Entrepreneurship" (n = 162), "Farmers" (n = 107), and "Association" (n = 89).

In relation to product names, "Pear" (n = 198) and "Apple" (n = 197) were the most frequent, followed by "Cantaloupe" (n = 108). It is important to note that Portugal has only one GI-certified pear but six different GI-certified apple varieties. Terms connected to the sensory aspects of products were present but not dominant: "Healthy" (n = 46), "Natural" (n = 35), "Flavour" (n = 29), "Taste" (n = 20), and "Fresh" (n = 16).

Geographical regions were highly represented in the corpus. "Loures" (n = 378), "Odivelas" (n = 306), and "Açores" (n = 280) stood out. "Loures" and "Odivelas" are part of the producer group name "Aecsclo", responsible for *Marmelada Branca de Odivelas*. Interestingly, despite the frequent mention of the producer and the region, the product term "Marmelada" did not appear in the corpus, reinforcing the tendency to promote the producer rather than the product itself. In the case of the "Açores", considerable mentions were also found for "Santa Maria" (n = 226) and "Graciosa" (n = 141), two islands of the archipelago. *Meloa de Santa Maria* is produced on Santa Maria island, while "Graciosa" is associated with *Alho da Graciosa*. However, although the location was strongly mentioned, the term "Alho" did not appear, once again highlighting the stronger promotion of the region and production site over the product itself.

Although several studies highlight that a consistent digital presence can generate significant benefits [6, 7, 23], the online promotion of Portuguese agricultural products with GI remains irregular, with 27% of the products showing no digital presence at all. Even on social media, most producers concentrate their efforts on Facebook (63%), a platform that has been losing relevance in the Portuguese market, where Instagram and TikTok have shown stronger growth in recent years [43]. Furthermore, many producer groups are responsible for more than one certified product but fail to promote each item adequately. In many cases, a single Facebook Page is used to represent all products, resulting in a lack of targeted communication that highlights the distinctive attributes of each GI product.

The data also shows a clear tendency to prioritise the use of producer group names. This reflects a community-oriented branding strategy, reinforcing collective reputation and market positioning by emphasising origin and authenticity [44, 45]. However, this approach overlooks the direct promotion of the product name itself, which, as a protected intellectual property, already represents a valuable brand asset and should be central to communication strategies.

Social media and digital platforms are increasingly used not only to promote products but also to inform and engage consumers on sustainability-related issues [3, 46]—an approach particularly relevant for GI products given their inherent link to sustainability. Nevertheless, the case study indicates that sustainability is not promoted in Facebook publications. In fact, few intrinsic attributes of GI products appear in the analysed corpus, with the notable exception of terms such as "Local" and "Regional," which were recurrent in the posts. This lack of focus represents a missed opportunity. GI products already possess an authentic and credible storytelling capacity rooted in sustainability, tradition, and territorial identity. Leveraging these attributes could align strongly with consumer expectations, especially considering that consumers prefer to buy from environmentally responsible companies, they are willing to pay more for eco-friendly products [1, 2, 11] and demand for transparent communication about added value [19, 20].

5 Conclusions and Implications

This study has examined, through a case study, the intersection between sustainability and digitalisation in the promotion of PDO and PGI agricultural products in Portugal. The results show that while digital tools—particularly websites and social media—are increasingly incorporated into the communication strategies of GI producer groups, their potential remains largely underexploited. The analysis revealed inconsistencies in digital engagement, a lack of strategic content planning, and limited use of tools such as e-commerce, despite their ability to expand market access and strengthen consumer proximity. Facebook emerged as the most widely used platform; however, it is often underutilised or poorly maintained. In addition, the emphasis on naming according to producer groups shifts attention away from the product itself and makes it more difficult for consumers to access information about the products, since many posts focus exclusively on the group's name—something that is not widely recognised by the public.

GI products are, by their very nature, ideally suited for storytelling and marketing. They combine attributes highly valued by contemporary consumers, including tradition, authenticity, cultural identity, and sustainable production practices. Yet these intrinsic features are not consistently or systematically communicated across digital platforms. This represents a missed opportunity: at a time when consumers increasingly demand transparency and purpose in their consumption choices, GI products already possess the credentials but often lack the communication.

Governments, policymakers, and retailers could collaborate with producers to construct a more coherent and sustainability-oriented vision for GI products. Such cooperation would not only enhance their market positioning but also reinforce an agri-food value chain that is inherently connected to sustainability and to the vitality of rural communities.

From a management and policy perspective, the findings highlight the need to strengthen digital communication and marketing skills among producer organisations. Targeted training and support—especially for small-scale and rural cooperatives—could transform fragmented efforts into strategic communication plans that improve product visibility and reinforce the territorial development goals of the GI system. Policy initiatives should also encourage the broader adoption of digital tools, moving beyond a basic online presence towards meaningful consumer engagement. In short, GI products already have a strong narrative; what is often missing is the digital strategy to communicate it effectively.

While this study offers relevant insights, it is not without limitations. The identification of producer groups relied on the GI View portal, which contained incomplete information for some products, requiring additional searches with limited accuracy. The analysis focused primarily on Facebook, excluding other platforms that may be more relevant for certain audiences. The study was quantitative and observational, without incorporating producers' perspectives, which limits the assessment of practical challenges in digital strategy. Finally, the dynamic nature of social media means that results may not be fully replicable, as producer activity and content evolve over time.

Future studies could go beyond presence analysis to explore qualitative dimensions of digital communication, such as positioning, storytelling, and audience interaction across platforms like Instagram, YouTube, or TikTok. Expanding the scope to include multiple platforms would reduce potential selection bias caused by focusing only on Facebook. In addition, integrating qualitative methods—such as interviews or participatory approaches—would provide deeper insight into producers' perspectives, strategies, and capacity-building needs, supporting the design of more effective and context-sensitive digital communication frameworks for GI promotion.

Acknowledgments All authors contributed to the study conception and design. All authors read and approved the final manuscript. The authors report there are no competing interests to declare. No AI-based software was used to generate or write the chapter content. Marisa R. Ferreira gratefully acknowledges financial support from FCT- Fundação para a Ciência e Tecnologia (Portugal), national funding through project UIDB/04728/2020.

Appendix

See Table 2.

Table 2 Portuguese PDO and PGI agricultural products and their producer groups

Product name	Producer groups	GI Type
Alho da Graciosa	AAIG—Associação de Agricultores da Ilha Graciosa	PGI
Ameixa d'Elvas	FRUTECO—Fruticultura Integrada, Lda.	PDO
Amêndoa Douro	Associação dos Produtores de Amêndoa do alto Douro	PDO
Ananás dos Açores / São Miguel	APROAN - Associação dos Produtores de Ananás dos Açores	PDO
Anona da Madeira	Associação dos Agricultores da Madeira	PDO
Arroz Carolino das Lezírias Ribatejanas	ORIVÁRZEA - Orizicultores da Várzea de Samora e Benavente, S.A.	PGI
Arroz Carolino do Baixo Mondego	AABM—Associação dos Agricultores do Baixo Mondego	PGI
Azeitona de conserva Negrinha de Freixo	ACFEC - Adega Cooperativa de Freixo de Espada à Cinta	PDO
Azeitona Galega da Beira Baixa	APABI - Associação de Produtores de Azeite da Beira interior	PDO
Azeitonas de Conserva de Elvas e Campo Maior	AGRODELTA—Indústrias Alimentares, Lda.	PDO
Batata de Trás-Os-Montes	Cooperativa Agrícola Norte Transmontano, C. R. L	PDO
Batata doce de Aljezur	Associação dos Produtores de Batata Doce de Aljezur	PDO
Castanha da Padrela	Associação regional dos Agricultores das Terras de Montenegro	PDO
Castanha da Terra Fria	Cooperativa Soutos os Cavaleiros	PDO
Castanha dos Soutos da Lapa	Bandarra - Cooperativa Agrícola do concelho de Trancoso, C. R. L.	PDO
Castanha Marvão-Portalegre	APAFNA - Agrupamento de Produtores Agrícolas e Florestais do Norte Alentejano	PDO
Cebola da Madeira	Agrupamento da Cebola da Madeira	PDO
Cereja da Cova da Beira	Cooperativa Agrícola dos Fruticultores da Cova da Beira, C. R. L	PGI
Cereja de São Julião-Portalegre	Producer group not found	PDO
Cereja do Fundão	CERFUNDÃO—Embalamento e Comercialização de Cerejas da Cova da Beira, Lda	PGI
Citrinos do Algarve	UNIPROFRUTAL - União dos Produtores Horto-fruticolas do Algarve	PGI

(continued)

Table 2 (continued)

Product name	Producer groups	GI Type
Ginja de Óbidos e Alcobaça	APMA - Associação dos Produtores de Maçã de Alcobaça	PGI
Maçã bravo de Esmolfe	FELBA - Promoção das Frutas e legumes da Beira Alta, ACE	PDO
Maçã da Beira Alta	FELBA - Promoção das Frutas e legumes da Beira Alta, ACE	PGI
Maçã da Cova da Beira	Cooperativa Agrícola dos Fruticultores da Cova da Beira, C. R. L	PGI
Maçã de Alcobaça	APMA - Associação dos Produtores de Maçã de Alcobaça	PGI
Maçã de Portalegre	APAFNA - Agrupamento de Produtores Agrícolas e Florestais do Norte Alentejano	PGI
Maçã Riscadinha de Palmela	Cooperativa Agrícola de Palmela, CRL	PDO
Maracujá dos Açores / S. Miguel	FRUTAÇOR - Cooperativa Agrícola Açoreana de Hortofruticultores, C. R. L	PDO
Marmelada Branca de Odivelas	Associação Empresarial de Comércio e Serviços dos Concelhos de Loures e Odivelas	PGI
Meloa de Santa Maria - Açores	Agromariensecoop- Cooperativa de Produtores agro - Pecuários da Ilha de Santa Maria, C. R. L	PGI
Pêra Rocha do Oeste	ANP - Associação Nacional de Produtores de Pêra Rocha	PDO
Pêssego da Cova da Beira	Cooperativa Agrícola dos Fruticultores da Cova da Beira, C. R. L	PGI

References

1. Iliopoulou, E., Koronaki, E., Vlachvei, A., & Notta, O. (2024). From knowledge to action: The power of green communication and Social media engagement in sustainable food consumption. *Sustainability, 16*(21). https://doi.org/10.3390/su16219202
2. Lees, N. J., Sivakumar, S., & Lucock, X. (2024). Agrifood sustainability transitions in firms and industry: A bibliographic analysis of research themes. *Sustainability, 16*(16). https://doi.org/10.3390/su16167079
3. Yu, L. (2024). Digital Marketing for Behavioral Change: Encouraging sustainable consumer practices to address environmental issues and support SDGs. *Journal of Lifestyle and SDGs Review, 5*(2). https://doi.org/10.47172/2965-730X.SDGsReview.v5.n02.pe03866
4. Parekh, V., & Svenfelt, Å. (2022). Taking sustainable eating practices from niche to mainstream: The perspectives of Swedish food-provisioning actors on barriers and potentials. *Sustainability: Science, Practice and Policy, 18*(1), 292–308. https://doi.org/10.1080/15487733.2022.2044197
5. Dias, C., & Mendes, L. (2018). Protected designation of origin (PDO), protected geographical indication (PGI) and traditional speciality guaranteed (TSG): A bibiliometric analysis. *Food Research International. Elsevier Ltd., 103*, 492–508. https://doi.org/10.1016/j.foodres.2017.09.059

6. Kumar, R., Pathak, S. K., & Khan, A. K. (2024). Digital marketing strategies for promoting sustainability in agro-based industries: Opportunities and challenges. *International Journal For Multidisciplinary Research*, 6. https://doi.org/10.36948/ijfmr.2024.v06i03.20774
7. Polenzani, B., Riganelli, C., & Marchini, A. (2023). Why do small firms implement web sales? The Italian olive oil case. *Journal of International Food & Agribusiness Marketing., 35*(4), 428–458. https://doi.org/10.1080/08974438.2021.2012547
8. Han, H., Xiong, J., & Zhao, K. (May 12, 2021). Digital inclusion in social media marketing adoption: The role of product suitability in the agriculture sector. *Information Systems and e-Business Management*, 1–27. https://doi.org/10.1007/S10257-021-00522-7/TABLES/9
9. Council of the European Union. (July 14, 1992). Council Regulation (EEC) No 2081/92 of 14 July 1992 on the protection of geographical indications and designations of origin for agricultural products and foodstuffs. *Official Journal of the European Communities, 35.* Published online. 1–8.
10. FAO. (2010). Linking People, Places and Products. A Guide for Promoting Quality Linked to Geographical Origin and Sustainable Geographical Indications. 2nd ed. (Vandecandelaere E, Arfini F, Belletti G, Marescotti A, eds.). Produced in collaboration with SINER-GI.
11. Widiastuti, E., Sukesi, S., & Sarsiti, S. (2024). Sustainable Marketing in the Digital age: A systematic review of the latest strategies and tactics. *International Journal of Economics, Business and Accounting Research (IJEBAR), 8*(1). https://doi.org/10.29040/ijebar.v8i1.12158
12. Hasimu, H., Marchesini, S., & Canavari, M. (2017). A concept mapping study on organic food consumers in Shanghai, China. *Appetite, 108*, 191–202. https://doi.org/10.1016/j.appet.2016.09.019
13. Stolz, H., Bodini, A., Stolze, M., Hamm, U., & Richter, T. (2009). Food quality from the consumer perspective - a synthesis of qualitative studies about consumers' perception and assessment of individual quality criteria of organic products. *Berichte über Landwirtschaft., 87*(1), 153–182.
14. Stolz, H., Stolze, M., Hamm, U., Janssen, M., & Ruto, E. (2011). Consumer attitudes towards organic versus conventional food with specific quality attributes. *NJAS: Wageningen Journal ofLife Sciences, 58*(3–4), 67–72. https://doi.org/10.1016/j.njas.2010.10.002
15. Borsellino, V., Schimmenti, E., & El Bilali, H. (2020). Agri-food markets towards sustainable patterns. *Sustainability, 12*(6), 2193. https://doi.org/10.3390/su12062193
16. Broad, G. M. (2020). Know your indoor farmer: Square roots, techno-local food, and transparency as publicity. *American Behavioral Scientist., 64*(11), 1588–1606. https://doi.org/10.1177/0002764220945349
17. Lioutas, E. D., & Charatsari, C. (2021). Enhancing the ability of agriculture to cope with major crises or disasters: What the experience of COVID-19 teaches us. *Agricultural Systems, 187*, 103023. https://doi.org/10.1016/j.agsy.2020.103023
18. Stampa, E., Zander, K., & Hamm, U. (2020). Insights into German consumers' perceptions of virtual fencing in grassland-based beef and dairy systems: Recommendations for communication. *Animals, 10*(12), 2267. https://doi.org/10.3390/ani10122267
19. Kühl, S., Schütz, A., & Busch, G. (March 1, 2024). Willingness to pay for a multi-level animal husbandry label: An analysis of German meat consumers. *British Food Journal.* https://doi.org/10.1108/BFJ-08-2023-0719
20. Michalke, A., Stein, L., Fichtner, R., Gaugler, T., & Stoll-Kleemann, S. (2022). True cost accounting in Agri-food networks: A German case study on informational campaigning and responsible implementation. *Sustainability Science, 17*(6), 2269–2285. https://doi.org/10.1007/s11625-022-01105-2
21. Cristobal-Fransi, E., Daries, N., Cardona, J. R., Del Río-Rama, M., & de la C. (2023). Challenges of digitization in the social economy in times of pandemic: The evolution of online presence and e-commerce in Agri-food cooperatives. *Agricultural and Food Economics., 11*(1), 49. https://doi.org/10.1186/s40100-023-00291-6
22. Altarturi, H. H. M., Nor, A. R. M., Jaafar, N. I., & Anuar, N. B. (January 24, 2023). A bibliometric and content analysis of technological advancement applications in agricultural e-commerce. *Electronic Commerce Research. Published online.* https://doi.org/10.1007/s10660-023-09670-z

23. Grijalba, M. A., Hernández, Y. B., Perez-Encinas, A., & Urda, B. S. (March 2, 2024). Does the use of digital tools improve a firm's performance? *Review of Managerial Science*, 1–18. https://doi.org/10.1007/S11846-024-00750-4/FIGURES/2
24. Borsellino, V., Pecoraro, N., & Schimmenti, E. (2023). Promoting dairy products through the web: The case of pecorino Siciliano PDO during the COVID-19 pandemic. *International Journal of Dairy Technology, 76*(2), 393–407. https://doi.org/10.1111/1471-0307.12933
25. Appel, G., Grewal, L., Hadi, R., & Stephen, A. T. (2020). The future of social media in marketing. *Journal of the Academy of Marketing Science, 48*(1), 79–95. https://doi.org/10.1007/S11747-019-00695-1/TABLES/2
26. Petril'ak M, Jansto E, Horska E. (2020). Communication of local farmers' products through Facebook: The case study of Nase-vase. *Communication Today., 11*(1), 150–163.
27. Kamruzzaman, M., Chowdhury, A., Odame, H. H., & Sarapura, S. (2019). How do extension agents of DAE use social media for strengthening agricultural innovation in Bangladesh? *Rural Extension And Innovation Systems Journal, 15*(1), 10–19.
28. Likoudis, Z., Sdrali, D., Costarelli, V., & Apostolopoulos, C. (2016). Consumers' intention to buy protected designation of origin and protected geographical indication foodstuffs: The case of Greece. *The International Journal of Consumer Studies, 40*, 283–289. https://doi.org/10.1111/ijcs.12253
29. Asif, S. (2025). The role of digital food Marketing in Shaping Consumer Behavior and Sustainability in emerging economies. In Z. Hussain, A. Albattat, F. Fakir, & Z. Yi (Eds.), *Innovative trends shaping food marketing and consumption* (pp. 57–70). IGI Global Scientific Publishing. https://doi.org/10.4018/979-8-3693-8542-5.ch003
30. Hashem, N. M., Hassanein, E. M., Hocquette, J. F., et al. (2021). Agro-livestock farming system sustainability during the COVID-19 era: A cross-sectional study on the role of information and communication technologies. *Sustainability, 13*(12). https://doi.org/10.3390/su13126521
31. Hennessy, T., Läpple, D., & Moran, B. (2016). The digital divide in farming: A problem of access or engagement? *Applied Economic Perspectives and Policy, 38*(3), 474–491. https://doi.org/10.1093/AEPP/PPW015
32. Li, X., Sarkar, A., Xia, X., & Memon, W. H. (2021). Village environment, capital endowment, and farmers' participation in E-commerce sales behavior: A demand observable bivariate Probit model approach. *Agriculture, 11*(9), 868. https://doi.org/10.3390/agriculture11090868
33. Opolot, H. N., Isubikalu, P., Obaa, B. B., & Ebanyat, P. (2018). Influence of university entrepreneurship training on farmers' competences for improved productivity and market access in Uganda. *Cogent Food & Agriculture, 4*(1), 1469211. https://doi.org/10.1080/23311932.2018.1469211
34. Prause, L., Hackfort, S., & Lindgren, M. (2021). Digitalization and the third food regime. *Agriculture and Human Values, 38*(3), 641–655. https://doi.org/10.1007/s10460-020-10161-2
35. Hartlieb, S., & Jones, B. (2009). Humanising business through ethical labelling: Progress and paradoxes in the UK. *Journal of Business Ethics., 88*(3), 583–600. https://doi.org/10.1007/s10551-009-0125-x
36. Marinescu, V., Fox, B., Cristea, D., Roventa-Frumusani, D., Marinache, R., & Branea, S. (2021). Talking about sustainability: How the media construct the Public's understanding of sustainable food in Romania. *Sustainability, 13*(9). https://doi.org/10.3390/su13094609
37. Deichmann, U., Goyal, A., & Mishra, D. (2016). Will digital technologies transform agriculture in developing countries? *Agricultural Economics, 47*(S1), 21–33. https://doi.org/10.1111/agec.12300
38. Aker, J. C., Ghosh, I., & Burrell, J. (2016). The promise (and pitfalls) of ICT for agriculture initiatives. *Agricultural Economics, 47*(S1), 35–48. https://doi.org/10.1111/agec.12301
39. Nakasone, E., & Torero, M. (2016). A text message away: ICTs as a tool to improve food security. *Agricultural Economics, 47*(S1), 49–59. https://doi.org/10.1111/agec.12314
40. Leat, P., & Revoredo-Giha, C. (2013). Risk and resilience in Agri-food supply chains: The case of the ASDA PorkLink supply chain in Scotland. *Supply Chain Management: An International Journal, 18*(2), 219–231. https://doi.org/10.1108/13598541311318845

41. Munz, J., & Doluschitz, R. (2021). Status quo of the adoption of digital technologies in agricultural retail and trade cooperatives - possibilities of a digital differentiation strategy. *Berichte Uber Landwirtschaft, 99*(2).
42. We Are Social, Meltwater. Digital 2024 Global Overview Report. https://wearesocial.com/uk/blog/2024/01/digital-2024-5-billion-social-media-users/
43. Gázquez-Abad, J. C., Jiménez-Castillo, D., & Marín-Carrillo, G. M. (2012). Differences in perceived image of fruit and vegetables from a specific origin: The moderating role of familiarity. *Journal of Food Products Marketing., 18*(5), 397–416. https://doi.org/10.1080/10454446.2012.685034
44. Tselempis, D., Karipidis, P., Tzimas, D., & Karypidou, I. (2019). Factors that impact farmers' engagement in local food brand development. *EuroMed Journal of Business, 15*(1), 86–101. https://doi.org/10.1108/EMJB-06-2019-0079
45. Alfonsi, R. M., Naidoo, M., & Gasparatos, A. (April 12, 2024). Stakeholder perspectives for information and communication technologies (ICTs) for sustainable food systems in South Africa: Challenges, opportunities and a proposed ICT framework. *Environment, Development and Sustainability*. https://doi.org/10.1007/s10668-024-04838-5
46. Borrero, J. D., Zabalo, A., Manzano-Agugliaro, F., & Salmerón-Manzano, E. (2021). Identification and analysis of strawberries' consumer opinions on twitter for marketing purposes. *Agronomy, 11*(4), 809. https://doi.org/10.3390/AGRONOMY11040809

Development of a Rehabilitation Action Plan to Promote Urban Sustainability in Existing Neighbourhoods

Jocelyn Erandi Reyes Nieto, Constança Rigueiro, Luís Simões da Silva, Vítor Murtinho, and Maria Alzira Pimenta Dinis

Abstract Urban degradation in existing neighbourhoods has become a pressing global issue, driven by demographic pressure, environmental degradation, and inefficient urban management practices. This chapter proposes a strategic framework for a sustainable rehabilitation action plan aimed at reversing these trends and promoting urban sustainability in already urbanised contexts. The framework addresses both the physical transformation of the built environment and the social dynamics required to foster long-term community engagement and resilience. Grounded in theories of sustainable urban development and participatory planning, the research employs a mixed-methods approach that integrates spatial analysis, evaluation, and policy review. This methodology supports the identification of key components necessary for effective urban rehabilitation, including diagnostic tools to assess neighbourhood conditions, guidelines for sustainable construction practices, and participatory mechanisms to ensure local ownership of interventions. The chapter's core contribution

J. E. R. Nieto
Escuela de Arquitectura, Arte y Diseño—EAAD, Tecnologico de Monterrey, Monterrey, Mexico
e-mail: jocelyn.reyes@tec.mx

C. Rigueiro (✉)
Institute for Sustainability and Innovation in Structural Engineering (ISISE), Instituto Politécnico de Castelo Branco, Coimbra, Portugal
e-mail: constanca@ipcb.pt

L. S. da Silva
ISISE, Department of Civil Engineering, University of Coimbra, Coimbra, Portugal

V. Murtinho
CES, Department of Architecture, University of Coimbra, Coimbra, Portugal

M. A. P. Dinis
Fernando Pessoa Research, Innovation and Development Institute (FP-I3ID), University Fernando Pessoa (UFP), Porto, Portugal

Marine and Environmental Sciences Centre (MARE), University of Coimbra, Coimbra, Portugal

M. A. P. Dinis
e-mail: madinis@ufp.edu.pt

L. Farinha et al. (eds.), *Sustainable Innovations in Mechanical Engineering and Management Sciences*, Lecture Notes on Multidisciplinary Industrial Engineering,
https://doi.org/10.1007/978-3-032-30372-1_13

lies in outlining a structured rehabilitation process that bridges regulatory planning frameworks with residents' demands and aspirations. It emphasises the need for local governance to take a proactive role in enabling sustainable change through legislation that is responsive to community needs. Aligning environmental, social, and economic objectives, the proposed action plan aims to not only improve the physical environment but also to enhance social cohesion, mitigate urban desertification, and foster a stronger sense of place and responsibility among residents. Ultimately, this work offers policymakers, urban planners, and researchers a practical and theoretically grounded proposal for urban regeneration that prioritises sustainability and inclusion in the rehabilitation of existing urban neighbourhoods.

Keywords Urban sustainability · Rehabilitation · Urban centers · Sustainable practices · Existing neighbourhoods · Urban assessment

1 Introduction

In recent years, significant urban degradation has been observed worldwide [11, 15, 31]. This phenomenon reflects the growing vulnerability of urban areas in the face of complex socio-environmental, economic, and spatial challenges. Among the main contributing factors are the overpopulation experienced by urban centres, which places increasing pressure on infrastructure [14], public services, and housing systems [28]; the environmental impacts generated by the built environment, which include high energy consumption, carbon emissions, and the overuse of natural resources [24]; and the unequal distribution of resources, which leads to social disparities and exacerbates territorial imbalances within cities [2, 25]. Furthermore, many urban areas lack comprehensive and effective processes for promoting rehabilitation and regeneration, often resulting in the neglect of ageing infrastructure and the deterioration of urban spaces [1]. As a result of these interrelated issues, the negative impact of the built environment on climate change has become increasingly pronounced [12, 13, 29], with urban areas now recognised as significant contributors to global carbon emissions and environmental degradation [10, 11, 27]. At the same time, the cumulative effects of these problems have brought to light critical urban phenomena such as uninhabitability, depopulation, desertification, and the fragmentation or marginalisation of certain districts, particularly those already vulnerable or historically underinvested [18]. In response, there has been a growing effort within the academic and professional communities to identify innovative solutions aimed at reversing these negative trends. This has led to increased interest in the development of intervention mechanisms that integrate technical, social, and environmental dimensions, support diverse interpretative frameworks, and encourage the adoption of sustainable practices across the construction and urban planning sectors [6, 7, 16, 17]. These mechanisms seek not only to rehabilitate the physical structure of urban areas but also to promote equity, resilience, and long-term sustainability in urban development. The research presented in this paper aligns with this approach, proposing a method for urban regeneration. It provides a comprehensive, theoretically grounded framework that integrates the physical transformation of the built environment with the social dynamics necessary for long-term community resilience. The methodology combines the Urban Integrated Sustainability Assessment for Existing Neighborhoods (UISA fEN) with the Conceptual Framework (CF) intervention model, a building ensuring targeted and adaptable solutions [20]. It includes the key components of a sustainable rehabilitation action plan to reverse the degradation of existing neighborhoods and promote their sustainable development.

2 Methods

The methodology integrates the UISA fEN with the Conceptual Framework (CF) intervention model to guide sustainable urban rehabilitation. Initially, neighbourhood with non-satisfied needs, is selected and analysed. The UISA fEN methodology is applied to assess its current sustainability status through a tailored set of indicators, including mandatory categories such as management, economic viability, waste management, infrastructure, energy, and water. Subsequently, the CF model is implemented, subdividing the urban area into subsectors and introducing multipurpose modular buildings that address the specific needs identified in the assessment, at both urban and building levels. After implementation, the UISA fEN assessment is repeated to assess the immediate effects of the intervention. Comparing pre and post-intervention results enables a clear understanding of the rehabilitation process's impact. This proactive approach, which we can call a rehabilitation action plan, supports ongoing monitoring and policy development, with the entire rehabilitation cycle recommended to be repeated approximately every 20 years, a medium-term in line with construction sector life cycle and refurbishment practices [9].

3 Rehabilitation Action Plan

The development of a rehabilitation action plan that advances urban sustainability in existing neighbourhoods emerges from the need to respond to the ongoing degradation affecting our urban areas. Its main objectives are, therefore, to address the key urban challenges faced by these neighbourhoods (Fig. 1), to foster their urban requalification and sustainable enhancement, and to improve the quality of life for their inhabitants.

The strategy adopted to develop this sustainable rehabilitation action plan is based on the systematic analysis and understanding of multi-issues concerning urban planning and the sustainability of the built environment. For this purpose, it is necessary to collect relevant information that justifies the need for a new rehabilitation process in a given urban area, as well as the application of sustainable building solutions.

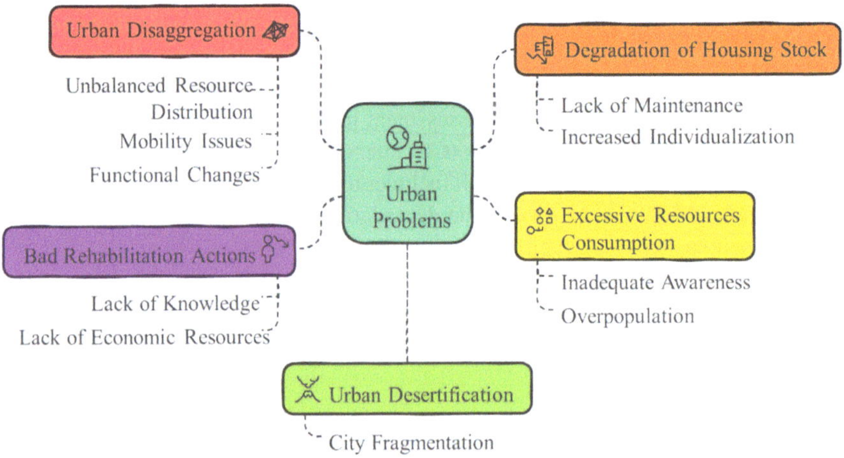

Fig. 1 Key urban challenges of existing neighbourhoods

These two approaches form the foundation of the rehabilitation process proposed in this chapter (Fig. 2).

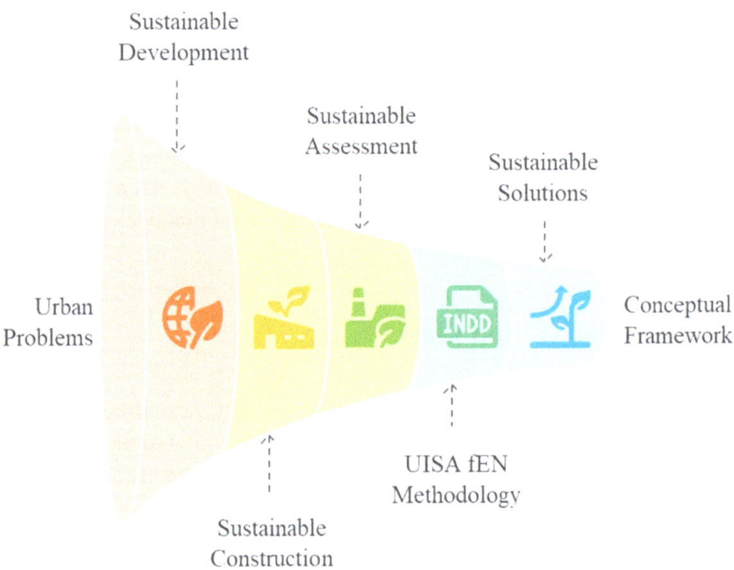

Fig. 2 Sustainable rehabilitation action plan from diagnosis to solution

3.1 Sustainable Solutions for Urban Rehabilitation

3.1.1 The Urban Integrated Sustainability Assessment for Existing Neighborhoods (UISA fEN) Methodology

The UISA fEN methodology is an urban sustainability assessment tool (USAT) designed to assist in identifying current urban challenges through the evaluation of categories (Urban morphology and Urban syntax), subcategories and indicators (key neighbourhood indicators) that should be analysed in any existing neighbourhood (Fig. 3). This methodology was developed to address the limitations observed in existing USAT [20] and to promote a sustainable framework for the rehabilitation action plan presented in this chapter. The UISA fEN methodology can be presented as a practical and accessible tool to support authorities in the elaboration of strategic plans for assessing and monitoring urban rehabilitation sustainably initiatives. It offers an innovative and holistic approach that seeks to understand the needs of a neighbourhood (through evaluation) and to measure its sustainability level through a systematic calculation process based on relevant indexes. UISA fEN includes a comprehensive set of indexes that facilitate urban sustainability assessments adaptable to the present and future needs of diverse contexts and cultures. To achieve this, it incorporates the management dimension, which enables and enhances the integration between the various sustainability dimensions [26]. Importantly, while the UISA fEN methodology may be regarded as an improvement upon existing USATs [20], it does not conflict with them. Instead, it was developed based on an analytical review of these tools and aims to complement their contributions.

The main innovations of this methodology come from its organizational framework and hierarchical design (Fig. 4). Unlike most current USATs, which are predominantly structured around the three core dimensions of sustainability, the UISA fEN methodology also draws from the complex socio-spatial realities of urban contexts [3, 5, 23]. Consequently, it is not conceived as a linear model but rather as a circular approach (Fig. 5) based on a top-down structure: the outer ring represents the main categories reflecting the core of an urban area (Urban morphology and Urban syntax categories); the intermediate ring embraces the essential themes for analysis (subcategories); and the innermost ring contains the specific indicators used to study the conditions of an existing neighbourhood. Additional inner rings may also be incorporated to represent the criteria and parameters associated with each indicator.

3.1.2 A Conceptual Framework (CF) for Improving Urban Areas

The second component of the sustainable solution proposed in this rehabilitation action plan refers to an urban-architectural model of decentralised management, which directly empowers neighbourhood residents and redefines lifestyles within existing urban centres in an innovative manner [18, 22]. The core concept of this CF is to serve as a feasible, sustainable, and flexible alternative aimed at the regeneration

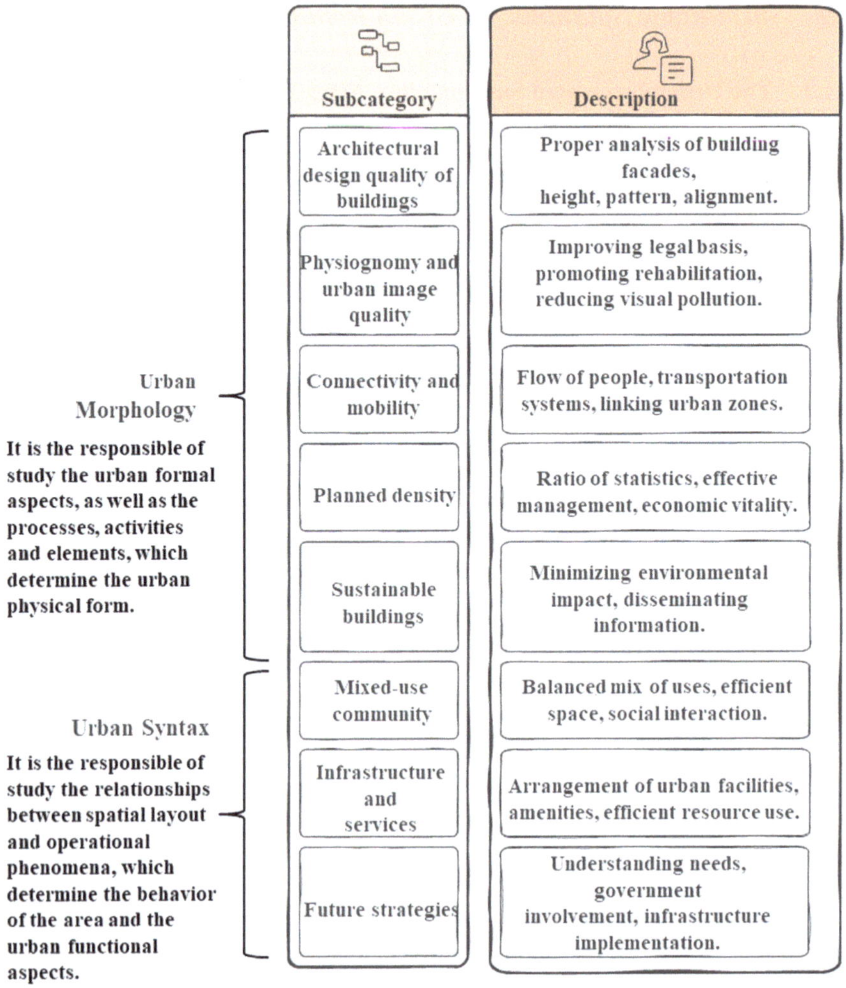

Fig. 3 UISA fEN methodology categories (Urban Morphology and Syntax) and subcategories, adapted from Reyes Nieto et al. [20]

and re-densification of existing neighbourhoods. As such, the CF introduces a novel framework for advancing sustainable urban management via the implementation of a two-level intervention process:

At the **urban level**, the intervention includes the reorganisation of the neighbourhood by dividing it into basic units, referred to as subsectors (Fig. 6). This subdivision enables a more precise characterisation of each urban area and facilitates the identification of the specific challenges affecting each subsector. These challenges are then addressed through targeted building-level interventions.

At the **building level**, the intervention entails the creation of a specialized architectural entity across each subsector. Commonly mentioned as a multipurpose building,

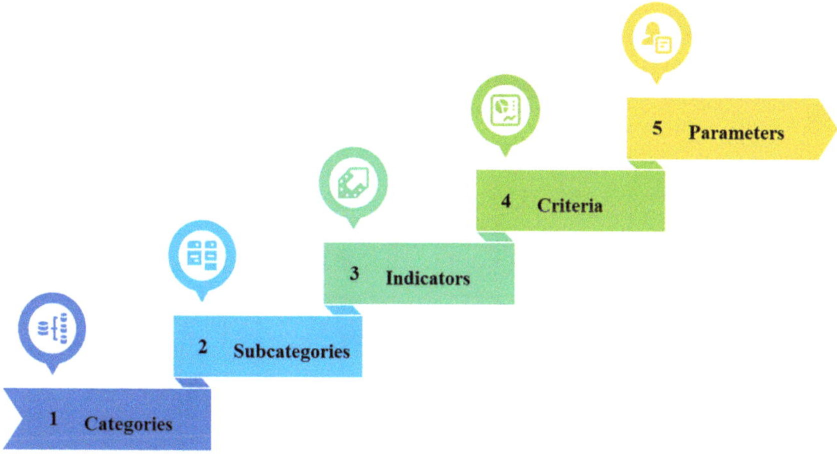

Fig. 4 UISA fEN methodology hierarchical system

Fig. 5 UISA fEN methodology three-level structure

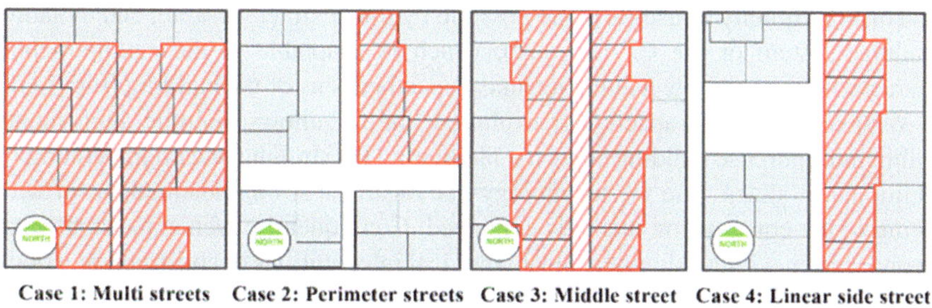

Fig. 6 UISA fEN subsectors typologies, Reyes Nieto [21]

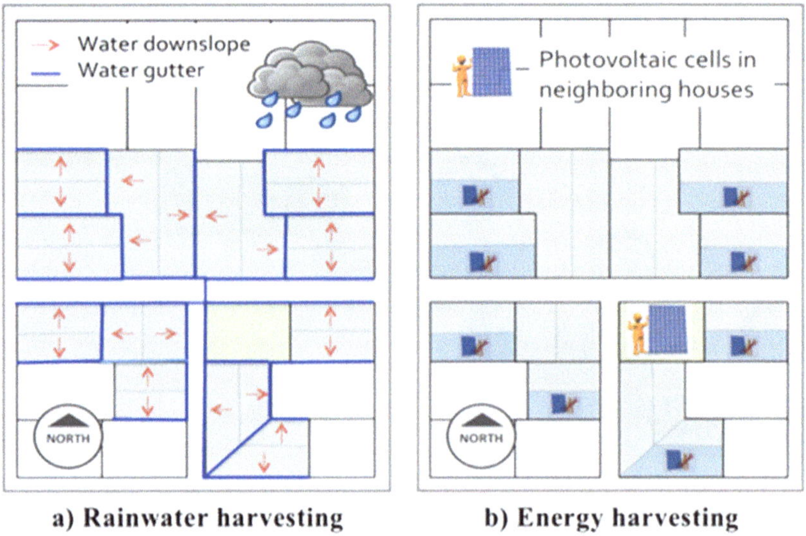

Fig. 7 Implementation of collective solutions within each subsector, Reyes Nieto [21]

this structure offers a wide range of facilities, services, amenities, and infrastructure designed to fulfil the residents' needs, support the efficient management of the subsector, and ultimately enhance the quality of life for the local population (Fig. 7).

The core innovations of this CF building are associated with its capacity for evolution over time [19]. Six potential uses (Fig. 8) were envisioned based on the needs identified in existing neighbourhoods and by considering the actions required to enhance their quality and habitability. These six potential uses (as automated parking, waste collection, water collection and storage, social activities and services, alternative transportation and production of electricity) give rise to twenty six distinct modular configurations which represent the range of facilities, amenities, infrastructure, and services that can be provided within the multipurpose buildings [20] (Fig. 9).

This multiplicity of uses is made possible by the flexible, versatile, and dynamic modular system of the CF (Fig. 10), which is adaptable to the environmental characteristics of the site and the technical specifications of the building (Fig. 11).

With regard to its capacity for evolution, the CF allows multiple adjustments within multipurpose building contexts, facilitating the straightforward integration of additional modules. The CF technology demonstrates strong potential for growth owing to the construction methods employed. Consequently, such expansions may be undertaken without significant technological interventions or construction works. Finally, in order to assess the efficiency of the CF, an economic feasibility analysis must be conducted—particularly for each multipurpose building. This economic study should take into account the various costs incurred throughout the buildings' life cycle [8].

Fig. 8 Potential uses of the multipurpose building concept, adapted from Reyes Nieto [21]

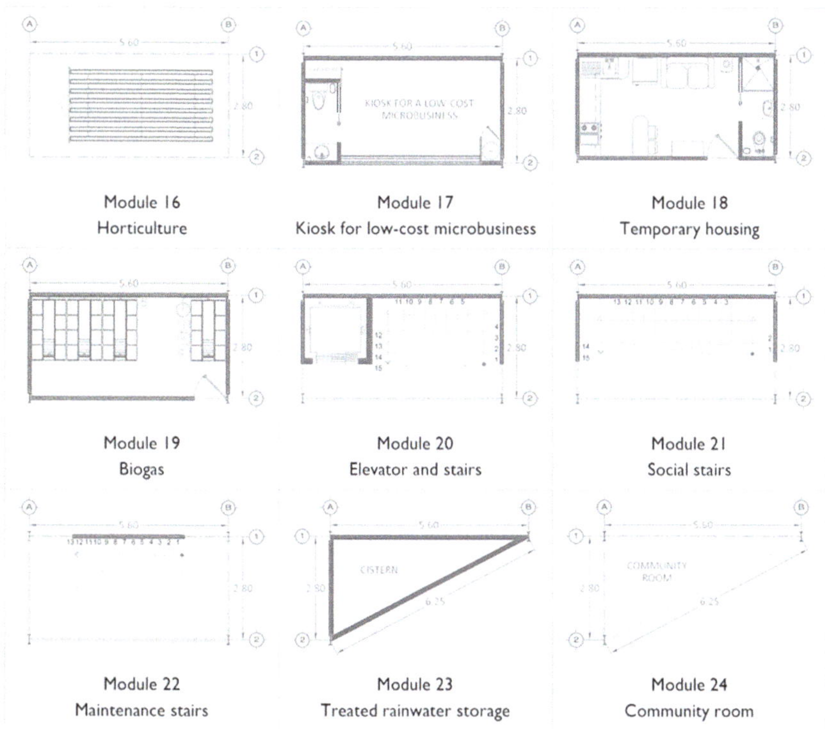

Fig. 9 Modular possibilities of the multipurpose building concept, Reyes Nieto [21]

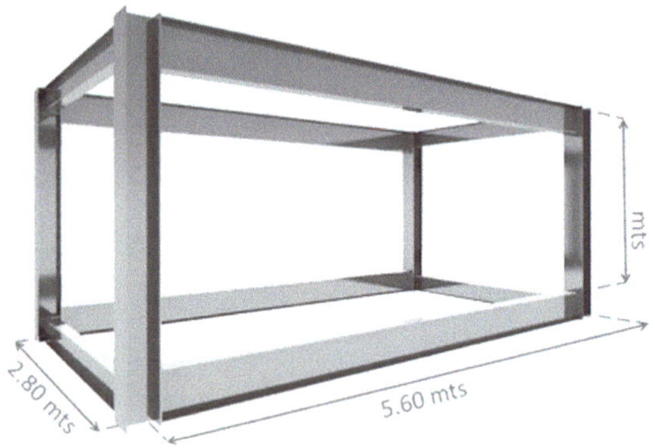

Fig. 10 Geometrical properties of the modular system, Reyes Nieto [21]

Fig. 11 Possible combinations of the modular system, Reyes Nieto [21]

3.2 Steps for the Implementation of the Rehabilitation Action Plan

Although each of the sustainable solutions presented above can be used independently, neither represents a complete rehabilitation process on its own, as they serve different objectives and functions. The implementation of the UISA fEN methodology assists in identifying the problematic areas within existing neighbourhoods, while the CF provides the necessary services to address these issues and restore the vitality and habitability of the neighbourhoods. Therefore, the combination of these two solutions is essential to create a comprehensive rehabilitation process. What

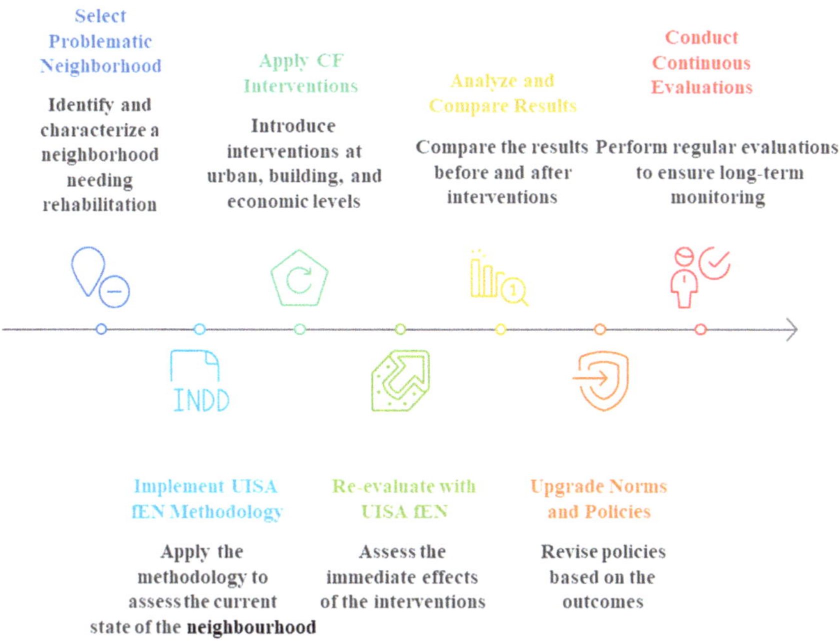

Fig. 12 Neighbourhood rehabilitation process

follows is the methodology for the rehabilitation action plan, which, through its seven-step structure, aims to restore existing neighbourhoods and promote sustainable development within them (Fig. 12).

The rehabilitation process begins with the careful selection and characterisation of a problematic neighbourhood. Once identified, the UISA fEN methodology is applied to assess the current situation of that neighbourhood. In this initial assessment, it is important to select the indicators most relevant to the study's objectives, depending on the specific purpose of the evaluation. However, certain mandatory indicators, such as Management, Economic viability, Waste management, Road infrastructure and equipment, Energy, and Water, must be included in every evaluation, as they are essential for achieving sustainable neighbourhoods [30]. Following this, the CF model is implemented within the selected neighbourhood. This intervention occurs at multiple levels: urban, building, and economic. After the CF has been applied, a subsequent evaluation using the UISA fEN methodology is conducted to assess the immediate effects of these interventions. The results from both evaluations, i.e., before and after the CF implementation, are then analysed and compared. This comparative assessment allows for the evaluation of real-world impacts and the robustness of the proposed sustainable solutions.

Based on these, existing norms and policies applied to the neighbourhood should be reviewed and updated accordingly. Furthermore, new action plans must be developed to address any remaining unsustainable areas within the neighbourhood. Finally, continuous evaluations are recommended at regular intervals, ideally every 20 years. This medium-term timescale aligns with life cycle assessment (LCA) frameworks mmonly used in the construction sector [4] and is well-suited for analysing and understanding the long-term metabolism of a neighbourhood.

4 Conclusions

Urban neighbourhoods worldwide face growing challenges related to degradation, declining habitability, and unsustainable resource use. In response, this chapter has presented a comprehensive rehabilitation action plan that integrates the UISA fEN methodology with a decentralised urban-architectural model, the CF building. Together, these tools offer a viable and adaptable approach to revitalising existing neighbourhoods, restoring their vibrancy, and fostering sustainable living environments. The application of the UISA fEN methodology enables a comprehensive diagnosis of the neighbourhood's contemporary urban challenge. Assessing a wide range of sustainability indicators, including mandatory elements such as management, economic viability, waste management, infrastructure, energy, and water, it facilitates the identification of priority areas for intervention. Subsequent evaluations using the UISA fEN methodology enable ongoing monitoring and comparison, facilitating the identification of effective mitigation measures and best practices across different contexts. This process supports the continuous evolution of the rehabilitation strategy and helps to consolidate a more resilient and sustainable urban environment. This targeted assessment ensures that the CF building can be tailored appropriately, selecting suitable locations and specifying the infrastructure, amenities, services, and facilities needed to meet the specific needs of residents and the community. Through its flexible and modular design, CF building supports improvements at the urban, building, and economic levels. Its deployment leads to enhanced neighbourhood management, optimised use of local resources, and increased awareness among residents regarding the efficient consumption of water and energy. Additionally, concerning outcomes, the promotion of renewable energy and alternative water sources contributes significantly to reducing the environmental footprint of the neighbourhood. Notably, the decentralised management of essential resources such as water, energy, and waste within the CF framework reduces transportation costs and enhances the overall sustainability of the neighbourhood.

Accordingly, the strategic combination of the UISA fEN methodology and the CF model offers a robust framework for neighbourhood requalification. It not only transforms the built environment but also promotes sustainable management practices and social engagement, thus laying the foundation for long-term urban regeneration and improved quality of life. Beyond physical and environmental improvements, this integrated rehabilitation plan empowers regulatory entities to develop tailored policies and legislation that respond directly to residents' needs. Moreover, fostering a sense of ownership and responsibility among inhabitants, it encourages active participation and strengthens community cohesion.

Although the proposed rehabilitation action plan is conceptually robust, some limitations must be acknowledged, namely: The study was only applied once in the city of Coimbra, so it needs to be validated in different urban contexts. Key performance indicators need to be tested and calibrated to ensure generalisation and adaptability to different socioeconomic conditions. For regulators and policymakers, the integrated UISA fEN–CF approach offers a structured basis for developing legislation and policies that promote sustainable urban regeneration while addressing local realities. It can serve as a tool for monitoring compliance with sustainability targets and as a decision-support tool for prioritising investment in the urban renewal that is so badly needed in Europe. For local technicians and authorities, the plan provides practical guidance for designing participatory and context-sensitive regeneration projects. Its modular and circular approach encourages continuous improvement, aligning the regulatory, technical and social dimensions of urban sustainability.

References

1. Alonso, S. (2011). *Problemas ambientais nos centros urbanos [Environmental problems in urban centers]*. Novos Diálogos.
2. Bergesen, J. D., Suh, S., Baynes, T. M., & Musango, J. K. (2017). Environmental and natural resource implications of sustainable urban infrastructure systems. Published by IOP publishing ltd. *Environmental Research Letters, 12*(12).
3. Cano Forrat, J. (2008). *Introducción a la historia del urbanismo [introduction to the history of urban planning]*. Universidad Politécnica de Valencia.
4. Curwell, S., Deakin, M., & Symes, M. (2005). *Sustainable urban development. Volume 1: The framework and protocols for environmental assessment. Routledge is an imprint of the Taylor & Francis Group* (pp. 105–108).
5. Double, N., Silvester, M., & Randall, M. (2009). *Urban structure and characterization study*. London Borough of Tower Hamlets (LBTH) council.
6. Fioretti, C., Pertoldi, M., Busti, M., & Van Heerden, S. (2020). *Handbook of sustainable urban development strategies, EUR 29990 EN*. Publications Office of the European Union, ISBN 978-92-76-13673-6. https://doi.org/10.2760/32842
7. Hosni, J., & Zumelzu, A. (2019). Assessing nodality in neighbourhoods in transformation: A concept of sustainable urban form. The case study of Rahue Bajo, Osorno, Chile. *Sustainable Development, 27*(2), 214–226. https://doi.org/10.1002/sd.1880
8. ISO 15686-5. (2017). *Building and constructed assets - service life planning: Part 5 - life cycle costing*. International Organization for Standardization.
9. ISO 20887. (2020). *Sustainability in buildings and civil engineering works — Design for disassembly and adaptability — Principles, requirements and guidance*. International Organization for Standardization.
10. Lazaroiu, G. C., & Roscia, M. (2012). Definition methodology for the smart cities model. *Energy, 47*, 326–332.
11. Leal Filho, W., Dinis, M. A. P., & Hassen, T. B. (2025**a**). Trade-offs among SDGs: How the pursuit of economic, food, and urban development goals may undermine climate and equity targets? *Sustainable Development*. https://doi.org/10.1002/sd.70029
12. Leal Filho, W., Kovaleva, M., Dinis, M. A. P., Luetz, J. M., Alves, F., Nagy, G. J., Yaffa, S., Ayal, D. Y., & Kalungu, J. (2025). Preface. In W. Leal Filho, M. Kovaleva, M. A. P. Dinis, J. M. Luetz, F. Alves, G. J. Nagy, S. Yaffa, D. Y. Ayal, & J. Kalungu (Eds.), *Climate change mitigation and adaptation in practice* (pp. 7–8). Springer Nature Switzerland AG.

13. Leal Filho, W., Vidal, D. G., & Dinis, M. A. P. (2023). Preface. In W. Leal Filho, D. G. Vidal, & M. A. P. Dinis (Eds.), *Climate change and health hazards: Addressing hazards to human and environmental health from a changing climate* (pp. 5–6). Springer Cham.
14. Leal Filho, W., Zuñiga, R. A. A., Sierra, J., Dinis, M. A. P., Corazza, L., Nagy, G. J., & Aina, Y. A. (2024). An assessment of priorities in handling climate change impacts on infrastructures. *Scientific Reports, 14*(1), 1–16., Article 14147. https://doi.org/10.1038/s41598-024-64606-3
15. Mateus, R. (2009). *Avaliação da Sustentabilidade da Construção. Propostas Para o Desenvolvimento de Edifícios Mais Sustentáveis [assessment of construction sustainability. Proposals for the development of more sustainable buildings].* Ph.D. dissertation,. University of Minho. Guimarães.
16. Natividade-Jesus, E., Almeida, A., Sousa, N., & Coutinho-Rodrigues, J. (2019). A case study driven integrated methodology to support sustainable urban regeneration planning and management. *Sustainability, 11*(15), 4129. https://doi.org/10.3390/su11154129
17. ONU. (1993). *Agenda 21.* Programme of actions for sustainable development.
18. Reyes Nieto, J., Simões da Silva, L., Murtinho, V., Rigueiro, C., & Gonçalves, A. (2016). Conceptual model for the sustainable rehabilitation of medium-size inner cities in Europe: Coimbra. *Portugal. Journal of Urban Planning and Development, 142*(3). https://doi.org/10.1061/(ASCE)UP.1943-5444.0000303
19. Reyes Nieto, J. E., Rigueiro, C., Simões da Silva, L., & Murtinho, V. (2017). A new urban management model applied to Baixa of Coimbra, Portugal. In L. Simões da Silva, P. Vila Real, P. Piloto, & J. P. Martins (Eds.), *XI Congresso de Construção Metálica e Mista* (pp. 127–136). CMM Press. Coimbra.
20. Reyes Nieto, J. E., Rigueiro, C., Simões da Silva, L., & Murtinho, V. (2018). Urban integrated sustainable assessment methodology for existing Neighborhoods (UISA fEN), a new approach for promoting sustainable development. *Sustainable Development, 26*(6), 564–587. https://doi.org/10.1002/sd.1720
21. Reyes Nieto, J. V. (2019). Development of sustainable solutions for urban rehabilitation. *Work developed to obtain a PhD degree in Steel and Composite Construction from the University of Coimbra.*
22. Reyes Nieto, J. E., Rigueiro, C., Simões da Silva, L., & Murtinho, V. (2024). Urban rehabilitation model: A blueprint for efficient, Livable, smart and sustainable urban environments. In *2024 IEEE international smart cities conference (ISC2)* (pp. 1–6). https://doi.org/10.1109/ISC260477.2024.11004267
23. Salat, S. (2011). *Cities and forms: On sustainable urbanism.* France. Editions Hermann.
24. Santos, P., Gervásio, H., Simões da Silva, L., & Gameiro, A. (2011). Influence of climate change on the energy efficiency of light-weight steel residential buildings. *Civil Engineering and Environmental Systems, 28*(4), 325–352. https://doi.org/10.1080/10286608.2011.637624
25. Sev, A. (2011). A comparative analysis of building environmental assessment tools and suggestions for regional adaptations. *Civil Engineering and Environmental Systems, 28*(3), 231–245. https://doi.org/10.1080/10286608.2011.588327
26. Spangenberg, J. (2002). Institutional sustainability indicators: An analysis of the institutions in agenda 21 and a draft set of indicators for monitoring their effectivity. *Sustainable Development, 10*(2), 103–115.
27. UNEP. (2015). *Sustainable buildings and climate initiative.* Promoting policies and practices for the built environment. UNEP.
28. United Nations. (2014). World populations prospects 2014..
29. Vidal, D. G., Lopes, H. S., Dinis, M. A. P., & Leal Filho, W. (2024). Lifting the veil of complexity, demanding for convergence: An introduction to the book "planetary health and climate change". In W. Leal Filho, D. G. Vidal, & M. A. P. Dinis (Eds.), *Planetary health and climate change: Understanding the impacts of climate change to the Well-being of our planet* (pp. 3–12). Springer Cham. https://doi.org/10.1007/978-3-031-72740-5_1

30. Zhao, P., Ali, Z. M., & Ahmad, Y. (2023). Developing indicators for sustainable urban regeneration in historic urban areas: Delphi method and analytic hierarchy process (AHP). *Sustainable Cities and Society, 99*, 104990., ISSN 2210-6707. https://doi.org/10.1016/j.scs.2023.104990
31. Zipperer, W., Northrop, R., & Andreu, M. (2020). Urban development and environmental degradation. *Oxford Research Encyclopedia of Environmental Science.*. Retrieved 26 Jun. 2025, from https://oxfordre.com/environmentalscience/view/10.1093/acrefore/9780199389414.001.0001/acrefore-9780199389414-e-97

MIX
Papier aus verantwortungsvollen Quellen
Paper from responsible sources
FSC® C105338

If you have any concerns about our products,
you can contact us on
ProductSafety@springernature.com

In case Publisher is established outside the EU,
the EU authorized representative is:
Springer Nature Customer Service Center GmbH
Europaplatz 3, 69115 Heidelberg, Germany

Printed by Libri Plureos GmbH
in Hamburg, Germany